# Yearbook of Astronomy 2026

Front Cover: The Einstein ring, named HerS J020941.1+001557, is made of light from a galaxy 19.5 billion light-years from Earth. In the very centre there is an elliptical galaxy with a brightly glowing core and a broad disc. A reddish warped ring of light, thicker at one side, surrounds its core, caused by gravitational lensing. A small galaxy intersects the ring as a bright dot. (ESA/Hubble & NASA, H. Nayyeri, L. Marchetti, J. Lowenthal)

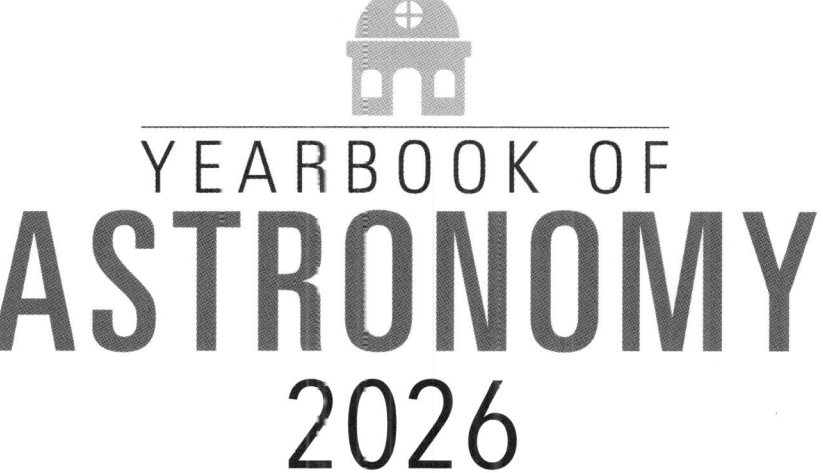

# YEARBOOK OF ASTRONOMY 2026

### EDITED BY
### Brian Jones

**WHITE OWL**
AN IMPRINT OF PEN & SWORD BOOKS LTD.
YORKSHIRE – PHILADELPHIA

First published in Great Britain in 2025 by
WHITE OWL
An imprint of
Pen & Sword Books Ltd
Yorkshire – Philadelphia

Copyright © Brian Jones, 2025

ISBN 978 1 03612 645 2

The right of Brian Jones to be identified as Author of this work has been asserted by him in accordance with the Copyright, Designs and Patents Act 1988.

A CIP catalogue record for this book is available from the British Library.

All rights reserved. No part of this book may be reproduced, transmitted, downloaded, decompiled or reverse engineered in any form or by any means, electronic or mechanical including photocopying, recording or by any information storage and retrieval system, without permission from the Publisher in writing. No part of this book may be used or reproduced in any manner for the purpose of training artificial intelligence technologies or systems.

Typeset by Mac Style

The Publisher's authorised representative in the EU for product safety is Authorised Rep Compliance Ltd., Ground Floor, 71 Lower Baggot Street, Dublin D02 P593, Ireland.
www.arccompliance.com

For a complete list of Pen & Sword titles please contact

PEN & SWORD BOOKS LIMITED
47 Church Street, Barnsley, South Yorkshire, S70 2AS, England
E-mail: enquiries@pen-and-sword.co.uk
Website: www.pen-and-sword.co.uk
or
PEN AND SWORD BOOKS
1950 Lawrence Road, Havertown, PA 19083, USA
E-mail: uspen-and-sword@casematepublishers.com
Website: www.penandswordbooks.com

# Contents

| | |
|---|---:|
| Editor's Foreword | 8 |
| Preface | 13 |
| About Time | 14 |
| Using the Yearbook of Astronomy as an Observing Guide | 16 |

**The Monthly Star Charts**

| | |
|---|---:|
| Northern Hemisphere Star Charts  *David Harper* | 27 |
| Southern Hemisphere Star Charts  *David Harper* | 53 |

| | |
|---|---:|
| The Planets in 2026  *Lynne Marie Stockman* | 78 |
| Mars finder chart – May 2026 to December 2026 | 83 |
| Jupiter finder chart – January 2026 to December 2026 | 84 |
| Saturn finder chart – January 2026 to December 2026 | 85 |
| Uranus finder chart – January 2026 to December 2026 | 86 |
| Neptune finder chart – January 2026 to December 2026 | 87 |

**Lunar Phenomena and Eclipses in 2026** — 88

| | |
|---|---:|
| Phases of the Moon | 88 |
| Apsides | 89 |
| Nodes | 90 |
| Lunar Occultations | 91 |
| Eclipses in 2026 | 92 |

**Monthly Sky Notes and Articles 2026**

| | |
|---|---:|
| *Monthly Sky Notes January* | 97 |
| Gone But Not Forgotten: Argo Navis  *Lynne Marie Stockman* | 100 |
| *Monthly Sky Notes February* | 106 |
| The Cosmic Microwave Background: Black Body Radiation  *David M. Harland* | 110 |
| *Monthly Sky Notes March* | 114 |
| Managing Healthcare in Space with an On Demand Pharmacy  *Martin Braddock* | 118 |

| | |
|---|---|
| *Monthly Sky Notes April* | 125 |
| Freedom Will Not Be Shackled  *Jonathan Powell* | 128 |
| *Monthly Sky Notes May* | 132 |
| Craters of Eternal Darkness  *Katrin Raynor* | 135 |
| *Monthly Sky Notes June* | 140 |
| Allan Rex Sandage  *David M. Harland* | 143 |
| *Monthly Sky Notes July* | 146 |
| Vega  *David M. Harland* | 149 |
| *Monthly Sky Notes August* | 153 |
| The Extraordinary George Eric Deacon Alcock: Return of the GEDA  *John McCue* | 156 |
| *Monthly Sky Notes September* | 160 |
| The Association between Comets and Meteor Showers  *Neil Norman* | 163 |
| *Monthly Sky Notes October* | 167 |
| Ussher Chronology: The Age of the Universe  *Jonathan Powell* | 171 |
| *Monthly Sky Notes November* | 174 |
| Eta Carinae  *David M. Harland* | 178 |
| *Monthly Sky Notes December* | 182 |
| Flammarion and the Comet: The Sorry Tail of the Downfall of an Honourable Man  *Neil Haggath* | 185 |
| Comets in 2026  *Neil Norman* | 188 |
| Minor Planets in 2026  *Neil Norman* | 193 |
| Meteor Showers in 2026  *Neil Norman* | 199 |

## Article Section

| | |
|---|---|
| Recent Advances in Astronomy  *Rod Hine* | 211 |
| Recent Advances in Solar System Exploration  *Peter Rea* | 216 |
| Anniversaries in 2026  *Neil Haggath* | 231 |
| The Astronomers' Stars: Taking It to Extremes  *Lynne Marie Stockman* | 238 |

| | |
|---|---|
| Hawking Stars  *Andrew D. Santarelli and Matthew E. Caplan* | 249 |
| Subrahmanyan Chandrasekhar and Professor A. S. Eddington   *David M. Harland* | 257 |
| Planetary Protection: Keeping the Planets Safe from Earthly Bacteria   *Peter Rea* | 270 |
| Nearby Worlds Out There: The Many Kinds of Exoplanet  *John McCue* | 282 |
| Comets and Literature in the Nineteenth Century  *Randall Stevenson* | 287 |
| On the Origin of NASA Names by Means of Imaginative Selection  *Peter Rea* | 296 |
| Mission to Mars: Countdown to Building a Brave New World: Pausing for Thought  *Martin Braddock* | 303 |
| A History of Observatory Designs: The Telescope Age from the Seventeenth to Nineteenth Centuries  *Katrin Raynor* | 312 |
| Sidewalk Astronomy: Cosmos to Kerbside  *Jonathan Powell* | 322 |

## Miscellaneous

| | |
|---|---|
| Some Interesting Variable Stars  *Tracie Heywood* | 335 |
| Some Interesting Double Stars  *Brian Jones* | 346 |
| Some Interesting Nebulae, Star Clusters and Galaxies  *Brian Jones* | 349 |
| Astronomical Organizations | 351 |
| Our Contributors | 357 |
| | |
| Society for the History of Astronomy (Advertisement) | 362 |
| Friends of the Royal Astronomical Society (Advertisement) | 363 |
| The Federation of Astronomical Societies / CAPCOM (Advertisements) | 364 |
| Space Oddities Live! (Advertisement) | 365 |
| British Astronomical Association (Advertisement) | 366 |
| Royal Astronomical Society of New Zealand (Advertisement) | 367 |
| Society for Popular Astronomy (Advertisement) | 368 |

# Editor's Foreword

The *Yearbook of Astronomy 2026* is the latest edition of what has long been an indispensable publication, the annual appearance of which has been eagerly anticipated by astronomers, both amateur and professional, for well over half a century. As ever, the *Yearbook* is aimed at both the armchair astronomer and the active backyard observer. Within its pages you will find a rich blend of information, star charts and guides to the night sky coupled with an interesting mixture of articles which collectively embrace a wide range of topics, ranging from the history of astronomy to the latest results of astronomical research; space exploration to observational astronomy; and our own celestial neighbourhood out to the farthest reaches of space.

The *Monthly Star Charts* have been compiled by David Harper and show the night sky as seen throughout the year. Two sets of twelve charts have been provided, one set for observers in the Northern Hemisphere and one for those in the Southern Hemisphere. Between them, each pair of charts depicts the entire sky as two semi-circular half-sky views, one looking north and the other looking south.

The *Monthly Star Charts* are followed by summaries of the observing conditions for each of the planets in *The Planets in 2026*. Apparition charts for all the major planetary members of our Solar System have been compiled by David Harper. Further details of these are given in the article *Using the Yearbook of Astronomy as an Observing Guide*. The finder charts for the superior planets follow the article *The Planets in 2026*, those depicting the morning and evening apparitions of Mercury and Venus being scattered throughout the *Monthly Sky Notes*.

A relatively new feature for the *Yearbook of Astronomy* is a set of pages called *Lunar Phenomena and Eclipses*, compiled by Lynne Marie Stockman and Brian Jones, which includes the phases and occultations as in previous editions but which also contains apsides and node crossings with references to eclipses. This section of the *Yearbook* concludes with details of *Eclipses in 2026*.

As with *The Planets in 2026*, the *Monthly Sky Notes* have been compiled by Lynne Marie Stockman and give details of the positions and visibility of the planets for each month throughout 2026. At the beginning of each of the monthly notes is a list of the significant solar system events occurring during that particular month, and which collectively replace the single list of events that has been a feature in previous editions of the *Yearbook of Astronomy*. Each section of the *Monthly Sky Notes* is accompanied by a short article, the range of which includes items on a variety of astronomy- and space-related topics including an interesting item by Lynne, in which we discover that, although constellations come and constellations go, some leave behind traces of their former existence. In the third instalment of her fascinating *Gone But Not Forgotten* series, Lynne examines the sprawling but interesting constellation Argo Navis (the Ship Argo), originally devised to commemorate the famous oared galley from Greek mythology which carried Jason and the Argonauts on their search for the Golden Fleece.

The *Monthly Sky Notes and Articles* section of the book concludes with a trio of articles penned by Neil Norman, these being *Comets in 2026*, *Minor Planets in 2026* and *Meteor Showers in 2026*, all three titles being fairly self-explanatory describing as they do the occurrence and visibility of examples of these three classes of object during and throughout the year.

In his article *Recent Advances in Astronomy*, regular contributor Rod Hine takes a look at the way that gravitational lensing is used to enhance the analysis of data from both the *Hubble* and the *James Webb* space telescopes, even as far as seeing individual stars and extremely young galaxies. He then reviews new results from the *Event Horizon Telescope* and an interesting new idea concerning some mysterious very long period repeating gamma-ray bursts. He rounds off with yet another superlative – the brightest quasar detected so far. So bright it was overlooked for several decades, incorrectly classified as a nearby star!

This is followed by *Recent Advances in Solar System Exploration* in which Peter Rea updates us on the progress of a number of planetary missions. The *Parker Solar Probe* is in the latter stages of its mission around the Sun and in the next year will come within six million kilometres of the Sun's surface. Still close to the Sun the European *BepiColombo* mission is about to enter orbit around Mercury. Further out from the Sun, Mars continues to be a major target for planetary missions, whilst two spacecraft are en-route to multiple asteroids, and Jupiter will soon have two more visitors from Earth for a detailed look at the large moons.

Something most astronomers take for granted, the equatorial mount, was invented by a man who died two centuries ago, and who made remarkable contributions to astronomy in a life tragically cut short. He was the renowned optician Joseph von Fraunhofer, who also discovered the absorption lines in the Sun's spectrum. In his article *Anniversaries in 2026*, Neil Haggath commemorates his life, together with those of Giuseppe Piazzi and Richard Carrington. He also remembers the Viking probes, which landed on Mars 50 years ago.

Some stars simply insist on Taking It to Extremes. In part five of her series of articles *The Astronomers' Stars*, in which she explores unusual stars named after the astronomers who discovered them, Lynne Marie Stockman examines some of the weirder members of the celestial zoo, including Babcock's Magnetic Star whose magnetic field is many times stronger than that of the Sun; the Becklin-Neugebauer Object, a recent escapee from the Trapezium; and Sanduleak's Star which is busily blasting its neighbours with the largest stellar jets yet discovered. Large or small, hot or cold, quiet or violent, the extreme stars are never dull.

In *Hawking Stars* by Andrew D. Santarelli and Matthew E. Caplan, we learn how black holes could be dark matter, and if they are, what would happen when the smallest of them end up inside the centres of stars like our Sun.

In his article *Subrahmanyan Chandrasekhar and Professor A. S. Eddington*, regular contributor David M. Harland looks at the work of the young Indian student Subrahmanyan Chandrasekhar whose study of conditions in white dwarf stars, taking into account the relativistic speed of the electrons, found there must be an upper limit to the mass of such a star. As we learn, this was rejected by Arthur Eddington, the leading astrophysicist of the day, but it later proved correct.

In his article *Planetary Protection*, Peter Rea reminds us of the insightful words of H.G. Wells when he states in his 1897 book *The War of the Worlds* that: "The Martians had no resistance to the bacteria in our atmosphere to which we have long since become immune. Once they had breathed our air, germs, which no longer affect us, began to kill them." Nothing could be more important today as we send our machines out in the solar system. We do not want to contaminate any of the solar system bodies with Earthly microbes, nor do we want to return any dangerous pathogens when we bring samples back to Earth. The recent COVID-19 pandemic showed us all the effects it has on our population when we have little or no resistance to a particularly virulent virus. We have a responsibility to safeguard our planet and others.

In his article *Nearby Worlds Out There: The Many Kinds of Exoplanet*, regular contributor John McCue looks at the main kinds of exoplanet, and describes some that are nearby so that readers may be able to see the parent stars in their own telescopes, and muse on the exoplanets that are there, but cannot be seen!

In his article *Comets and Literature in the Nineteenth Century*, Randall Stevenson assesses the impact of the nineteenth century's unusually abundant bright comets – and of advances in scientific understanding of their nature – on contemporary authors such as Tolstoy, Tennyson, Edgar Allan Poe and Thomas Hardy, as well as on science fiction writers including Jules Verne, Camille Flammarion and H.G. Wells.

NASA loves acronyms, proof of that being evident in the mission names of the last decade or so, the mission to the asteroid 101955 Bennu called OSIRIS-REx (*Origins, Spectral Interpretation, Resource Identification, Security-Regolith Explorer*) not exactly running off the tongue. In his article *On the Origin of NASA Names by Means of Imaginative Selection*, Peter Rea reveals a time when NASA used names that inspired a generation, with not an acronym in sight. Planetary missions such as *Mariner 9* or *Voyager 2* and lunar missions like *Surveyor 7* come from a time when NASA had a naming convention. That convention is now long forgotten, the names of more recent missions resorting to acronyms. Whilst some are quite clever, the mission to Mercury called *MESSENGER (MErcury Surface, Space ENvironment, GEochemistry and Ranging)* seem a little contrived, and irritates one particular author.

This is followed by *Mission to Mars: Countdown to Building a Brave New World - Pausing for Thought* by Martin Braddock, the sixth in a series of articles originally scheduled to appear in the *Yearbook of Astronomy* throughout the 2020s with a view to keeping the reader fully up to date with the ongoing preparations geared towards sending a manned mission to Mars at or around the turn of the decade. This article considers the repercussions of NASA's announcement in early 2024 that the Mars Sample Return (MSR) project is to be put on hold. Given this decision it is now appropriate for us to suspend our series of articles describing a future for Mars colonisation, review why the decision was made and consider a way forward.

With the invention of the telescope came the need for observatories, not only to house these delicate optical instruments but to establish academic research facilities for scientists to study the subject of astronomy. In the second part of her series of articles *A History of Observatory Designs*, regular contributor Katrin Raynor examines the period from the seventeenth to the nineteenth centuries – a time that may justifiably be called the 'Telescope Age' – describing a select few of the interesting, innovative and unique observatories that were established during the time following the global spread of telescope design and construction.

In *Sidewalk Astronomy: Cosmos to Kerbside*, regular contributor Jonathan Powell examines how one man's devotion to astronomy, coupled with a passion for telescope making, brought the wonders of the cosmos to the pavements of towns and cities worldwide. John Dobson's vision of bringing astronomy to a much wider audience evolved to incorporate a dedicated band of sidewalk astronomers that would offer members of the public a chance to experience courtesy of a telescope eyepiece, images of the universe that they may never have otherwise witnessed.

The final section of the book starts off with *Some Interesting Variable Stars* by Tracie Heywood which contains useful information on variables as well as predictions for timings of minimum brightness of the famous eclipsing binary Algol for 2026. *Some Interesting Double Stars* and *Some Interesting Nebulae, Star Clusters and Galaxies* present a selection of objects for you to seek out in the night sky. The lists included here are by no means definitive and may well omit your favourite celestial targets. If this is the case, please let us know and we will endeavour to include these in future editions of the *Yearbook of Astronomy*.

Next we have a selection of *Astronomical Organizations*, which lists organizations and associations across the world through which you can further pursue your interest and participation in astronomy (if there are any that we have omitted please let us know) and *Our Contributors*, which contains brief background details of the numerous writers who have contributed to this edition of the *Yearbook*.

New topics and themes are occasionally introduced into the *Yearbook of Astronomy*, allowing it to keep pace with the increasing range of skills, techniques and observing methods now open to amateur astronomers, this in addition to articles relating to our rapidly-expanding knowledge of the Universe in which we live. There is always an interesting mix, some articles written at a level which will appeal to the casual reader and some of what may be loosely described as at a more academic level. The intention is to fully maintain and continually increase the usefulness and relevance of the *Yearbook of Astronomy* to the interests of the readership who are, without doubt, the most important aspect of the *Yearbook* and the reason it exists in the first place. With this in mind, suggestions from readers for further improvements and additions to the *Yearbook* content are welcomed. All thoughts and comments can be sent via the *Yearbook of Astronomy* website at **yearbookofastronomy.com**. After all, the book is written for you …

As ever, grateful thanks are extended to those individuals who have contributed a great deal of time and effort to the *Yearbook of Astronomy 2026*, including David Harper who, as well as compiling all the planetary apparition charts featured in the book, has also provided the *Monthly Star Charts*, which were generated specifically for what has been described as the new generation of the *Yearbook of Astronomy*. Equally important are the efforts of Lynne Marie Stockman who has put together the *Monthly Sky Notes*. Their combined efforts have produced what can justifiably be described as the backbone of the *Yearbook of Astronomy*.

Thanks are due to Neil Haggath for supplying the (short) definition for *saros* which appears in the *Lunar Phenomena and Eclipses* section as well as the fuller definition which can be found in the downloadable *Glossary* available on the *Yearbook of Astronomy* website at **yearbookofastronomy.com**

Also worthy of mention is Mat Blurton, who has done an excellent job typesetting the *Yearbook*, and David M. Harland who has provided valuable assistance with several images used in this edition. My thanks are also due to Jonathan Wright, Charlotte Mitchell, Lori Jones, Janet Brookes, Paul Wilkinson, Charlie Simpson and Rosie Crofts of Pen & Sword Books Ltd for their efforts in producing and promoting the *Yearbook of Astronomy 2026*, the latest edition of this much-loved and iconic publication.

Brian Jones - Editor
Bradford, West Riding of Yorkshire
September 2024

# Preface

The information given in this edition of the *Yearbook of Astronomy* is in narrative form. The positions of the planets given in the *Monthly Sky Notes* often refer to the constellations in which they lie at the time. These can be found on the star charts which collectively show the whole sky via two charts depicting the northern and southern circumpolar stars and forty-eight charts depicting the main stars and constellations for each month of the year. The northern and southern circumpolar charts show the stars that are within 45° of the two celestial poles, while the monthly charts depict the stars and constellations that are visible throughout the year from Europe and North America or from Australia and New Zealand. The monthly charts overlap the circumpolar charts. Wherever you are on the Earth, you will be able to locate and identify the stars depicted on the appropriate areas of the chart(s).

There are numerous star atlases available that offer more detailed information, such as *Sky & Telescope's POCKET SKY ATLAS* and *Norton's STAR ATLAS and Reference Handbook* to name but a couple. In addition, more precise information relating to planetary positions and so on can be found in a number of publications, a good example of which is *The Handbook of the British Astronomical Association*, as well as many of the popular astronomy magazines such as the British monthly periodicals *Sky at Night* and *Astronomy Now* and the American monthly magazines *Astronomy* and *Sky & Telescope*.

# About Time

Before the late eighteenth century the biggest problem affecting mariners sailing the seas was finding their position. Latitude was easily determined by observing the altitude of the pole star above the northern horizon. Longitude, however, was far more difficult to measure. The inability of mariners to determine their longitude often led to them getting lost, and on many occasions shipwrecked. To address this problem King Charles II established the Royal Observatory at Greenwich in 1675 and from here, Astronomers Royal began the process of measuring and cataloguing the stars as they passed due south across the Greenwich meridian.

Now mariners only needed an accurate timepiece (the chronometer invented by Yorkshire-born clockmaker John Harrison) to display GMT (Greenwich Mean Time). Working out the local standard time onboard ship and subtracting this from GMT gave the ship's longitude (west or east) from the Greenwich meridian. Therefore mariners always knew where they were at sea and the longitude problem was solved.

Astronomers use a time scale called Universal Time (UT). This is equivalent to Greenwich Mean Time and is defined by the rotation of the Earth. The *Yearbook of Astronomy* gives all times in UT rather than in the local time for a particular city or country. Times are expressed using the 24-hour clock, with the day beginning at midnight, denoted by 00:00. Universal Time (UT) is related to local mean time by the formula:

Local Mean Time = UT − west longitude

In practice, small differences in longitude are ignored and the observer will use local clock time which will be the appropriate Standard (or Zone) Time. As the formula indicates, places in west longitude will have a Standard Time slow on UT, while those in east longitude will have a Standard Time fast on UT. As examples we have:

Standard Time in

| | |
|---|---|
| New Zealand | UT + 12 hours |
| Victoria, NSW | UT + 10 hours |
| Japan | UT + 9 hours |
| Western Australia | UT + 8 hours |
| India | UT + 5 hours 30 minutes |
| Pakistan | UT + 5 hours |
| Kenya | UT + 3 hours |
| South Africa | UT + 2 hours |
| British Isles | UT |
| Newfoundland Standard Time | UT − 3 hours 30 minutes |
| Atlantic Standard Time | UT − 4 hours |
| Eastern Standard Time | UT − 5 hours |
| Central Standard Time | UT − 6 hours |
| Mountain Standard Time | UT − 7 hours |
| Pacific Standard Time | UT − 8 hours |
| Alaska Standard Time | UT − 9 hours |
| Hawaii-Aleutian Standard Time | UT − 10 hours |

During the periods when Summer Time (also called Daylight Saving Time) is in use, one hour must be added to Standard Time to obtain the appropriate Summer/Daylight Saving Time. For example, Pacific Daylight Time is UT −7 hours.

# Using the Yearbook of Astronomy as an Observing Guide

## Notes on the Monthly Star Charts

The star charts on the following pages show the night sky throughout the year. There are two sets of charts, one for use by observers in the Northern Hemisphere and one for those in the Southern Hemisphere. The first set is drawn for latitude 52°N and can be used by observers in Europe, Canada and most of the United States. The second set is drawn for latitude 35°S and show the stars as seen from Australia and New Zealand. Twelve pairs of charts are provided for each of these latitudes.

Each pair of charts shows the entire sky as two semi-circular half-sky views, one looking north and the other looking south. A given pair of charts can be used at different times of year. For example, chart 1 shows the night sky at midnight on 21 December, but also at 2am on 21 January, 4am on 21 February and so forth. The accompanying table will enable you to select the correct chart for a given month and time of night. The caption next to each chart also lists the dates and times of night for which it is valid.

The charts are intended to help you find the more prominent constellations and other objects of interest mentioned in the monthly observing notes. To avoid the charts becoming too crowded, only stars of magnitude 4.5 or brighter are shown. This corresponds to stars that are bright enough to be seen from any dark suburban garden on a night when the Moon is not too close to full phase.

Each constellation is depicted by joining selected stars with lines to form a pattern. There is no official standard for these patterns, so you may occasionally find different patterns used in other popular astronomy books for some of the constellations.

Any map projection from a sphere onto a flat page will by necessity contain some distortions. This is true of star charts as well as maps of the Earth. The distortion on the half-sky charts is greatest near the semi-circular boundary of each chart, where it may appear to stretch constellation patterns out of shape.

The charts also show selected deep-sky objects such as galaxies, nebulae and star clusters. Many of these objects are too faint to be seen with the naked eye, and you will need binoculars or a telescope to observe them. Please refer to the table of deep-sky objects for more information.

## Planetary Apparition Diagrams

The diagrams of the apparitions of Mercury and Venus show the position of the respective planet in the sky at the moment of sunrise or sunset throughout the entire apparition. Two sets of positions are plotted on each chart: for latitude 52° North (blue line) and for latitude 35° South (red line). A thin dotted line denotes the portion of the apparition which falls outside the year covered by this edition of the *Yearbook*. A white dot indicates the position of Venus on the first day of each month, or of Mercury on the first, eleventh and 21st of the month. The day of greatest elongation (GE) is also marked by a white dot. Note that the dots do NOT indicate the magnitude of the planet.

Mars is in direct motion throughout the whole of 2026, conjunction of the planet occurring at the beginning of the year. The finder chart for Mars shows the path of the red planet from 1 May through to the end of December. The ecliptic is denoted by the dotted line running across the centre of the chart from left to right. The position of Mars is indicated on the 1st of each month as well as at opposition and at stationary points in Right Ascension. Note that the dots do NOT indicate the magnitude of Mars.

The finder charts for Jupiter, Saturn, Uranus and Neptune show the paths of the planets throughout the year. The position of each planet is indicated at opposition and at stationary points, as well as the start and end of the year and on the 1st of each month (1st of April, July and October only for Uranus and Neptune) where these dates do not fall too close to an event that is already marked. Stars are shown to magnitude 5.5 on the charts for Jupiter and Saturn. On the Uranus chart, stars are shown to magnitude 8; on the Neptune chart, the limiting magnitude is 10. In both cases, this is approximately two magnitudes fainter than the planet itself. Right Ascension and Declination scales are shown for the epoch J2000 to allow comparison with modern star charts. Note that the sizes of the dots denoting the planets do NOT indicate their magnitudes.

## Selecting the Correct Charts

The table below shows which of the charts to use for particular dates and times throughout the year and will help you to select the correct pair of half-sky charts for any combination of month and time of night.

The Earth takes 23 hours 56 minutes (and 4 seconds) to rotate once around its axis with respect to the fixed stars. Because this is around four minutes shorter than a full 24 hours, the stars appear to rise and set about 4 minutes earlier on each successive day, or around an hour earlier each fortnight. Therefore, as well as showing the stars at 10pm (22h in 24-hour notation) on 21 January, chart 1 also depicts the sky at 9pm (21h) on 6 February, 8pm (20h) on 21 February and 7pm (19h) on 6 March.

The times listed do not include summer time (daylight saving time), so if summer time is in force you must subtract one hour to obtain standard time (GMT if you are in the United Kingdom) before referring to the chart. For example, to find the correct chart for mid-September in the northern hemisphere at 3am summer time, first of all subtract one hour to obtain 2am (2h) standard time. Then you can consult the table, where you will find that you should use chart 11.

The table does not indicate sunrise, sunset or twilight. In northern temperate latitudes, the sky is still light at 18h and 6h from April to September, and still light at 20h and 4h from May to August. In Australia and New Zealand, the sky is still light at 18h and 6h from October to March, and in twilight (with only bright stars visible) at 20h and 04h from November to January.

| Local Time | 18h | 20h | 22h | 0h | 2h | 4h | 6h |
|---|---|---|---|---|---|---|---|
| January | 11 | 12 | 1 | 2 | 3 | 4 | 5 |
| February | 12 | 1 | 2 | 3 | 4 | 5 | 6 |
| March | 1 | 2 | 3 | 4 | 5 | 6 | 7 |
| April | 2 | 3 | 4 | 5 | 6 | 7 | 8 |
| May | 3 | 4 | 5 | 6 | 7 | 8 | 9 |
| June | 4 | 5 | 6 | 7 | 8 | 9 | 10 |
| July | 5 | 6 | 7 | 8 | 9 | 10 | 11 |
| August | 6 | 7 | 8 | 9 | 10 | 11 | 12 |
| September | 7 | 8 | 9 | 10 | 11 | 12 | 1 |
| October | 8 | 9 | 10 | 11 | 12 | 1 | 2 |
| November | 9 | 10 | 11 | 12 | 1 | 2 | 3 |
| December | 10 | 11 | 12 | 1 | 2 | 3 | 4 |

## Legend to the Star Charts

| STARS | | DEEP-SKY OBJECTS | |
|---|---|---|---|
| Symbol | Magnitude | Symbol | Type of object |
| • | 0 or brighter | ✳ | Open star cluster |
| • | 1 | ◦ | Globular star cluster |
| • | 2 | □ | Nebula |
| • | 3 | ▦ | Cluster with nebula |
| · | 4 | ○ | Planetary nebula |
| · | 5 | ◌ | Galaxy |
| ✦ | Double star |  | Magellanic Clouds |
| ⊙ | Variable star | | |

## Star Names

There are over 200 stars with proper names, most of which are of Roman, Greek or Arabic origin although only a couple of dozen or so of these names are used regularly. Examples include Arcturus in Boötes, Castor and Pollux in Gemini and Rigel in Orion.

A system whereby Greek letters were assigned to stars was introduced by the German astronomer and celestial cartographer Johann Bayer in his star atlas *Uranometria*, published in 1603. Bayer's system is applied to the brighter stars within any particular constellation, which are given a letter from the Greek alphabet followed by the genitive case of the constellation in which the star is located. This genitive case is simply the Latin form meaning 'of' the constellation. Examples are the stars Alpha Boötis and Beta Centauri which translate literally as 'Alpha of Boötes' and 'Beta of the Centaur'.

As a general rule, the brightest star in a constellation is labelled Alpha ($\alpha$), the second brightest Beta ($\beta$), and the third brightest Gamma ($\gamma$) and so on, although there are some constellations where the system falls down. An example is Gemini where the principal star (Pollux) is designated Beta Geminorum, the second brightest (Castor) being known as Alpha Geminorum.

There are only 24 letters in the Greek alphabet, the consequence of which was that the fainter naked eye stars needed an alternative system of classification. The system in popular use is that devised by the first Astronomer Royal John Flamsteed in which the stars in each constellation are listed numerically in order from west to east. Although many of the brighter stars within any particular constellation will have both Greek letters and Flamsteed numbers, the latter are generally used only when a star does not have a Greek letter.

## The Greek Alphabet

| α | Alpha   | ι | Iota    | ρ | Rho     |
|---|---------|---|---------|---|---------|
| β | Beta    | κ | Kappa   | σ | Sigma   |
| γ | Gamma   | λ | Lambda  | τ | Tau     |
| δ | Delta   | μ | Mu      | υ | Upsilon |
| ε | Epsilon | ν | Nu      | φ | Phi     |
| ζ | Zeta    | ξ | Xi      | χ | Chi     |
| η | Eta     | ο | Omicron | ψ | Psi     |
| θ | Theta   | π | Pi      | ω | Omega   |

## The Names of the Constellations

On clear, dark, moonless nights, the sky seems to teem with stars although in reality you can never see more than a couple of thousand or so at any one time when looking with the unaided eye. Each and every one of these stars belongs to a particular constellation, although the constellations that we see in the sky, and which grace the pages of star atlases, are nothing more than chance alignments. The stars that make up the constellations are often situated at vastly differing distances from us and only appear close to each other, and form the patterns that we see, because they lie in more or less the same direction as each other as seen from Earth.

A large number of the constellations are named after mythological characters, and were given their names thousands of years ago. However, those star groups lying close to the south celestial pole were discovered by Europeans only during the last few centuries, many of these by explorers and astronomers who mapped the stars during their journeys to lands under southern skies. This resulted in many of the newer constellations having modern-sounding names, such as Octans (the Octant) and Microscopium (the Microscope), both of which were devised by the French astronomer Nicolas Louis De La Caille during the early 1750s.

Over the centuries, many different suggestions for new constellations have been put forward by astronomers who, for one reason or another, felt the need to add new groupings to star charts and to fill gaps between the traditional constellations. Astronomers drew up their own charts of the sky, incorporating their new groups into them. A number of these new constellations had cumbersome names, notable examples including Officina Typographica (the Printing Shop) introduced by the German astronomer Johann Bode in 1801; Sceptrum Brandenburgicum (the Sceptre of Brandenburg) introduced by the German astronomer Gottfried Kirch in 1688; Taurus Poniatovii (Poniatowski's Bull) introduced by the Polish-Lithuanian astronomer Martin Odlanicky Poczobut in 1777; and Quadrans Muralis (the Mural Quadrant) devised by the French astronomer Joseph-Jerôme de Lalande in 1795. Although these have long since been rejected, the latter has been immortalised by the annual Quadrantid meteor shower, the radiant of which lies in an area of sky formerly occupied by Quadrans Muralis.

During the 1920s the International Astronomical Union (IAU) systemised matters by adopting an official list of 88 accepted constellations, each with official spellings and abbreviations. Precise boundaries for each constellation were then drawn up so that every point in the sky belonged to a particular constellation.

The abbreviations devised by the IAU each have three letters which in the majority of cases are the first three letters of the constellation name, such as AND for Andromeda, EQU for Equuleus, HER for Hercules, ORI for Orion and so on. This trend is not strictly adhered to in cases where confusion may arise. This happens with the two constellations Leo (abbreviated LEO) and Leo Minor (abbreviated LMI). Similarly, because Triangulum (TRI) may be mistaken for Triangulum Australe, the latter is abbreviated TRA. Other instances occur with Sagitta (SGE) and Sagittarius (SGR) and with Canis Major (CMA) and Canis Minor (CMI) where the first two letters from the second names of the constellations are used. This is also the case with Corona Australis (CRA) and Corona Borealis (CRB) where the first letter of the second name of each constellation is incorporated. Finally, mention must be made of Crater (CRT) which has been abbreviated in such a way as to avoid confusion with the aforementioned CRA (Corona Australis).

The table shown on the following pages contains the name of each of the 88 constellations together with the translation and abbreviation of the constellation name. The constellations depicted on the monthly star charts are identified with their abbreviations rather than the full constellation names.

# The Constellations

| Andromeda | Andromeda | AND |
|---|---|---|
| Antlia | The Air Pump | ANT |
| Apus | The Bird of Paradise | APS |
| Aquarius | The Water Carrier | AQR |
| Aquila | The Eagle | AQL |
| Ara | The Altar | ARA |
| Aries | The Ram | ARI |
| Auriga | The Charioteer | AUR |
| Boötes | The Herdsman | BOO |
| Caelum | The Graving Tool | CAE |
| Camelopardalis | The Giraffe | CAM |
| Cancer | The Crab | CNC |
| Canes Venatici | The Hunting Dogs | CVN |
| Canis Major | The Great Dog | CMA |
| Canis Minor | The Little Dog | CMI |
| Capricornus | The Goat | CAP |
| Carina | The Keel | CAR |
| Cassiopeia | Cassiopeia | CAS |
| Centaurus | The Centaur | CEN |
| Cepheus | Cepheus | CEP |
| Cetus | The Whale | CET |
| Chamaeleon | The Chameleon | CHA |
| Circinus | The Pair of Compasses | CIR |
| Columba | The Dove | COL |
| Coma Berenices | Berenice's Hair | COM |
| Corona Australis | The Southern Crown | CRA |
| Corona Borealis | The Northern Crown | CRB |
| Corvus | The Crow | CRV |
| Crater | The Cup | CRT |
| Crux | The Cross | CRU |
| Cygnus | The Swan | CYG |
| Delphinus | The Dolphin | DEL |
| Dorado | The Goldfish | DOR |
| Draco | The Dragon | DRA |
| Equuleus | The Foal | EQU |
| Eridanus | The River | ERI |
| Fornax | The Furnace | FOR |
| Gemini | The Twins | GEM |
| Grus | The Crane | GRU |
| Hercules | Hercules | HER |
| Horologium | The Pendulum Clock | HOR |
| Hydra | The Water Snake | HYA |
| Hydrus | The Lesser Water Snake | HYI |
| Indus | The Indian | IND |
| Lacerta | The Lizard | LAC |
| Leo | The Lion | LEO |
| Leo Minor | The Lesser Lion | LMI |
| Lepus | The Hare | LEP |
| Libra | The Scales | LIB |
| Lupus | The Wolf | LUP |
| Lynx | The Lynx | LYN |
| Lyra | The Lyre | LYR |
| Mensa | The Table Mountain | MEN |
| Microscopium | The Microscope | MIC |
| Monoceros | The Unicorn | MON |
| Musca | The Fly | MUS |
| Norma | The Level | NOR |
| Octans | The Octant | OCT |
| Ophiuchus | The Serpent Bearer | OPH |
| Orion | Orion | ORI |
| Pavo | The Peacock | PAV |
| Pegasus | Pegasus | PEG |
| Perseus | Perseus | PER |

| | | | | | | |
|---|---|---|---|---|---|---|
| Phoenix | The Phoenix | PHE | Sextans | The Sextant | SEX |
| Pictor | The Painter's Easel | PIC | Taurus | The Bull | TAU |
| Pisces | The Fish | PSC | Telescopium | The Telescope | TEL |
| Piscis Austrinus | The Southern Fish | PSA | Triangulum | The Triangle | TRI |
| Puppis | The Stern | PUP | Triangulum Australe | The Southern Triangle | TRA |
| Pyxis | The Mariner's Compass | PYX | Tucana | The Toucan | TUC |
| Reticulum | The Net | RET | Ursa Major | The Great Bear | UMA |
| Sagitta | The Arrow | SGE | Ursa Minor | The Little Bear | UMI |
| Sagittarius | The Archer | SGR | Vela | The Sail | VEL |
| Scorpius | The Scorpion | SCO | Virgo | The Virgin | VIR |
| Sculptor | The Sculptor | SCL | Volans | The Flying Fish | VOL |
| Scutum | The Shield | SCT | Vulpecula | The Fox | VUL |
| Serpens Caput and Cauda | The Serpent | SER | | | |

# The Monthly Star Charts

# Northern Hemisphere Star Charts

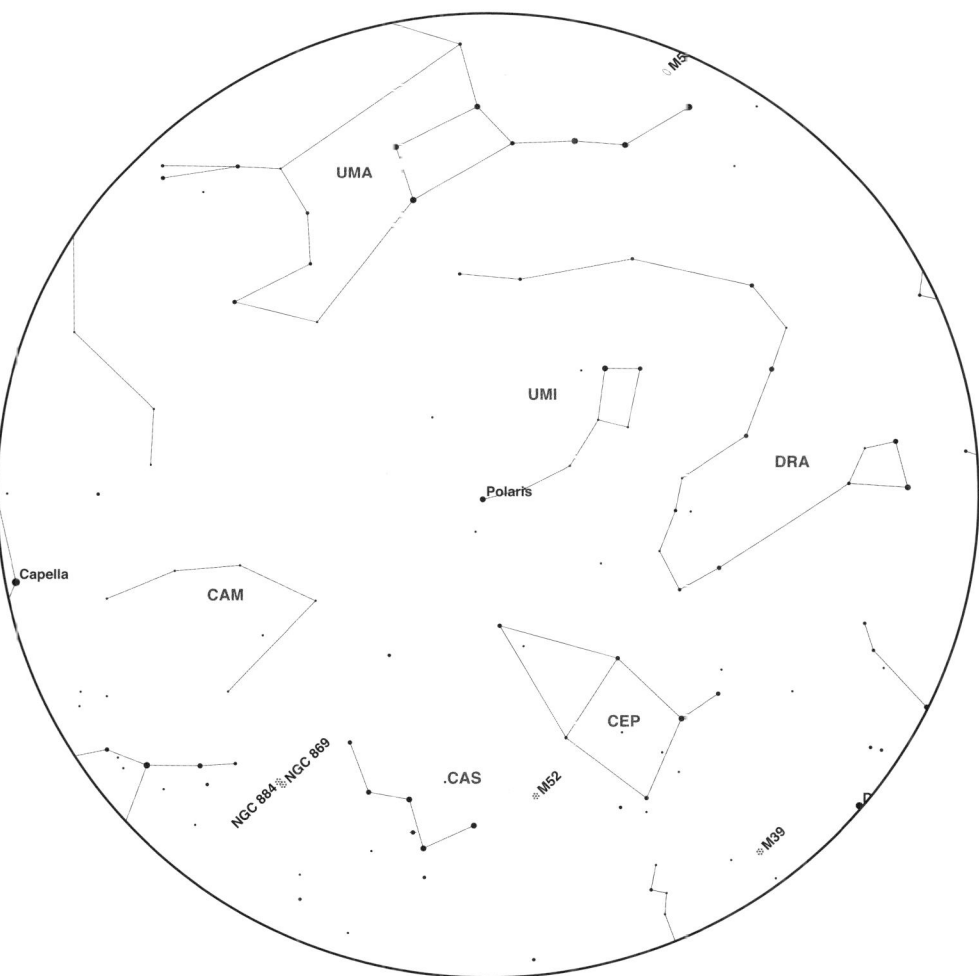

This chart shows stars lying at declinations between +45 and +90 degrees. These constellations are circumpolar for observers in Europe and North America.

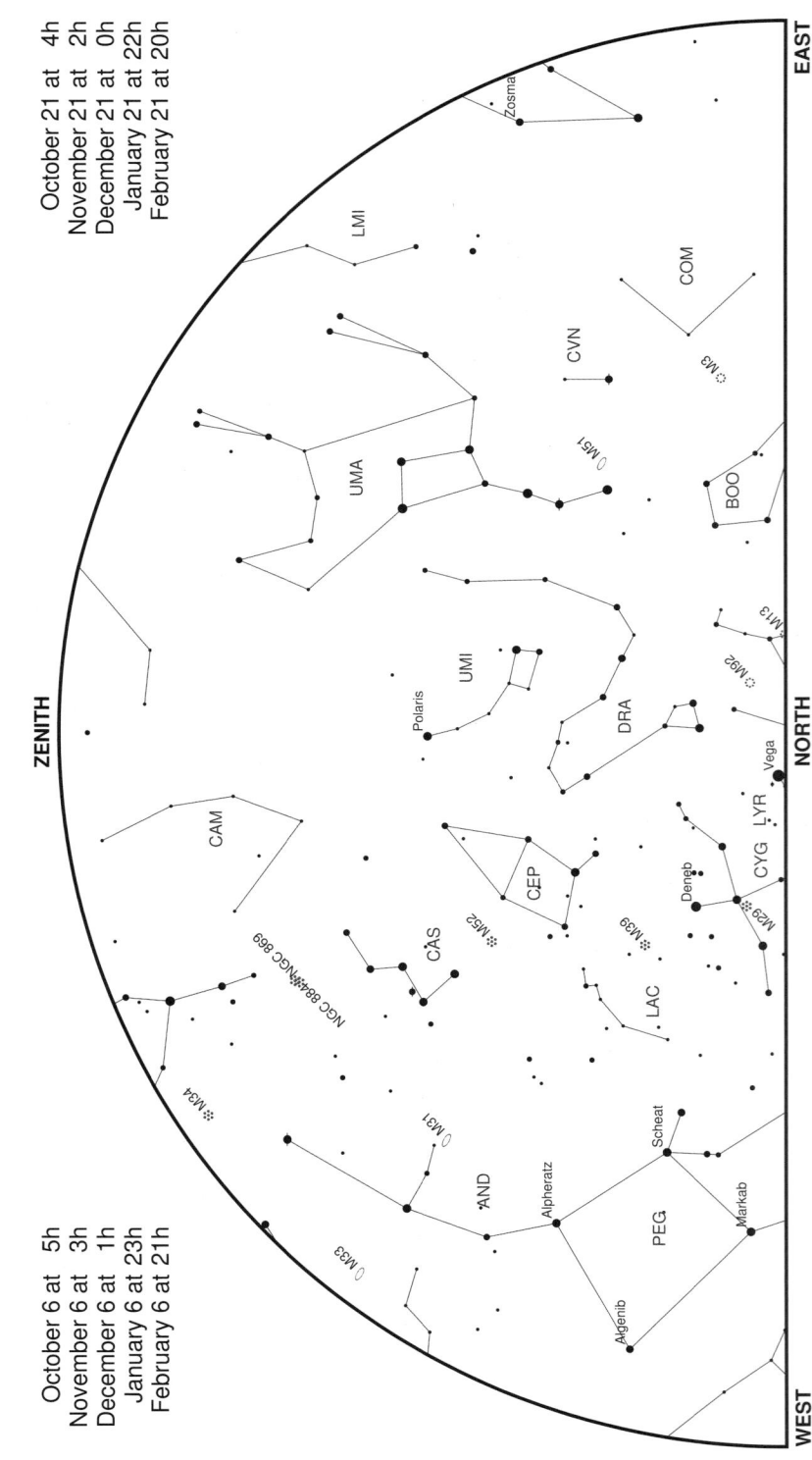

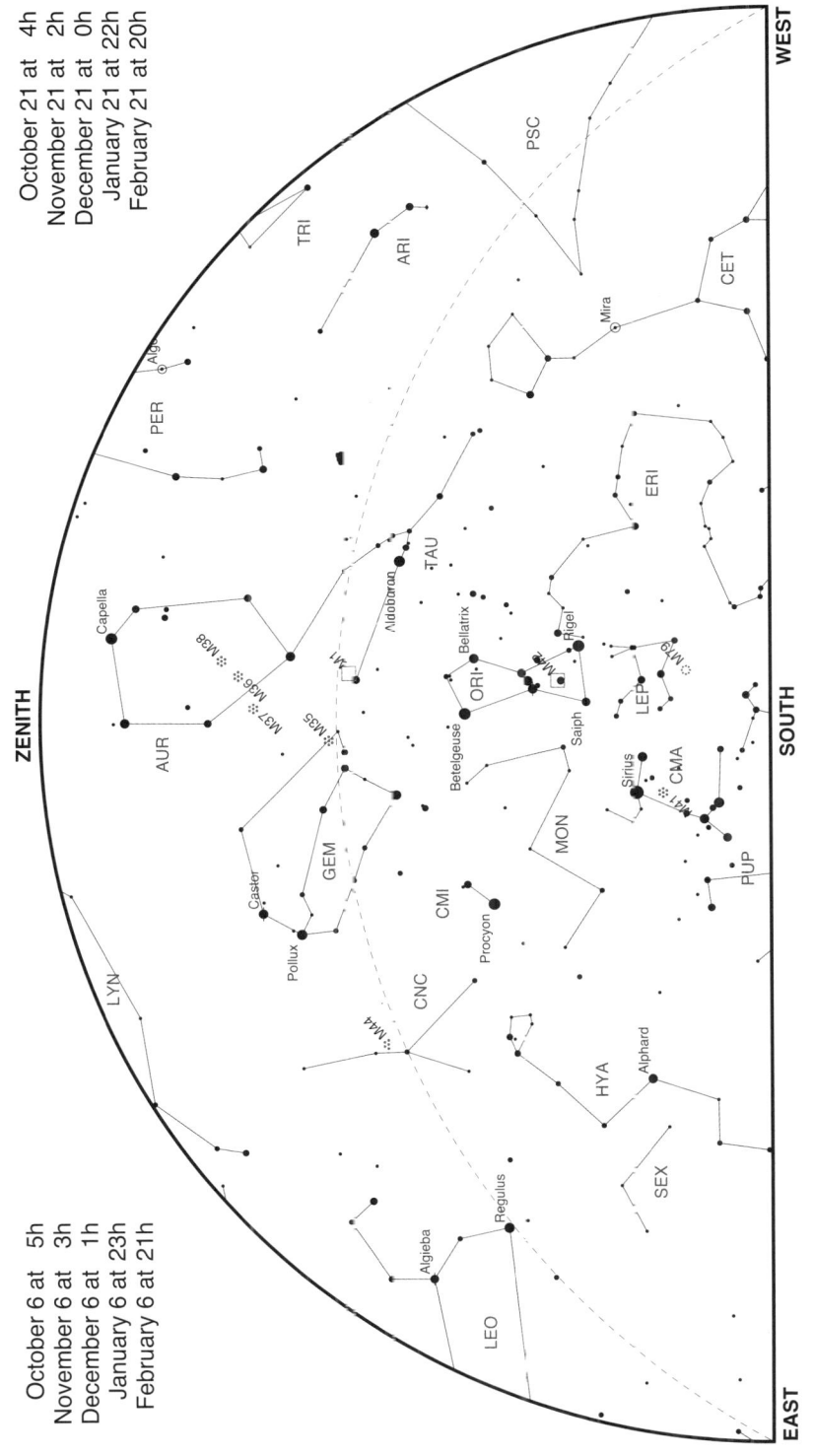

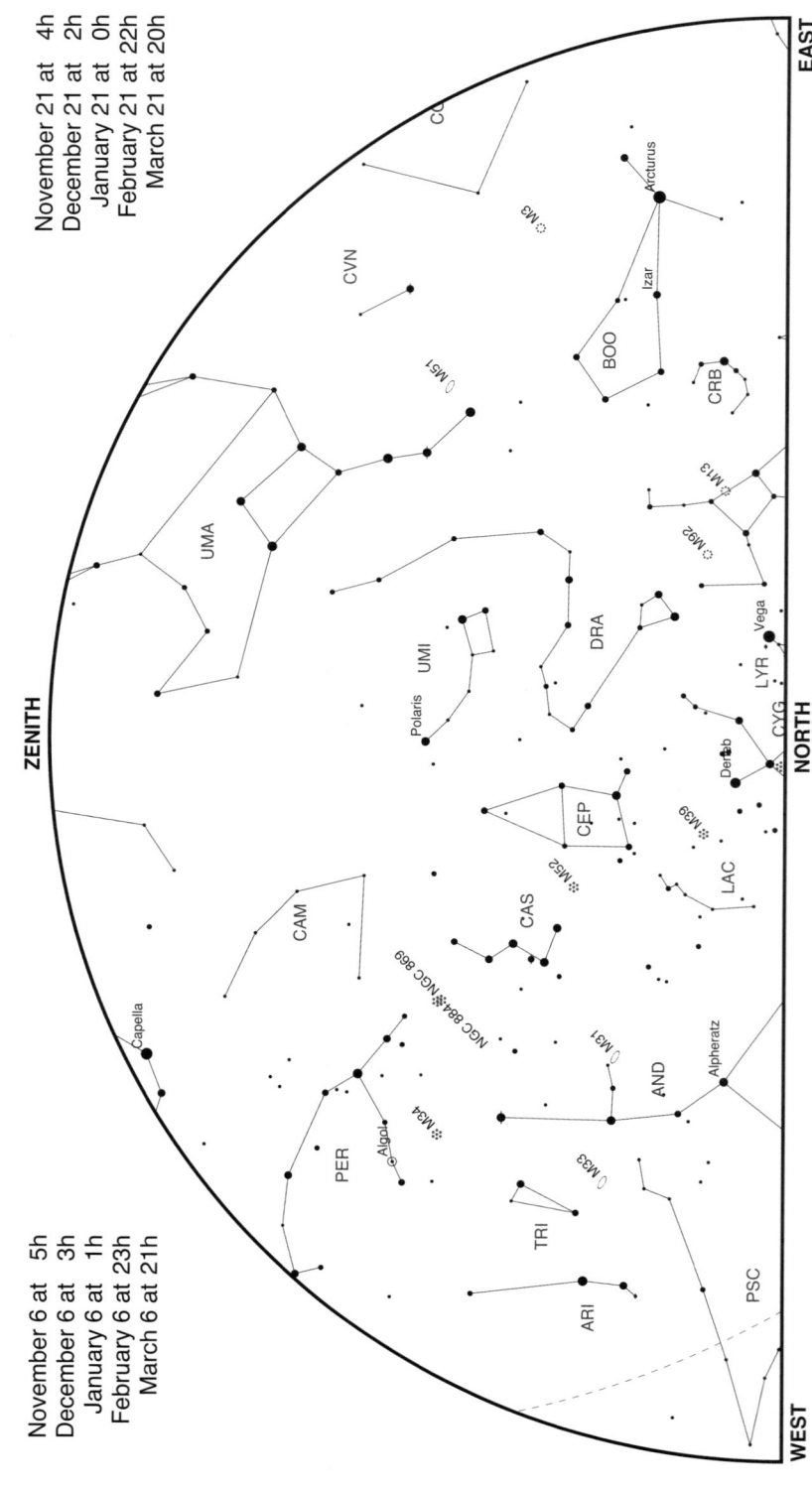

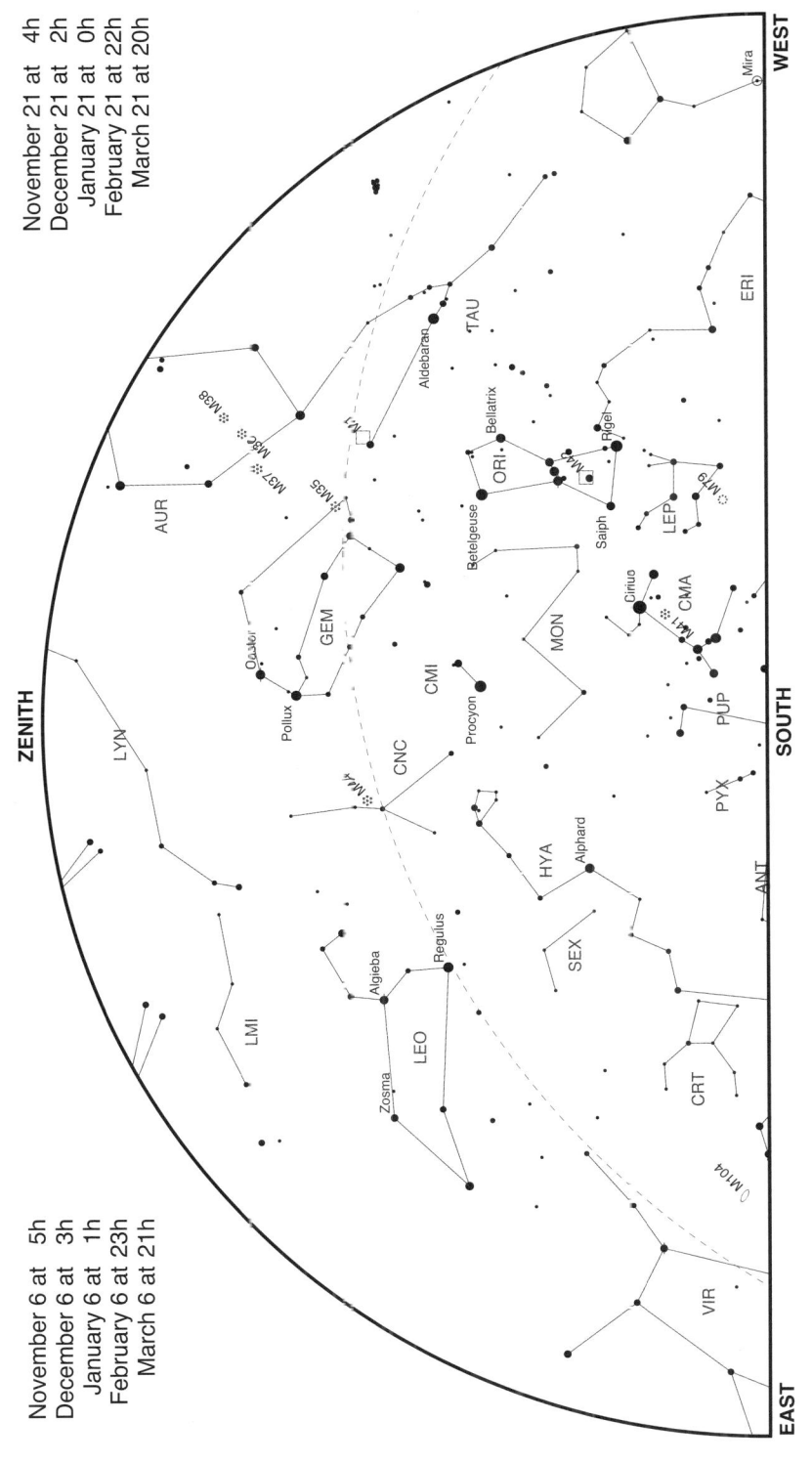

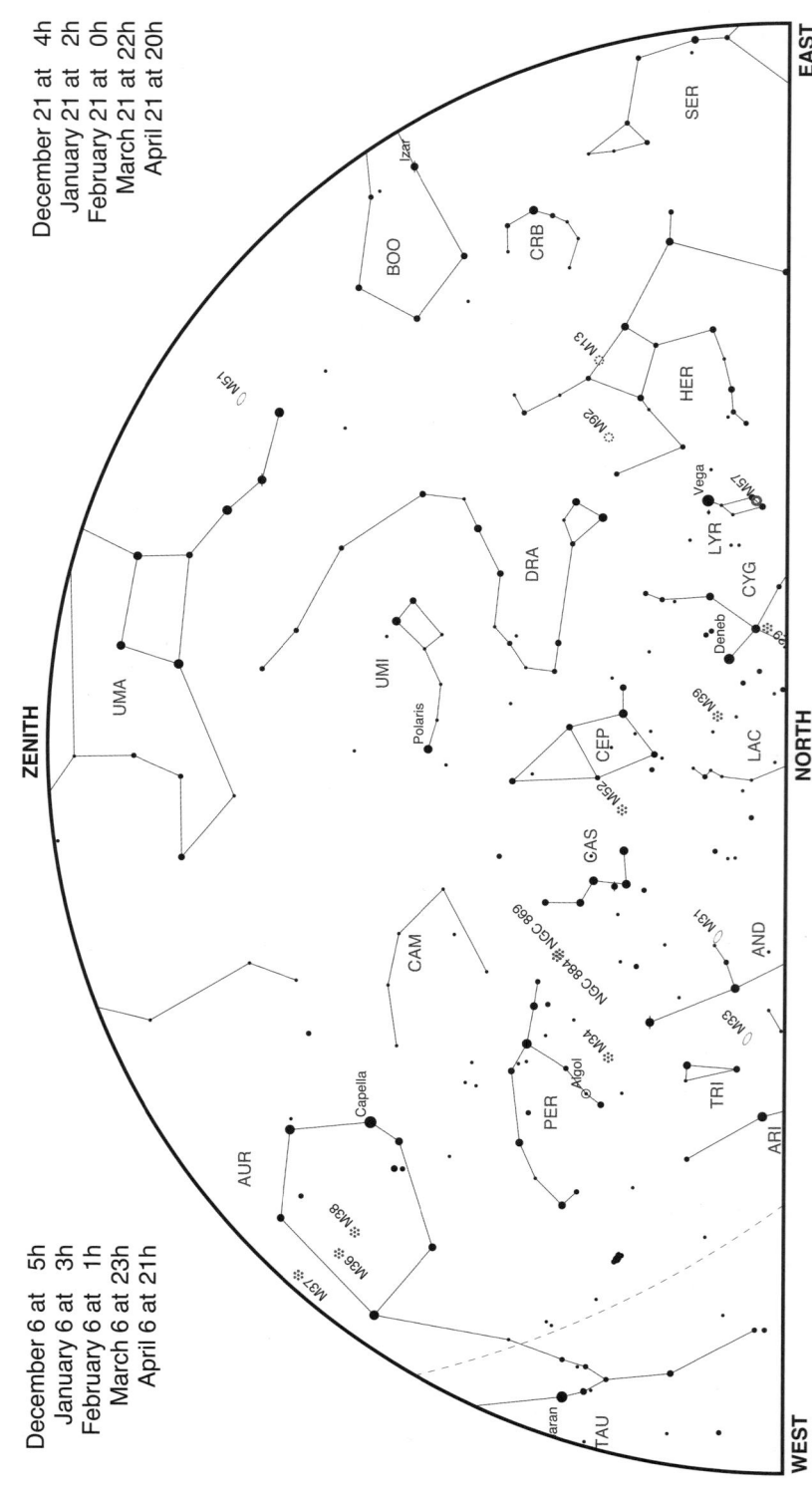

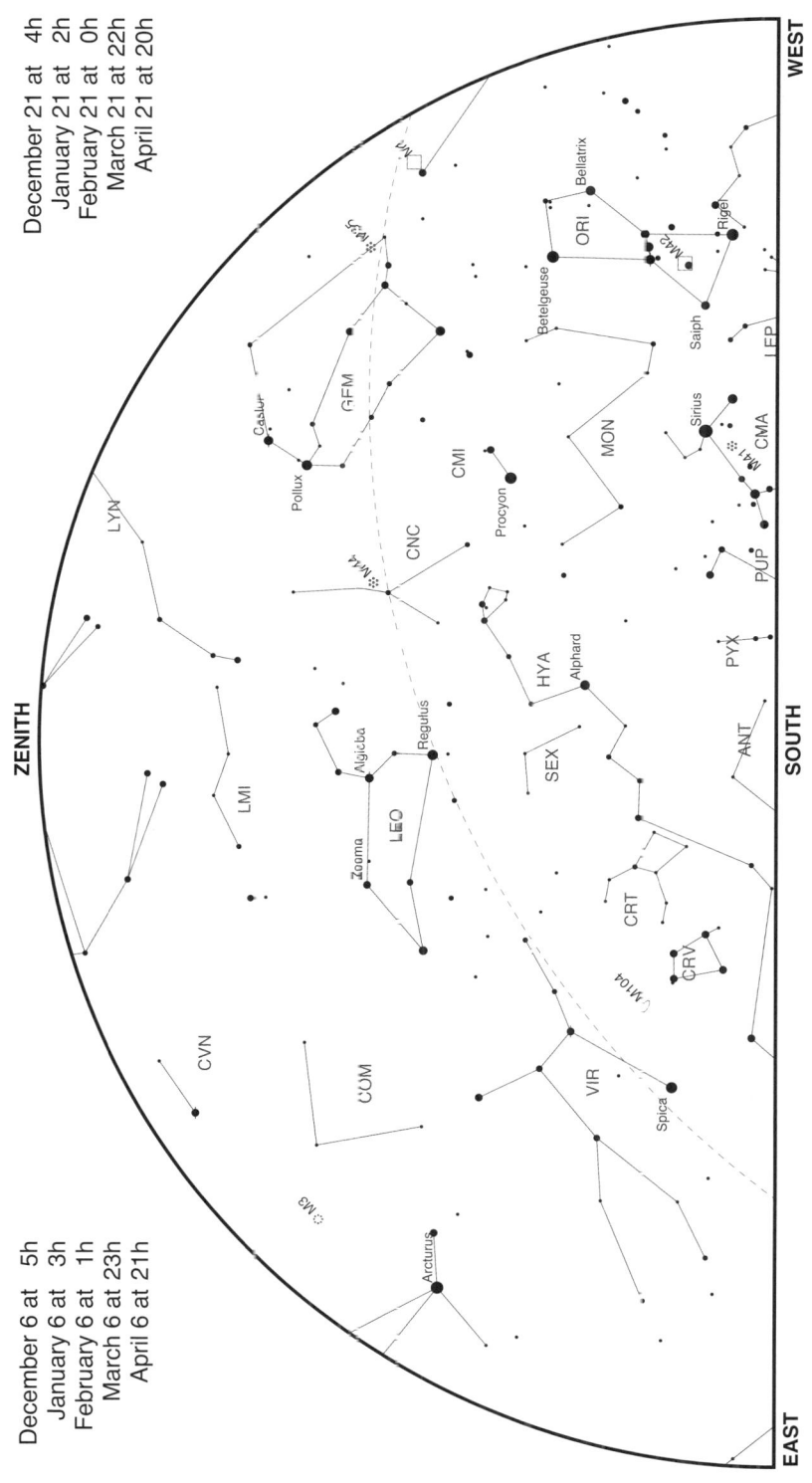

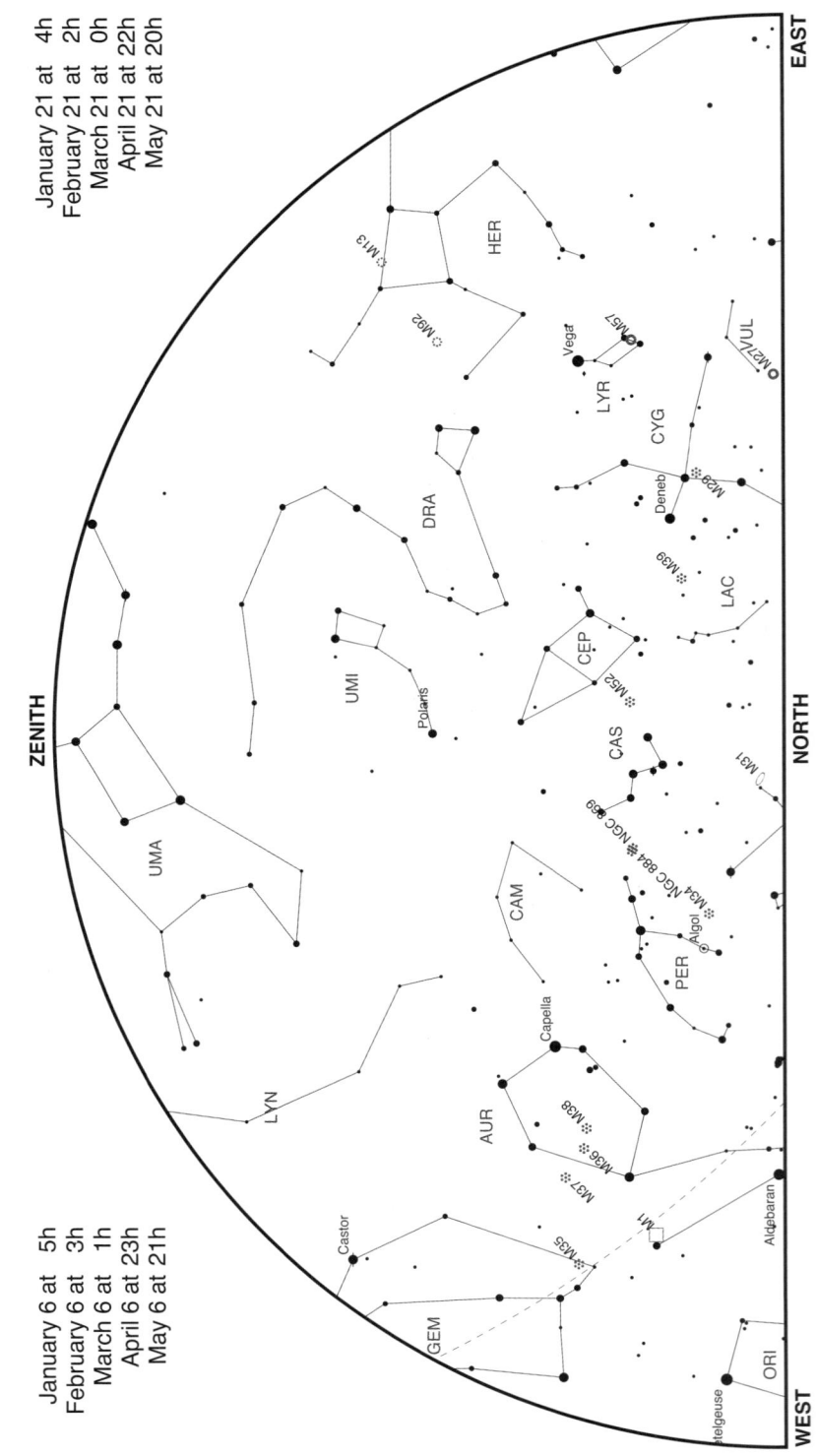

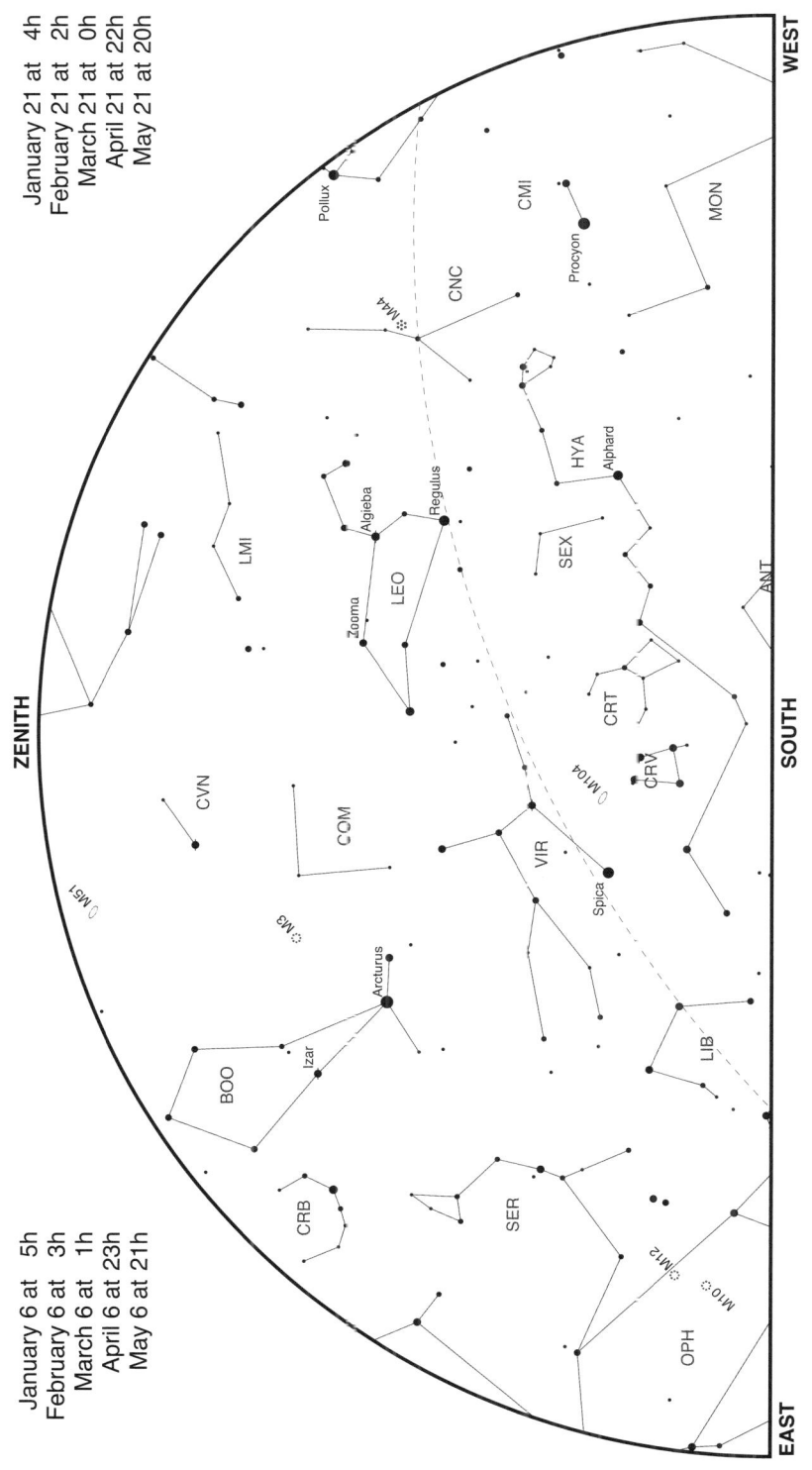

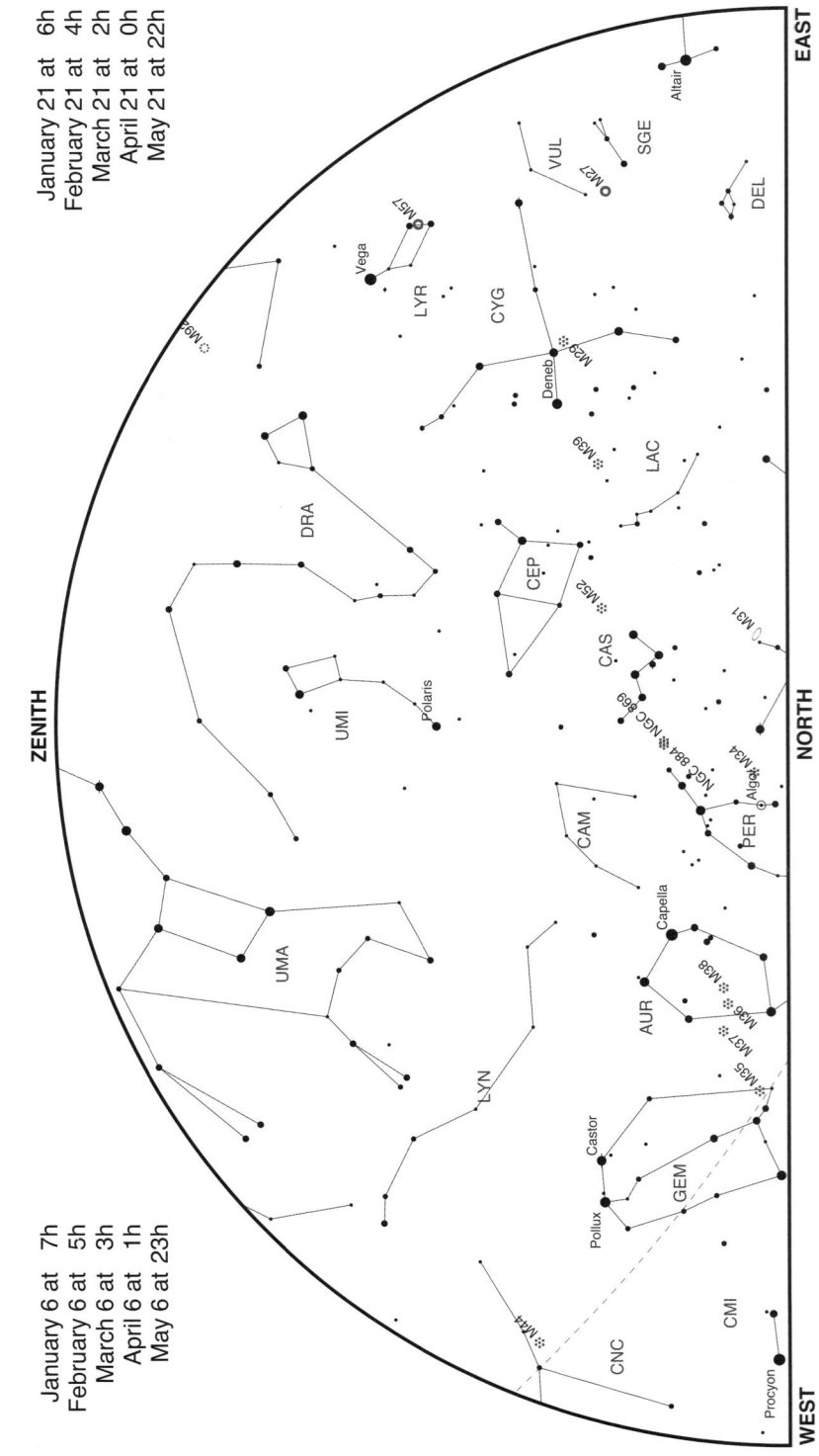

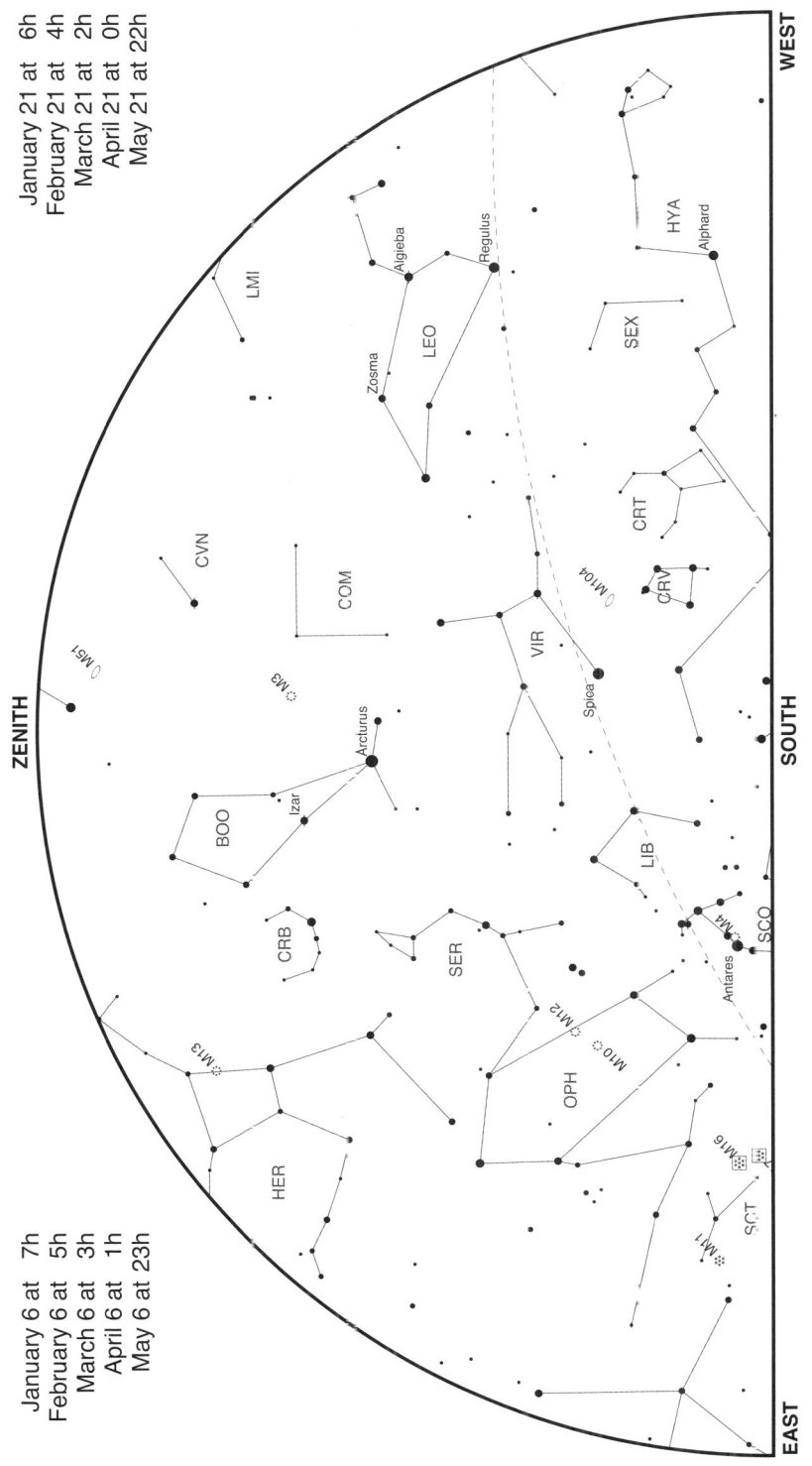

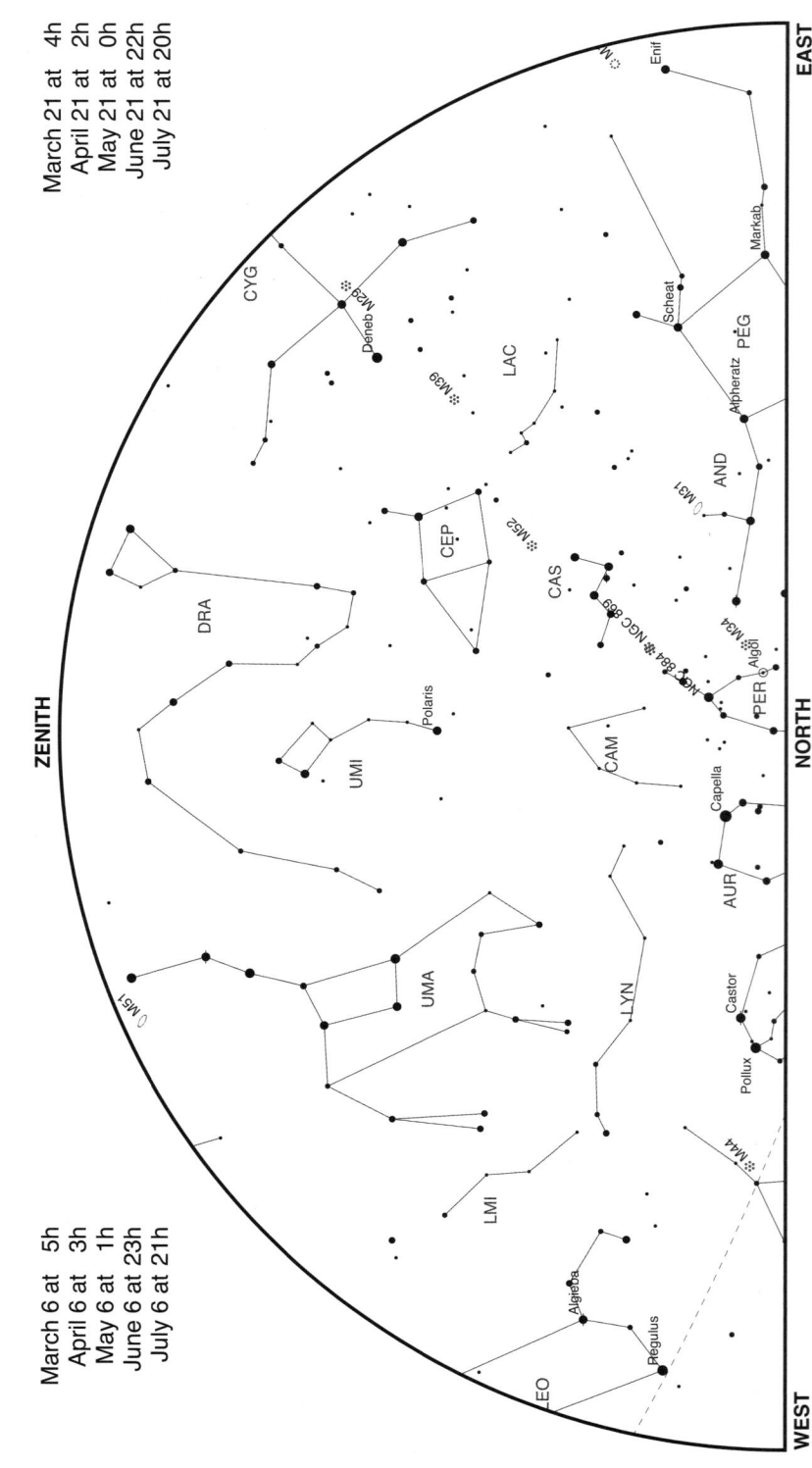

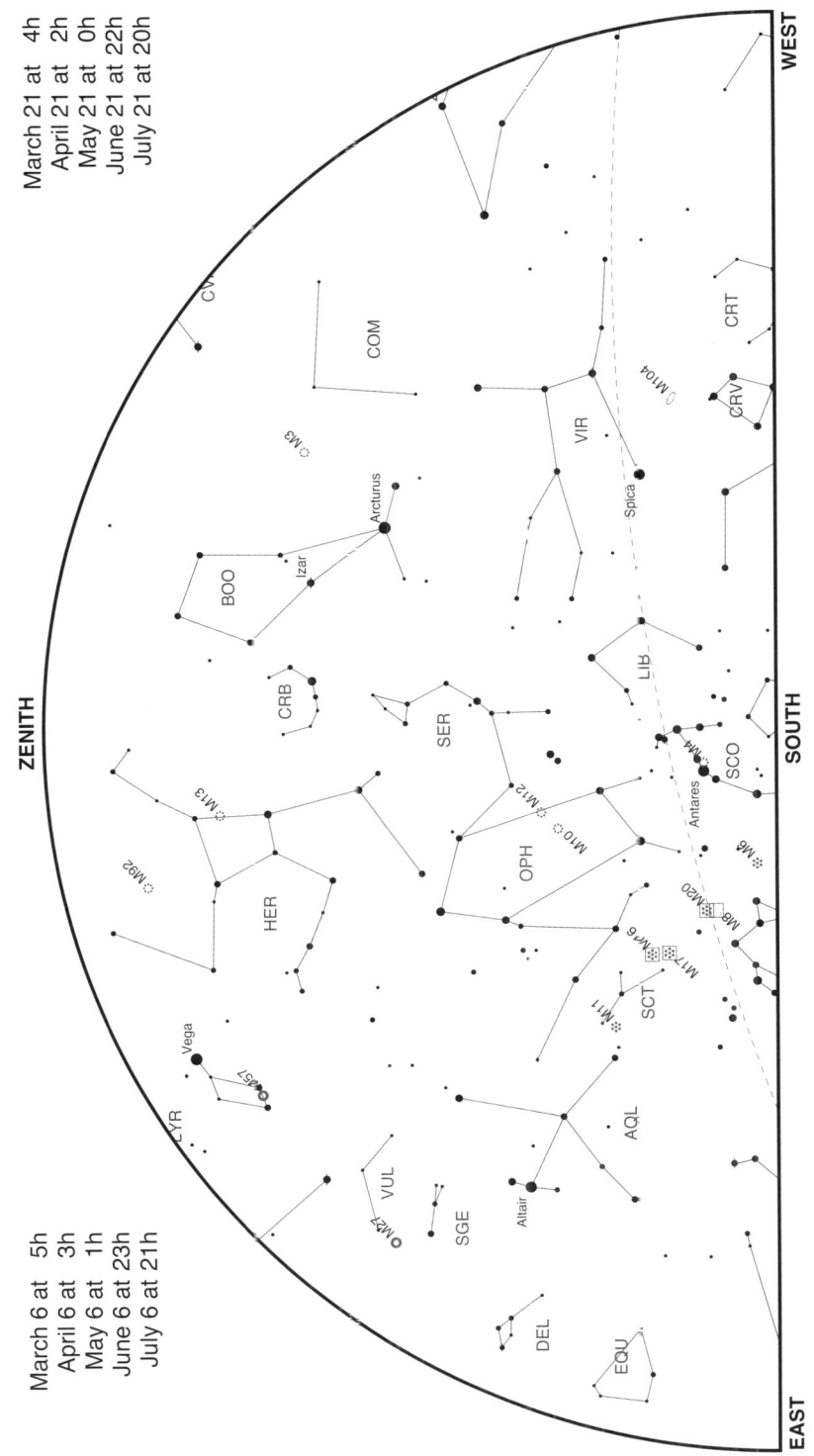

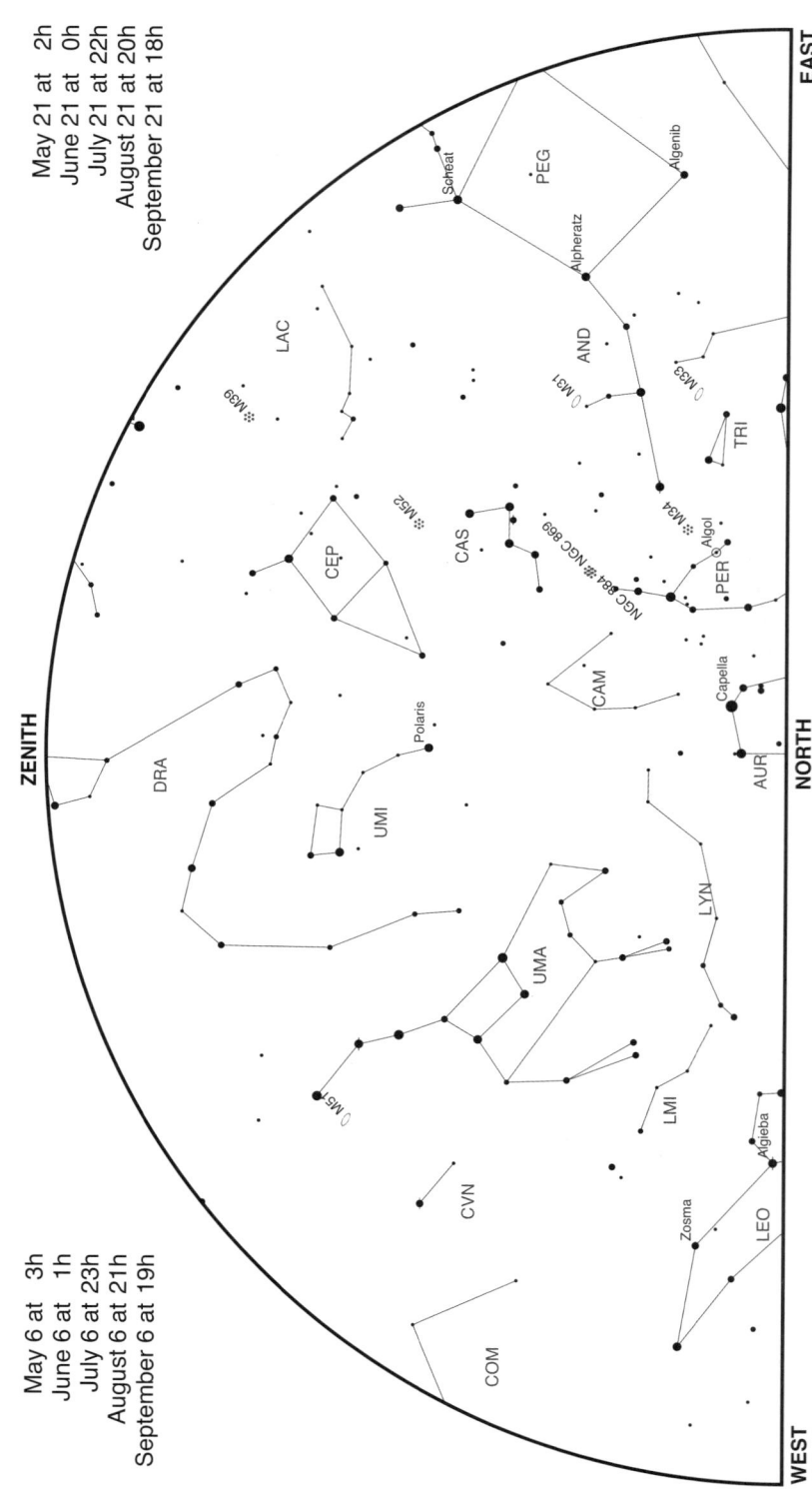

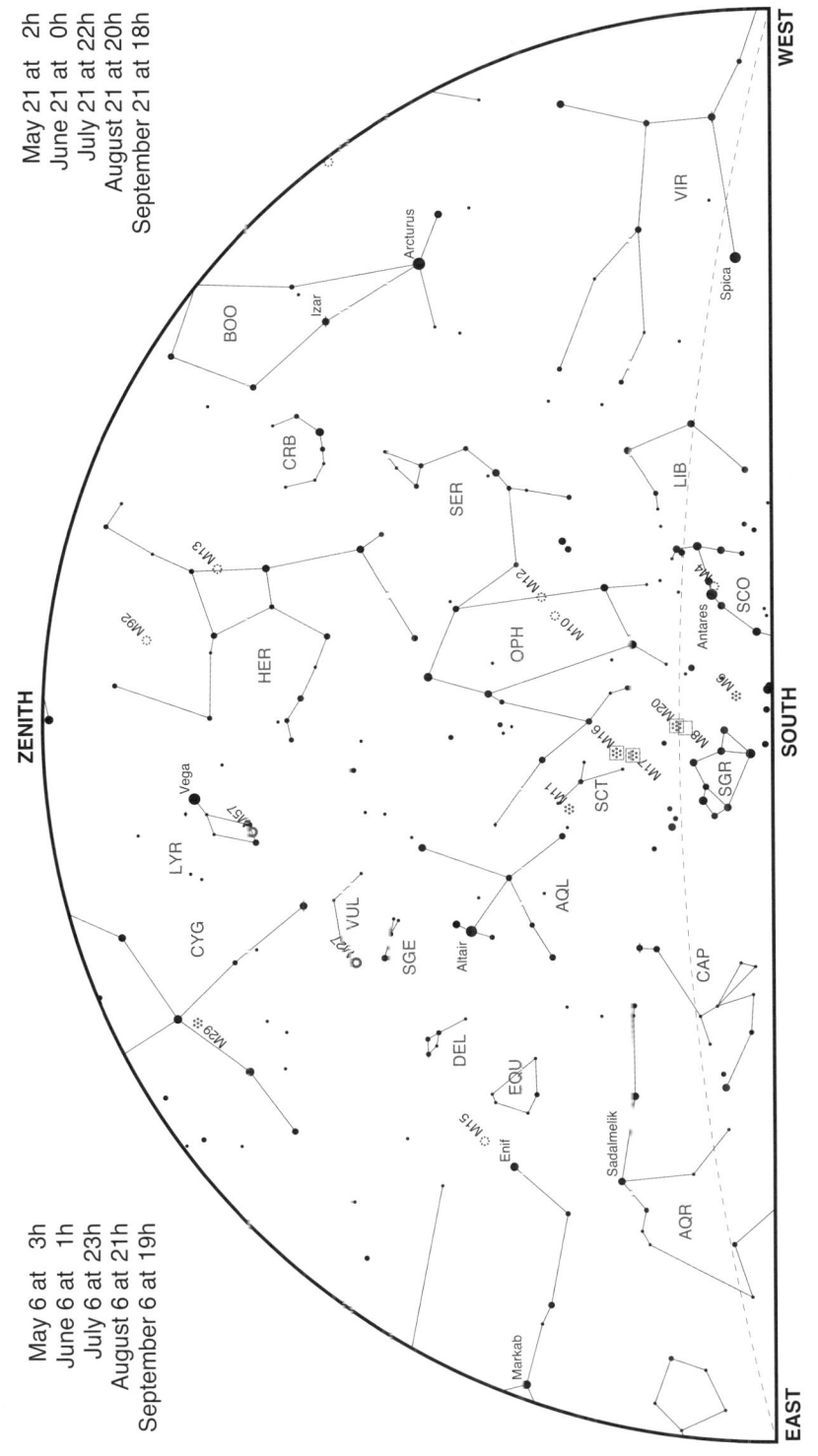

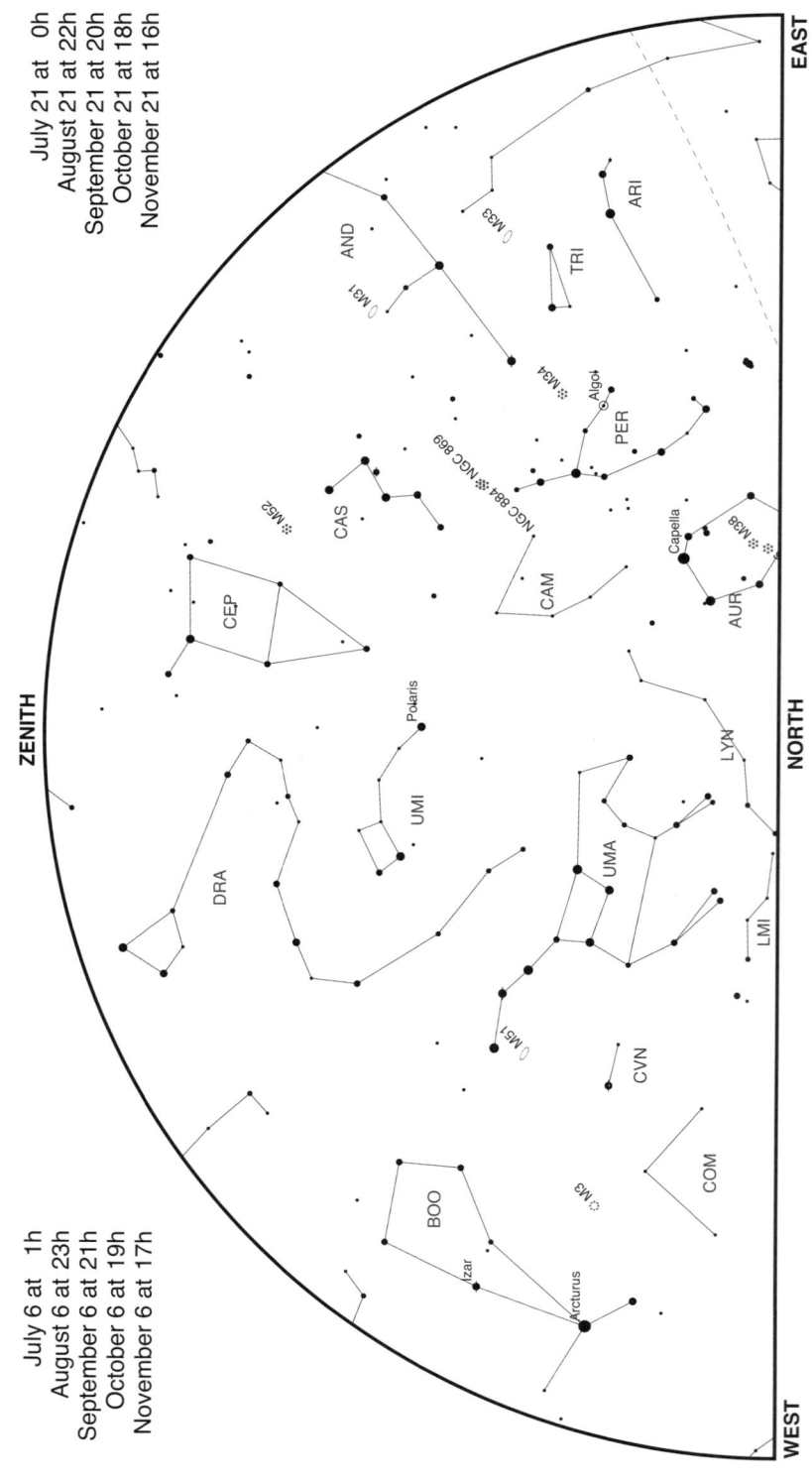

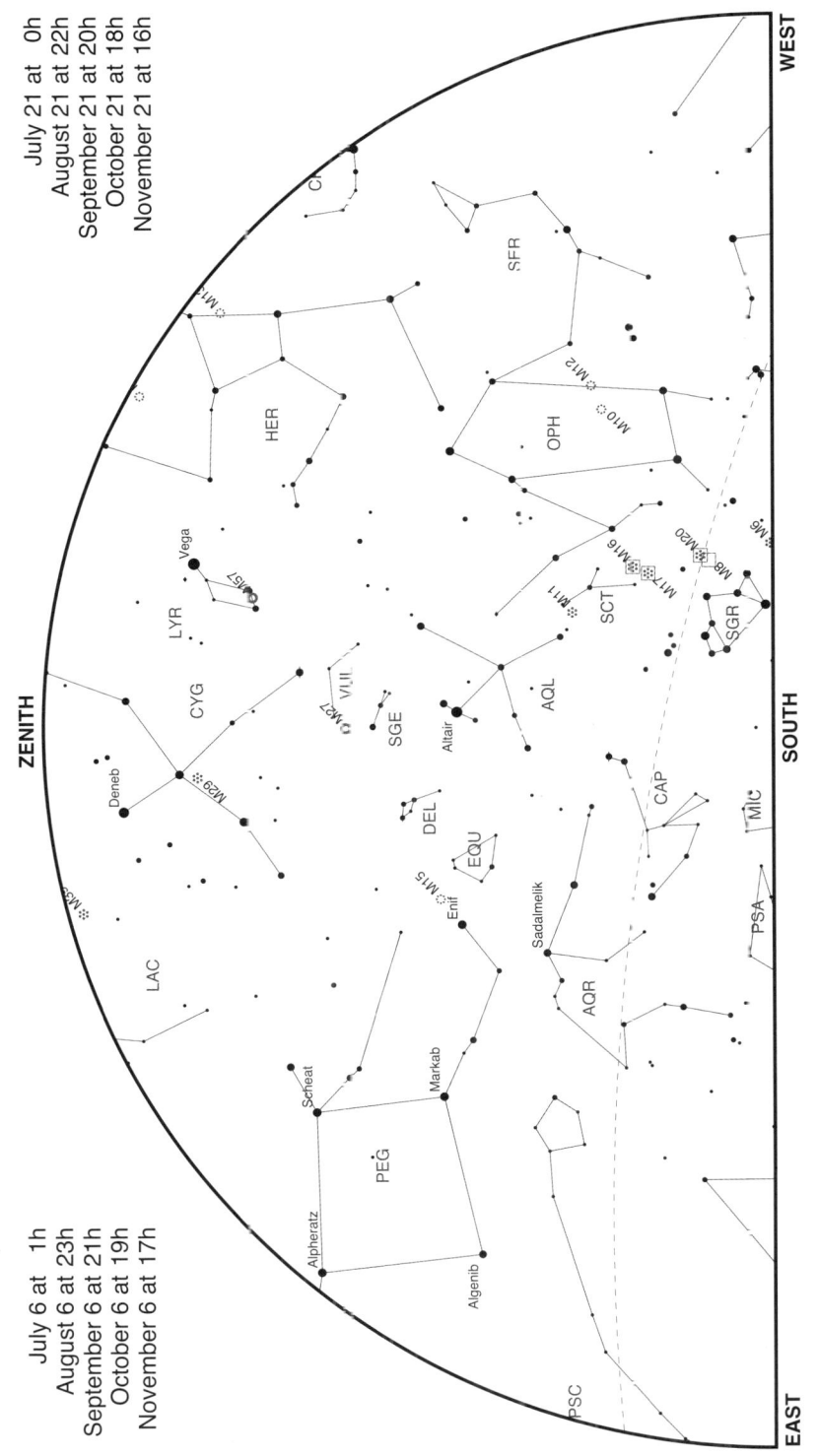

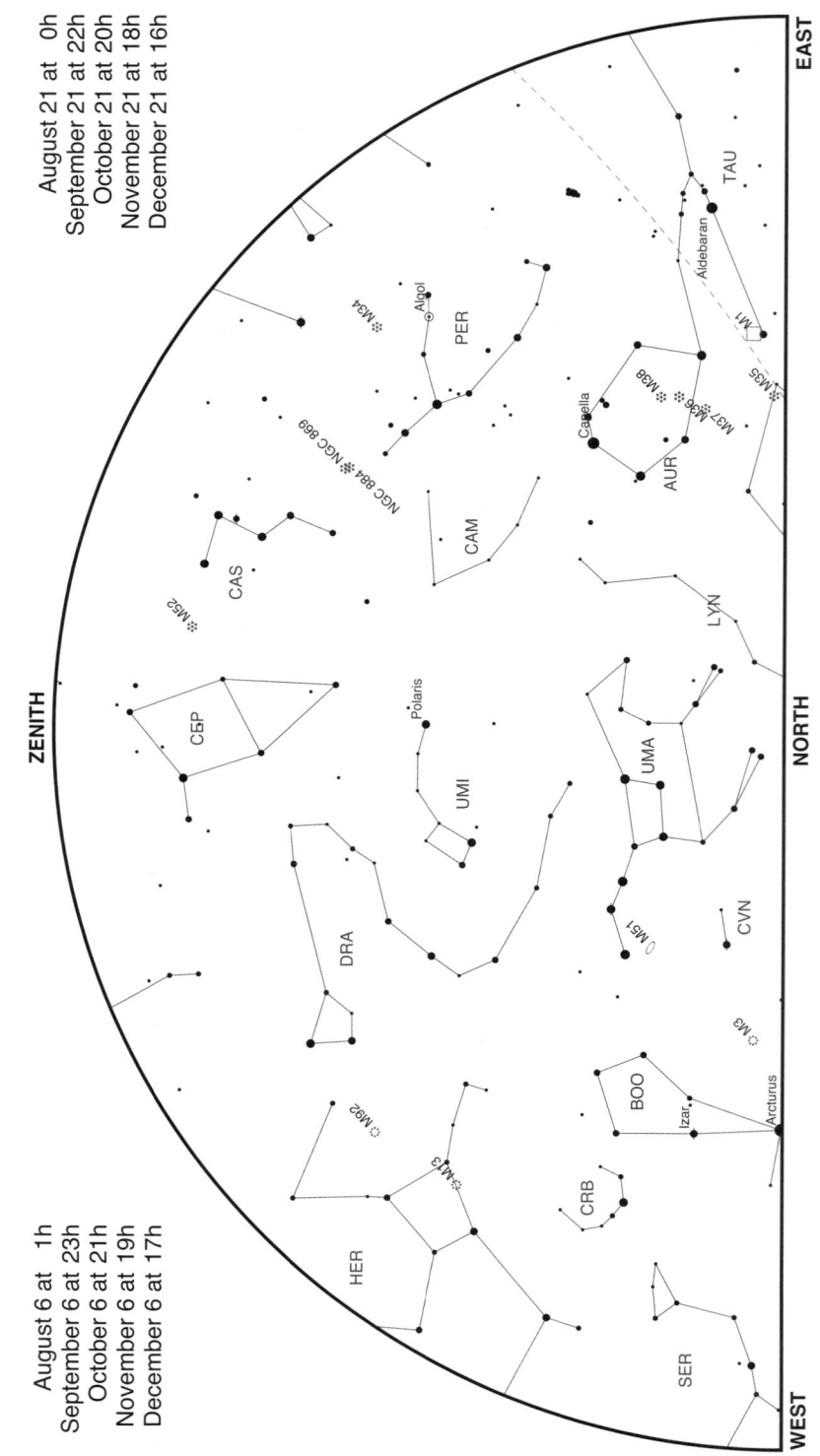

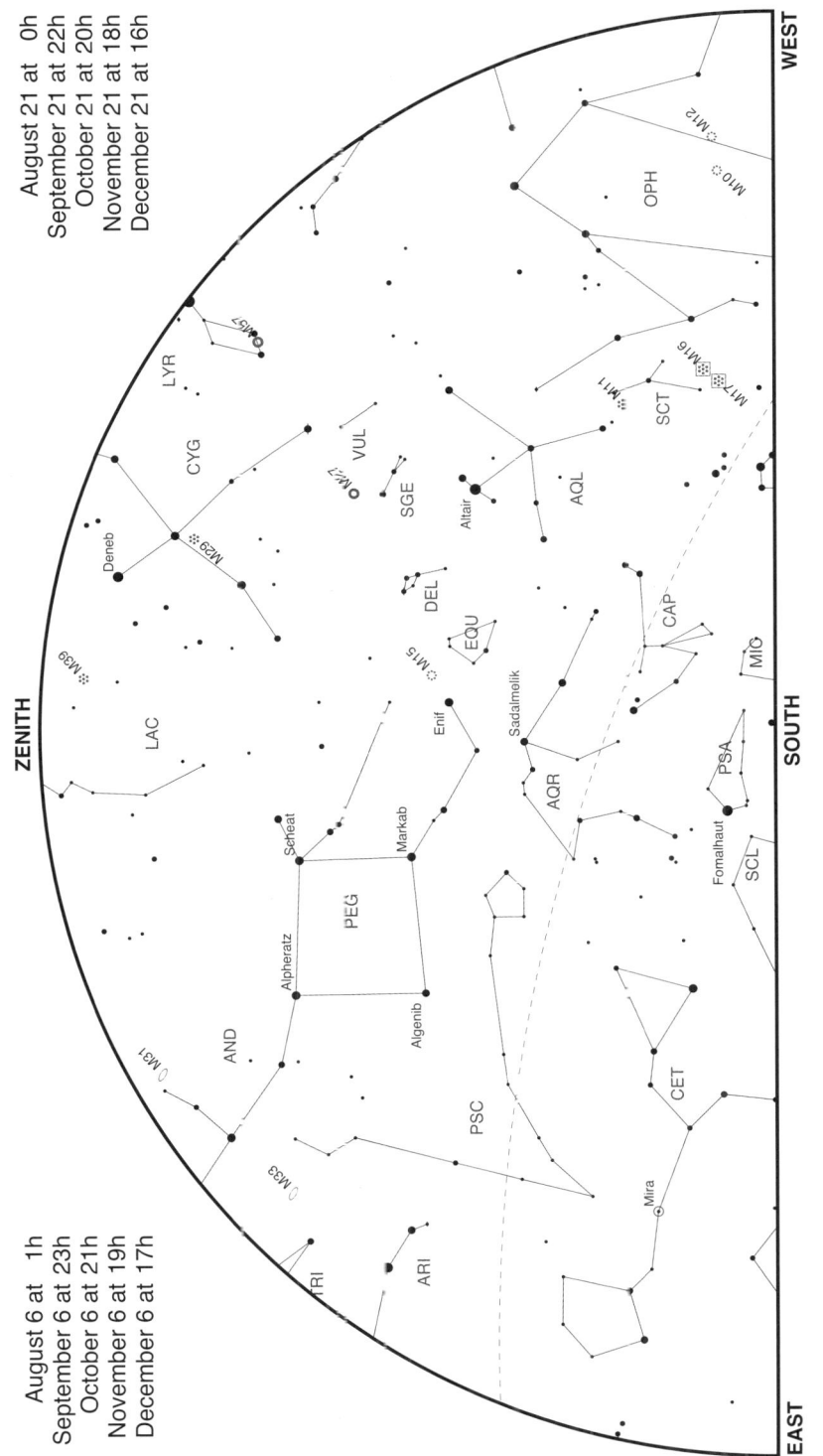

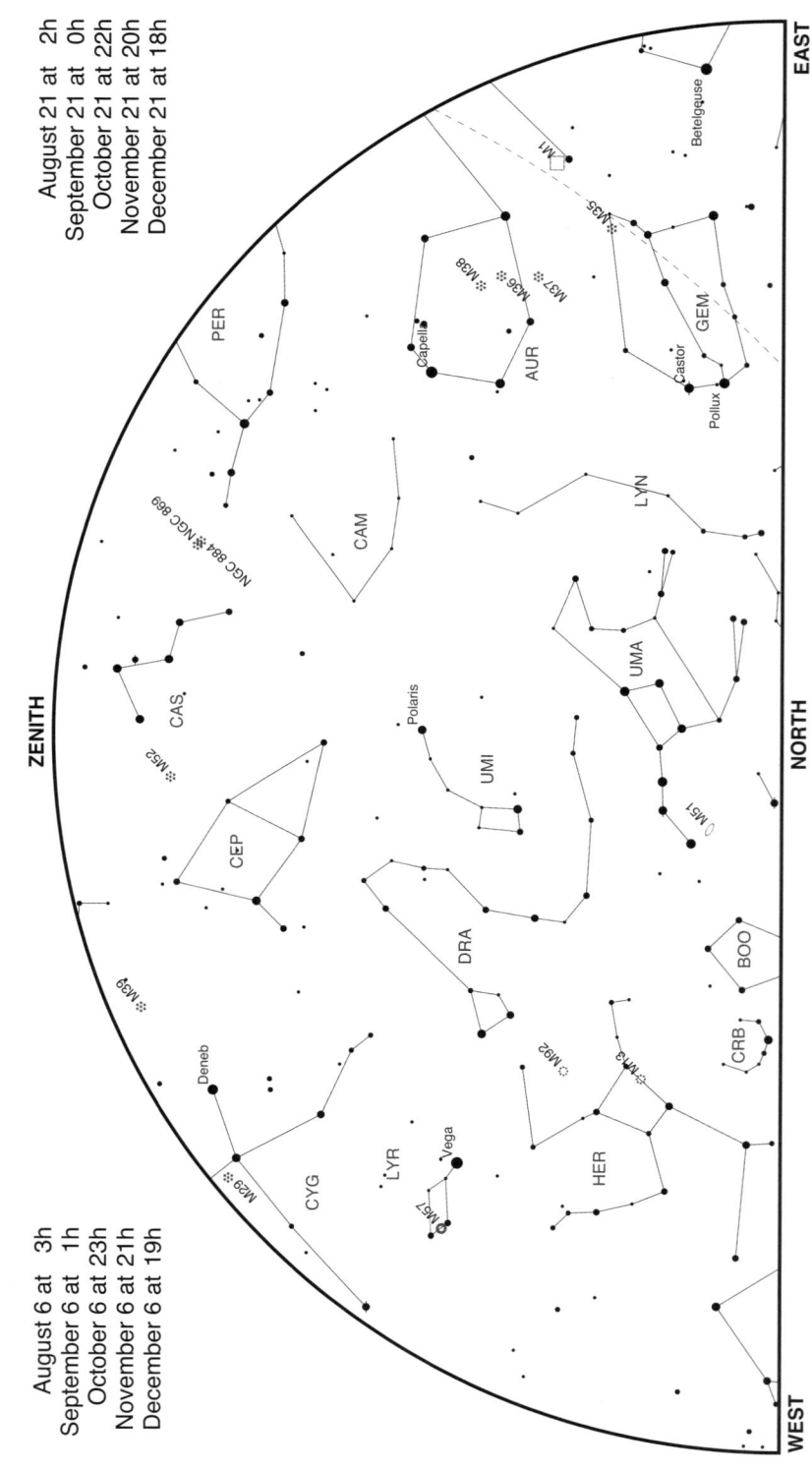

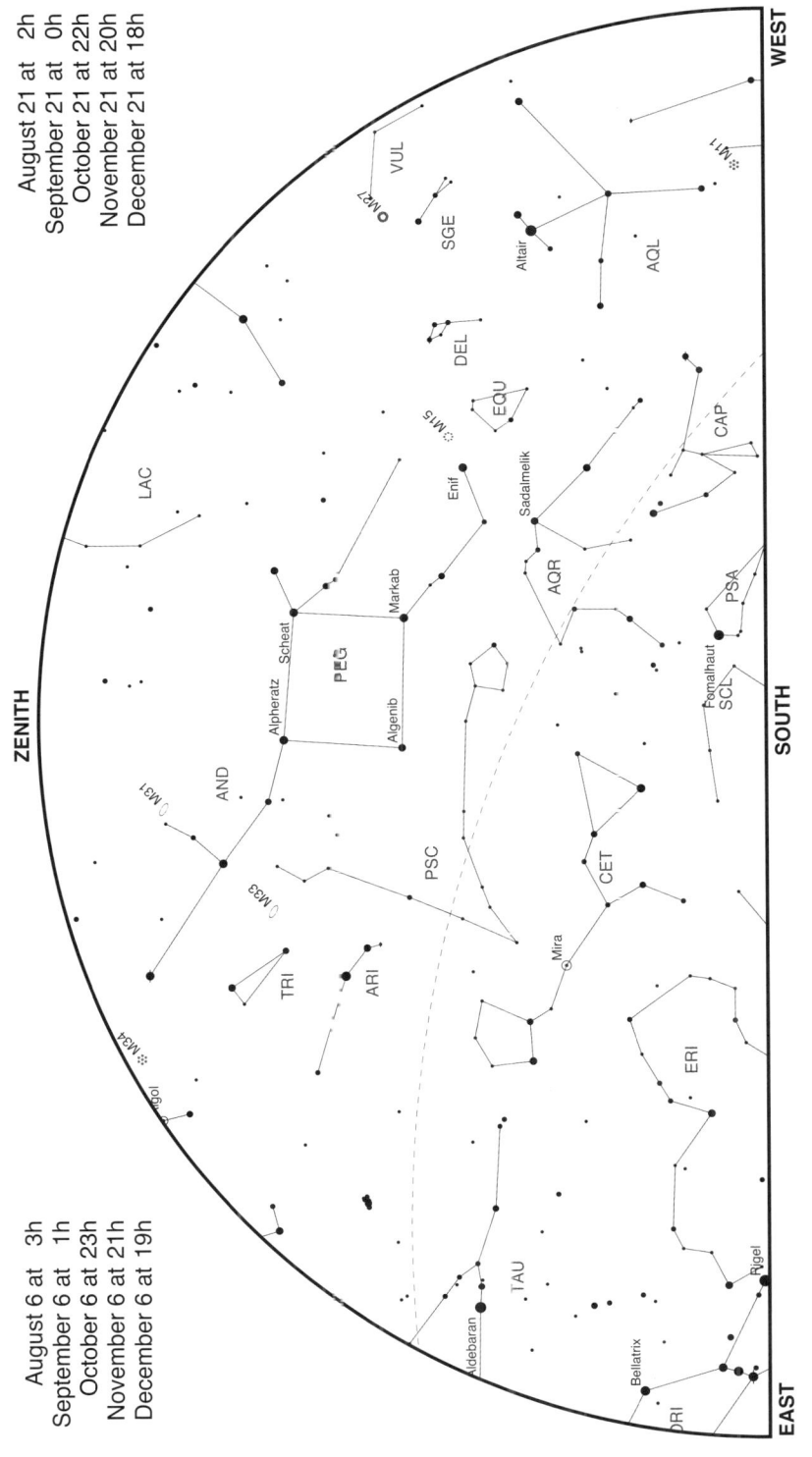

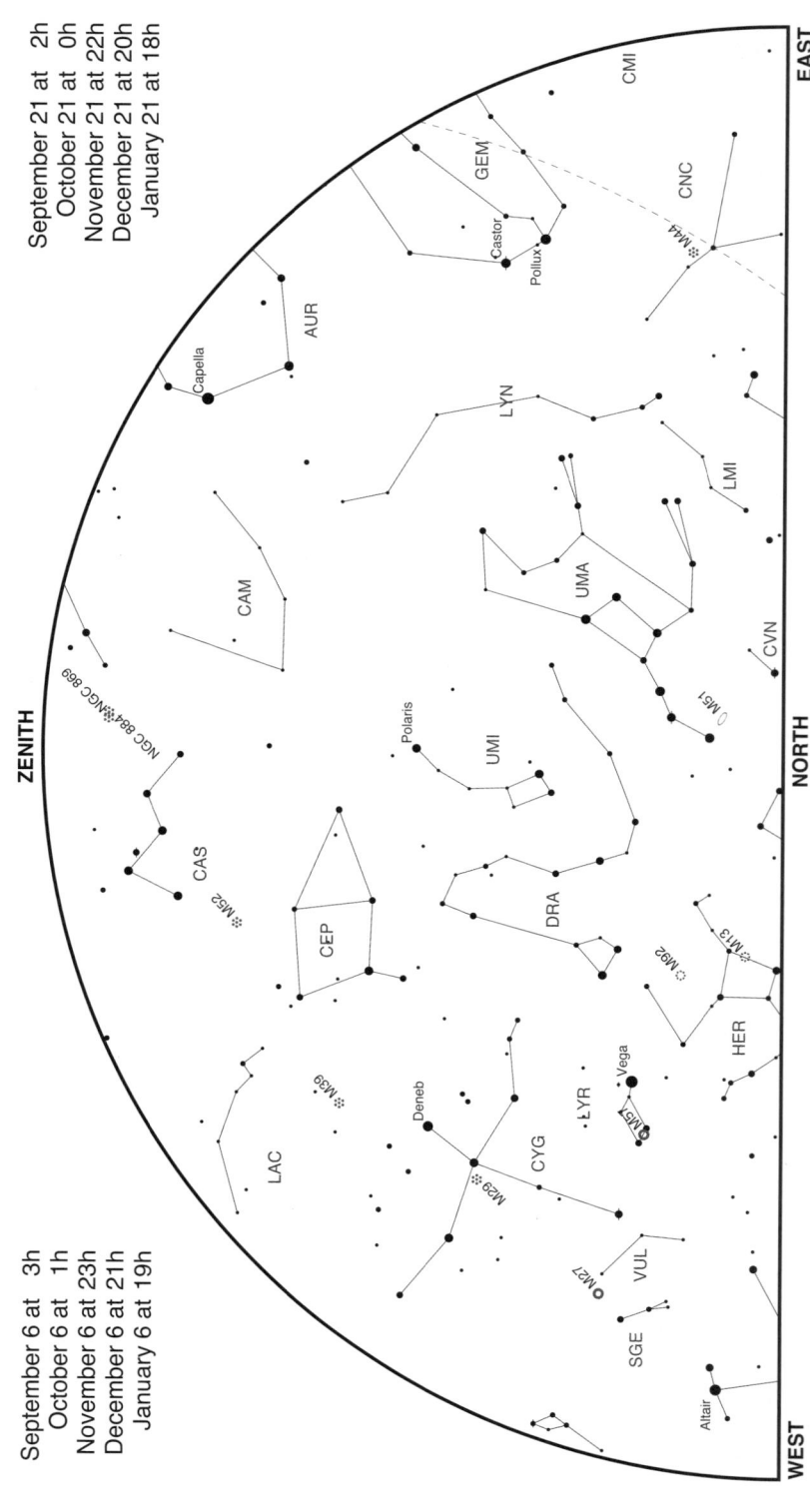

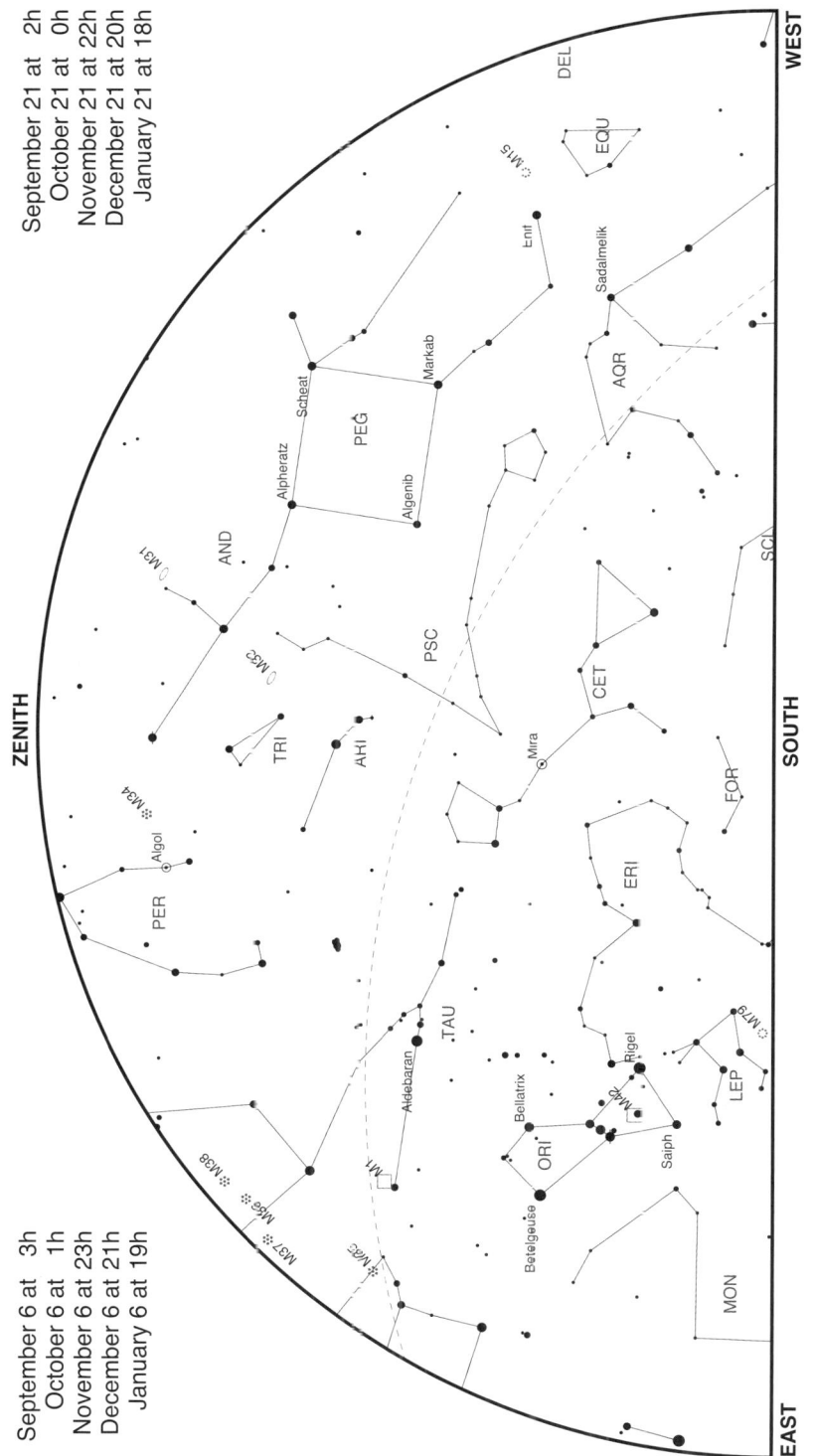

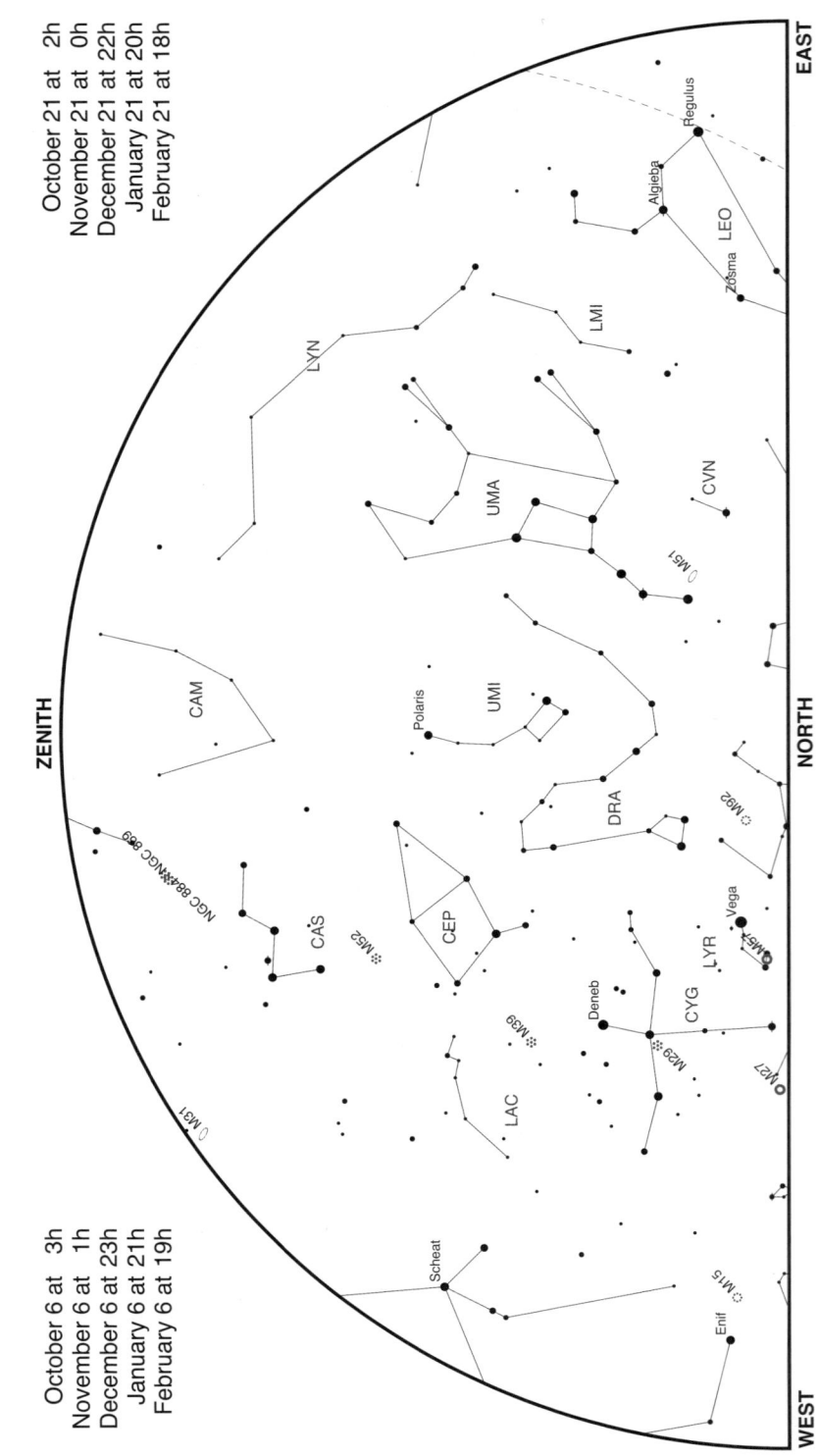

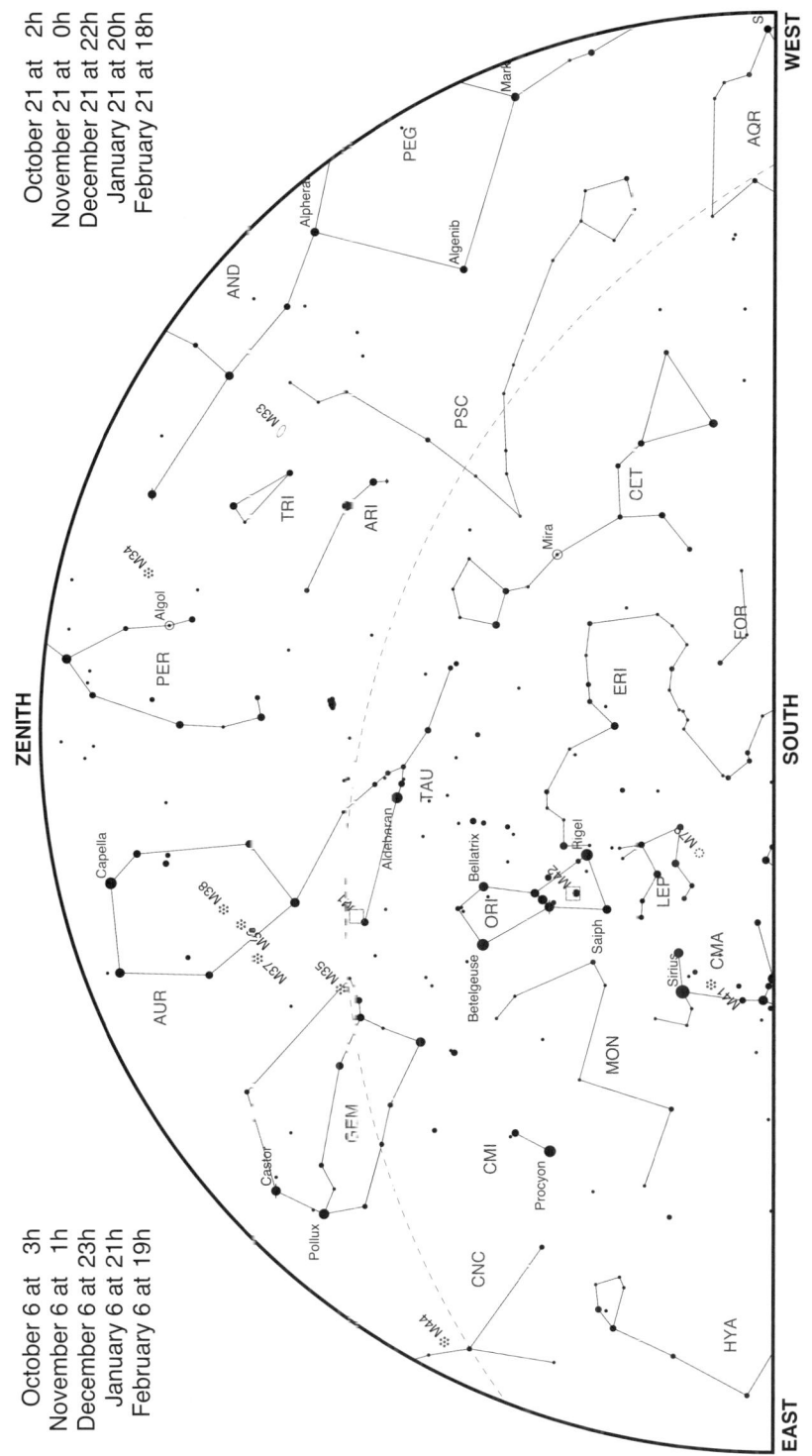

# Southern Hemisphere Star Charts

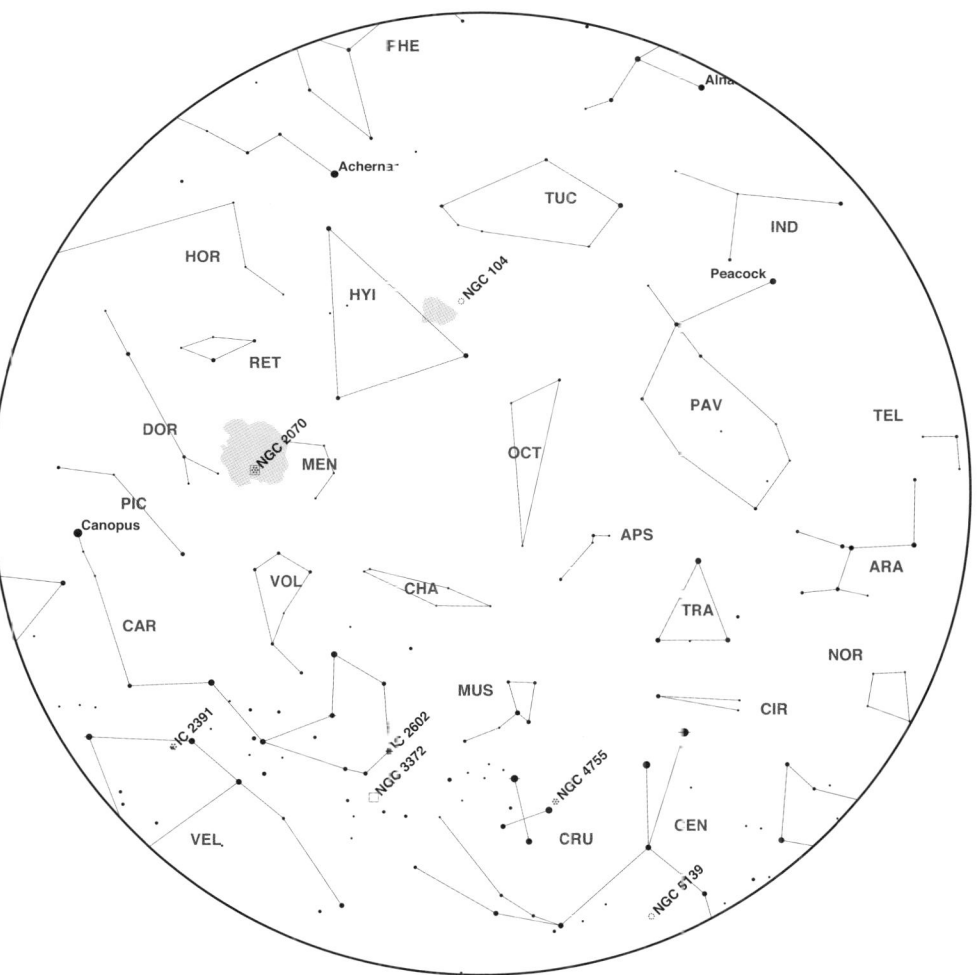

This chart shows stars lying at declinations between − 45 and − 90 degrees. These constellations are circumpolar for observers in Australia and New Zealand.

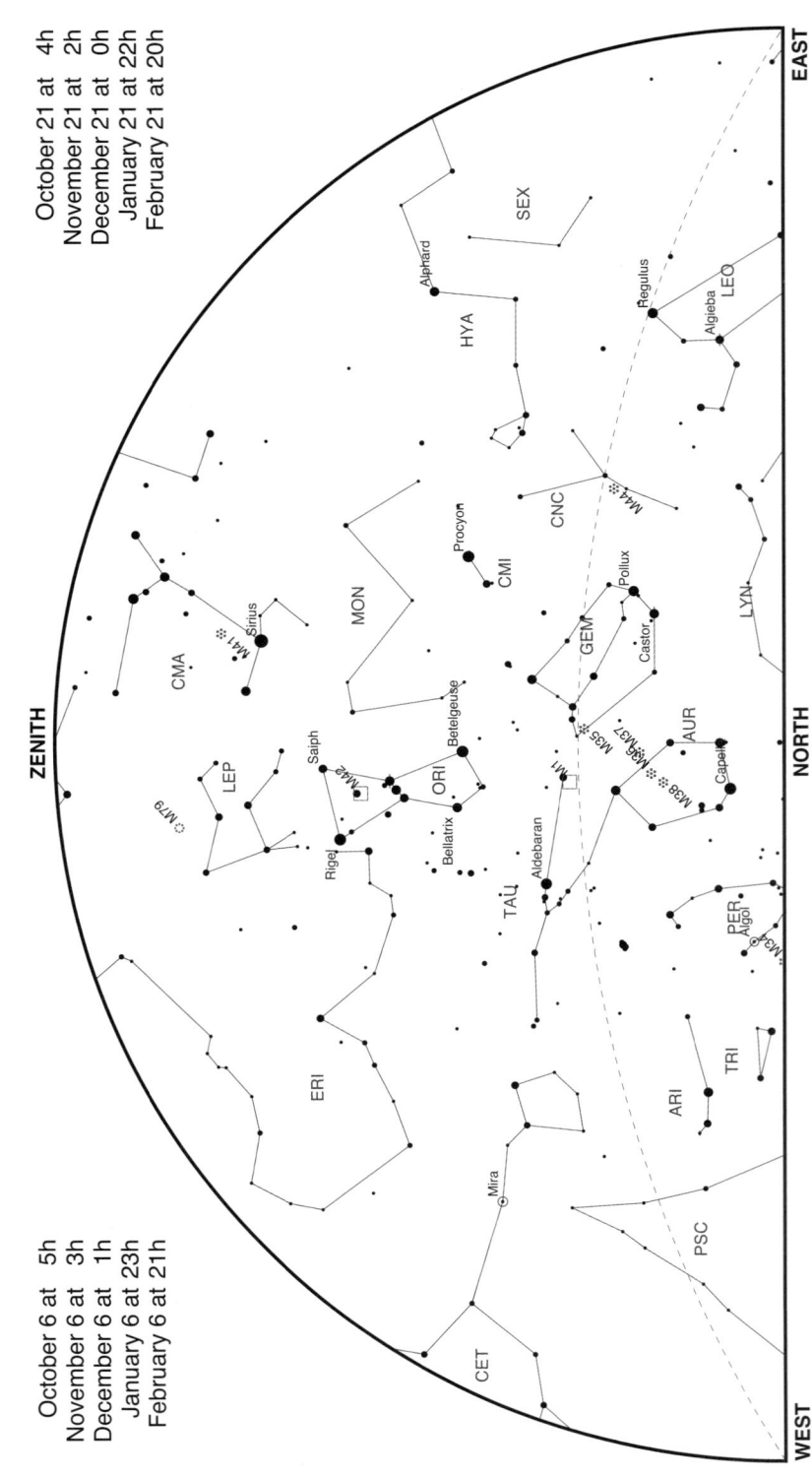

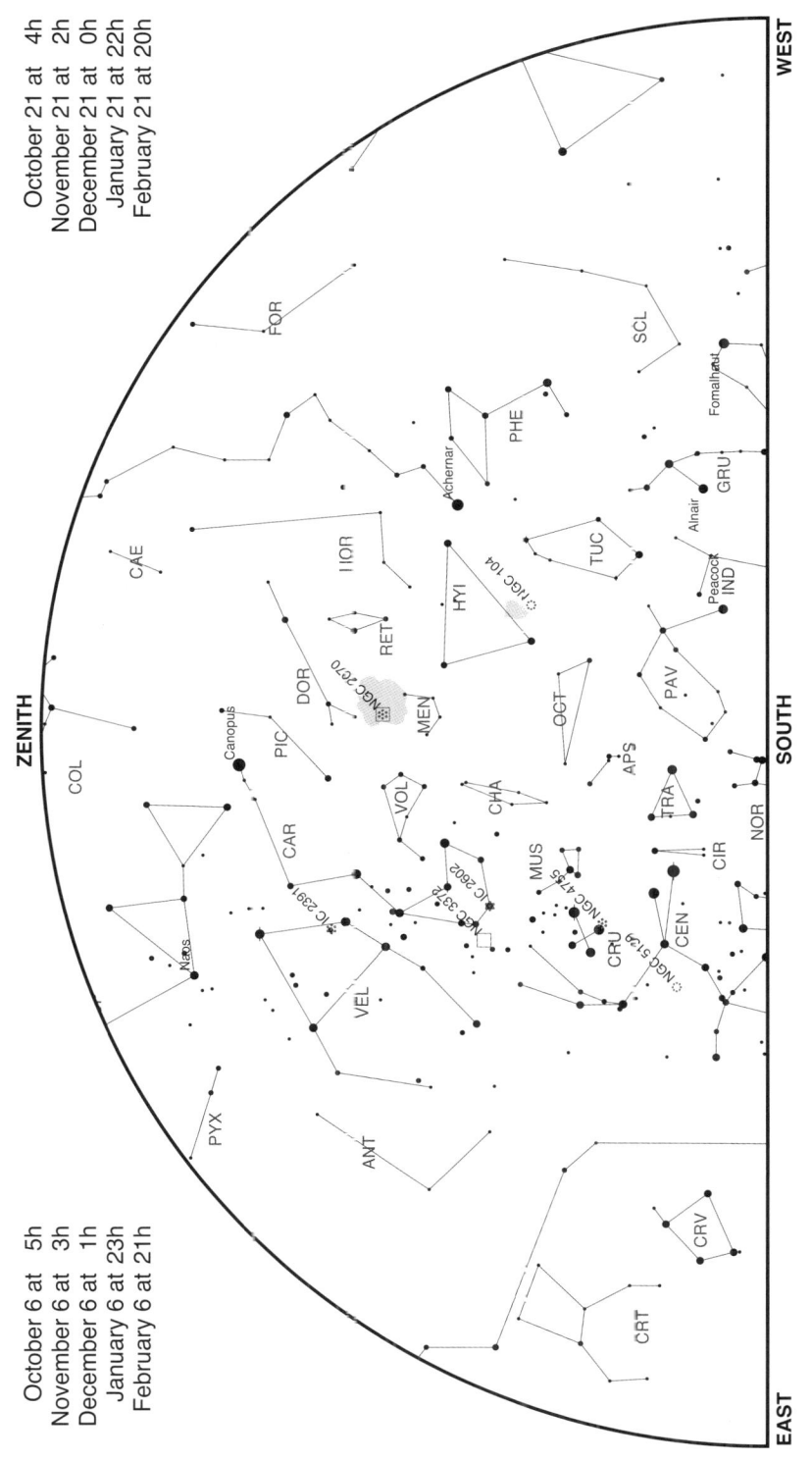

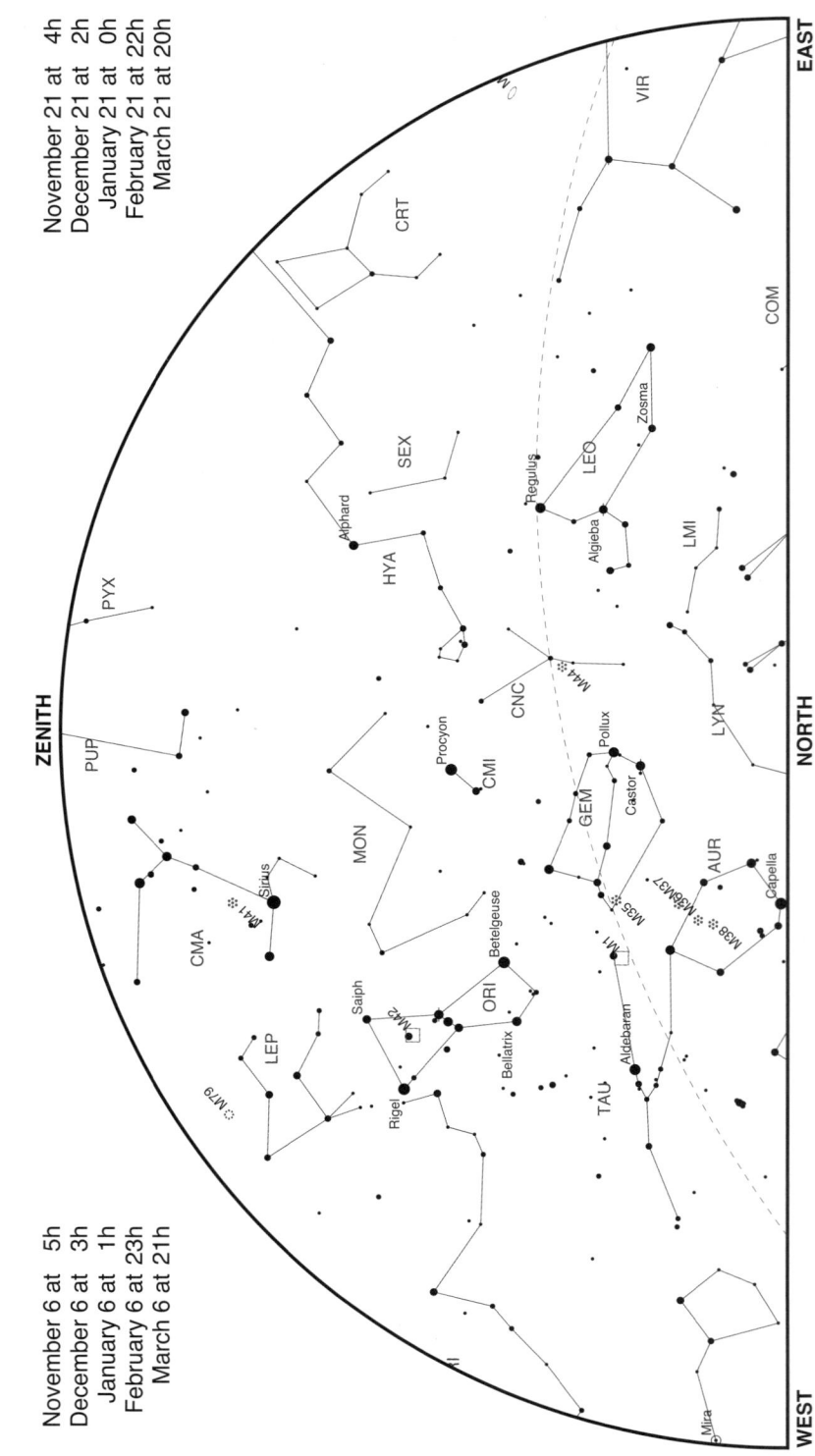

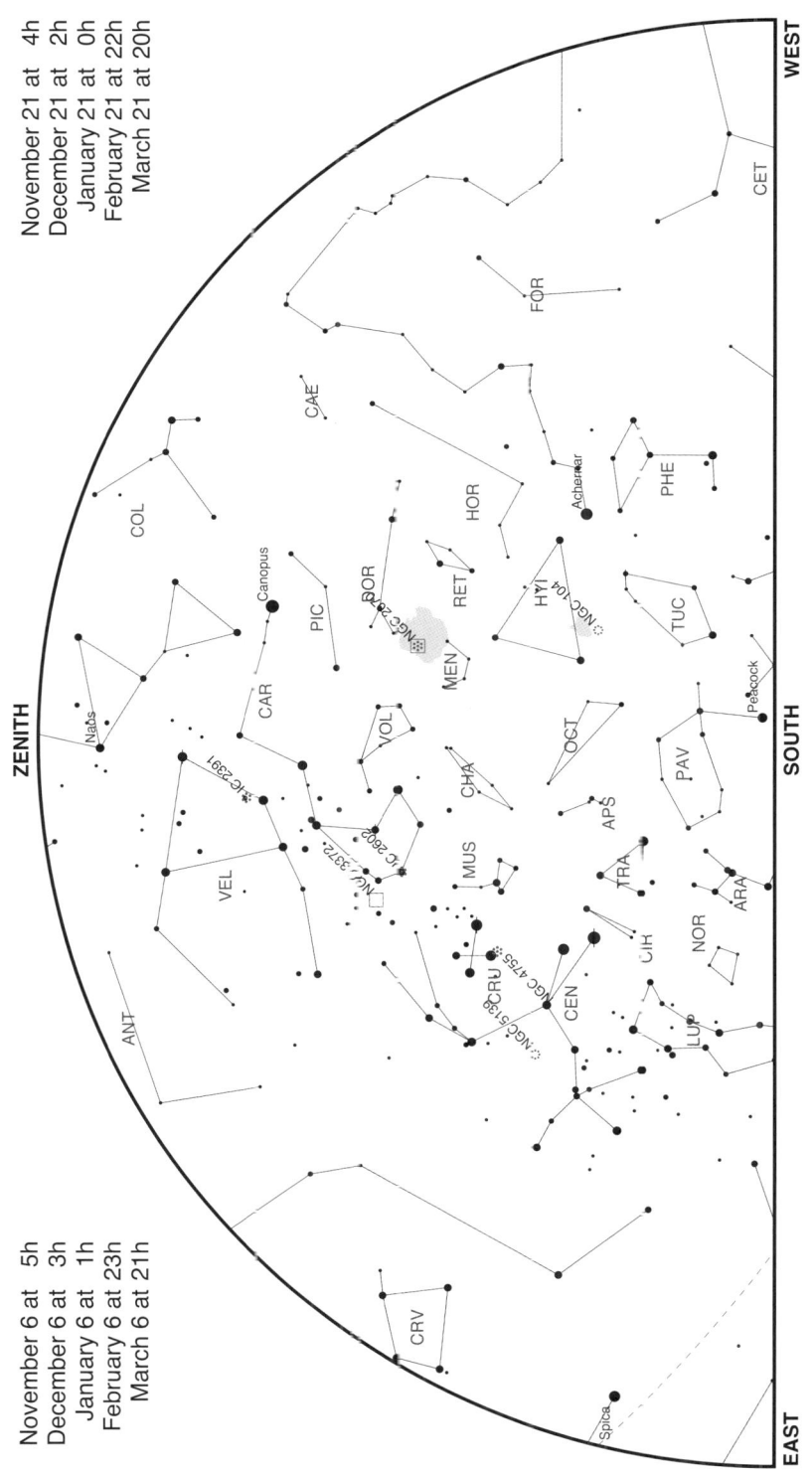

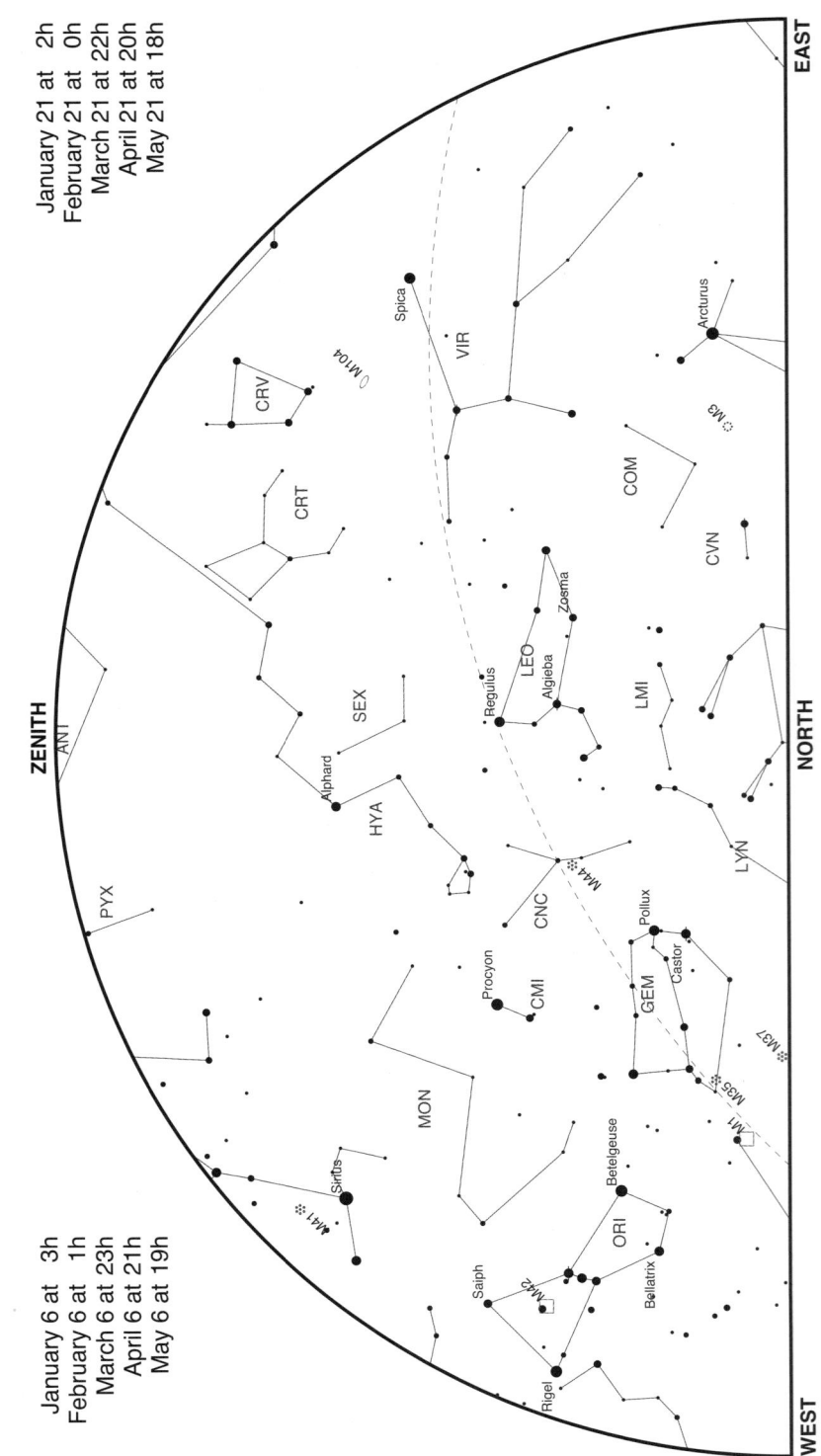

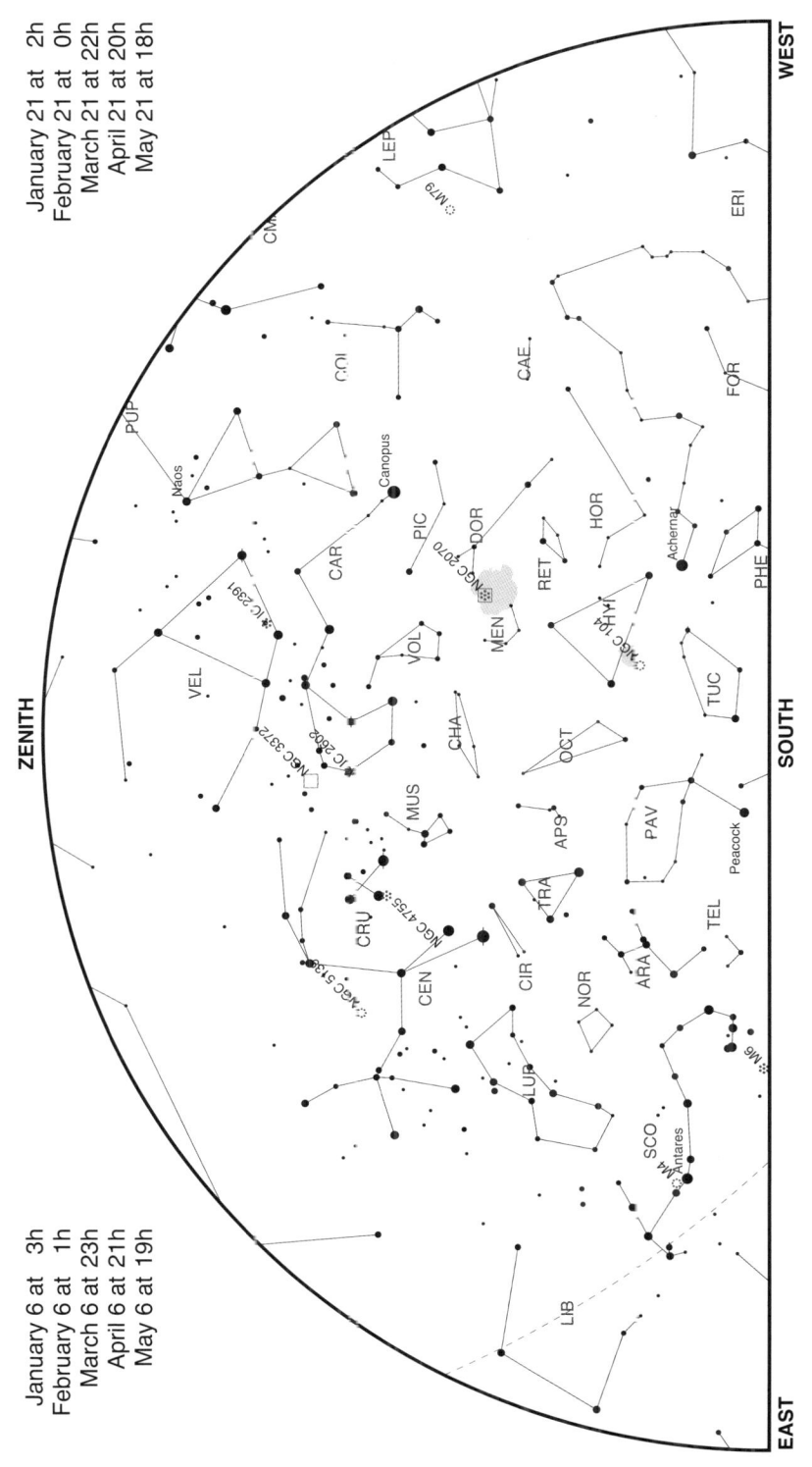

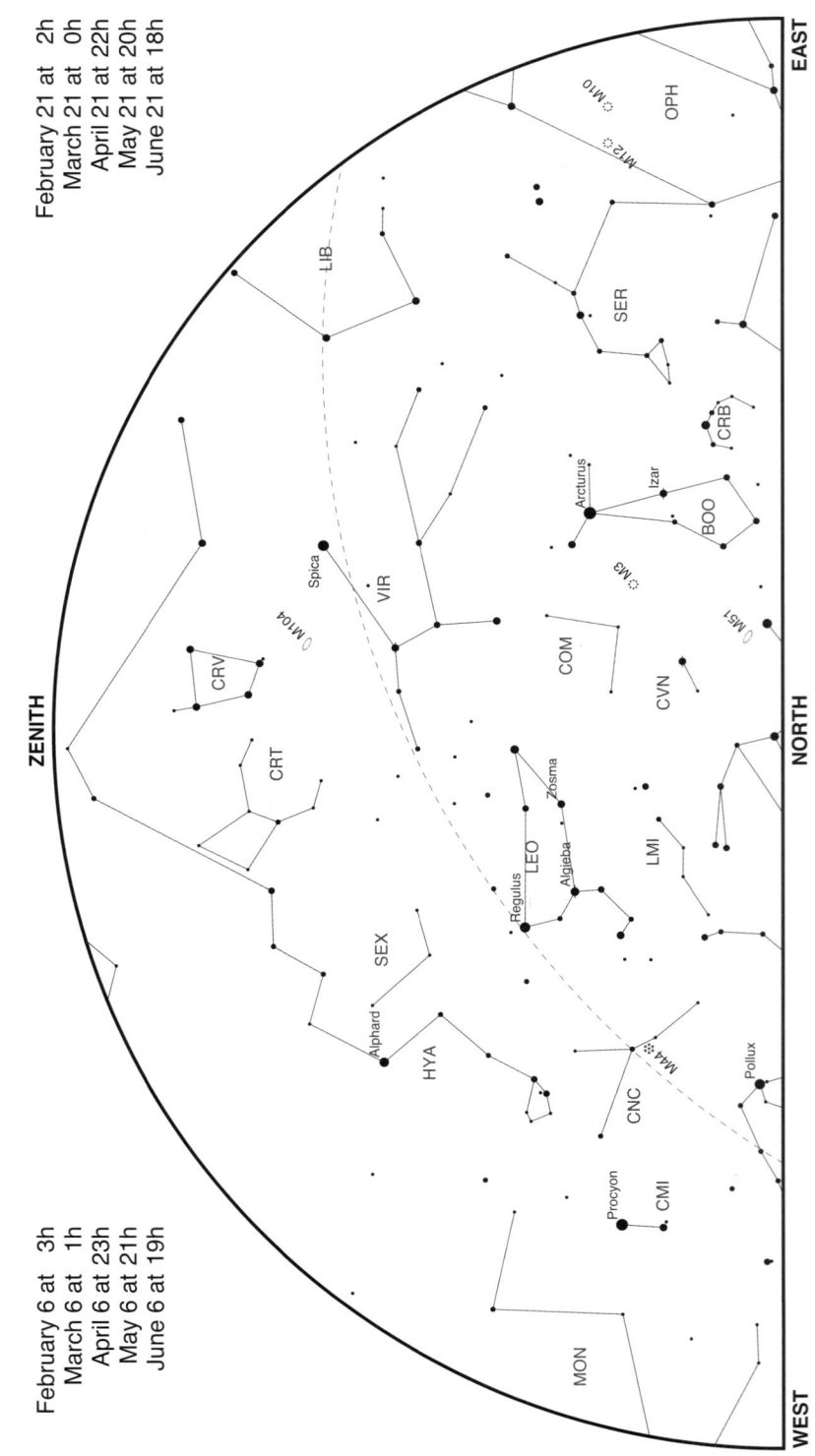

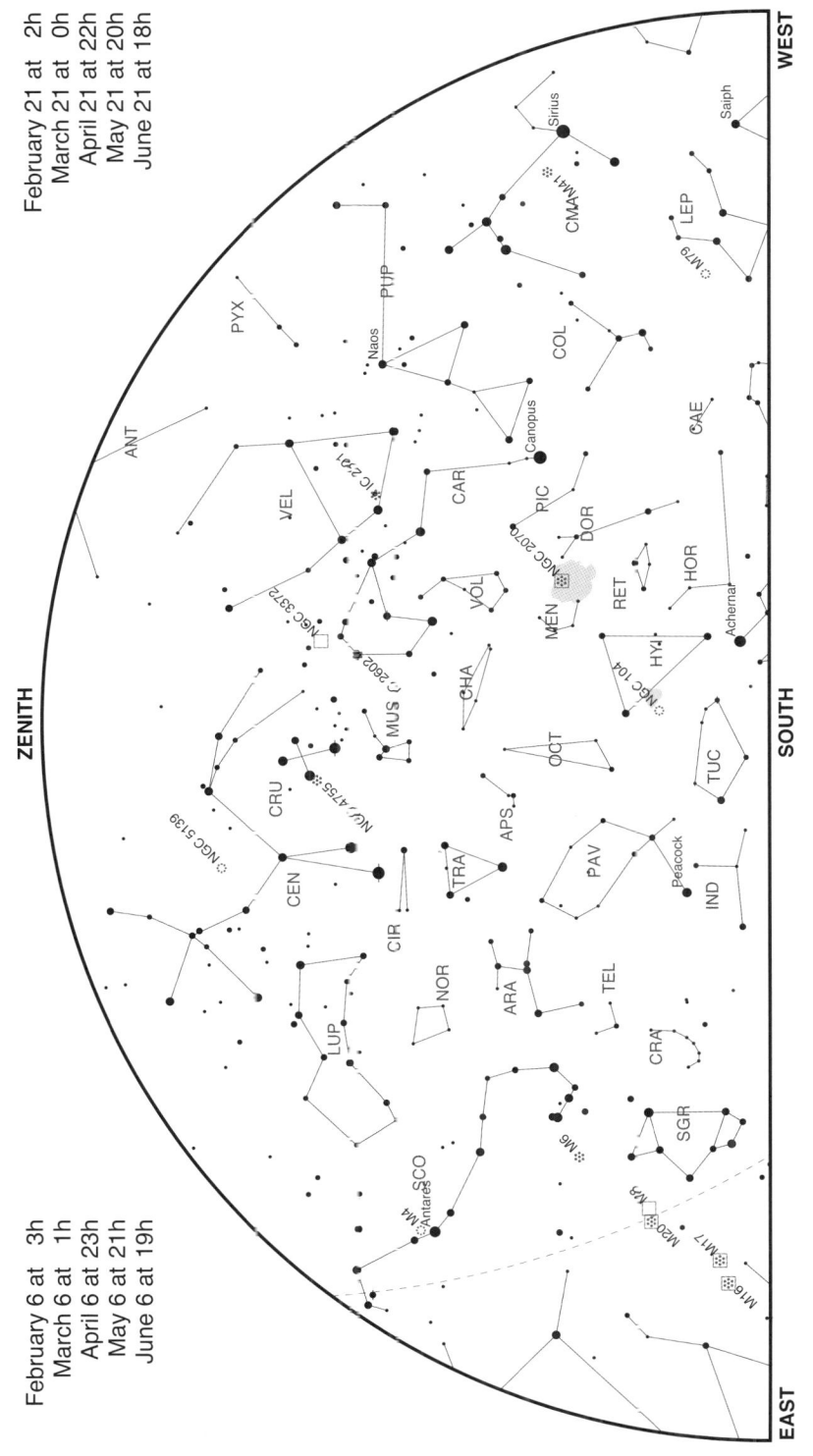

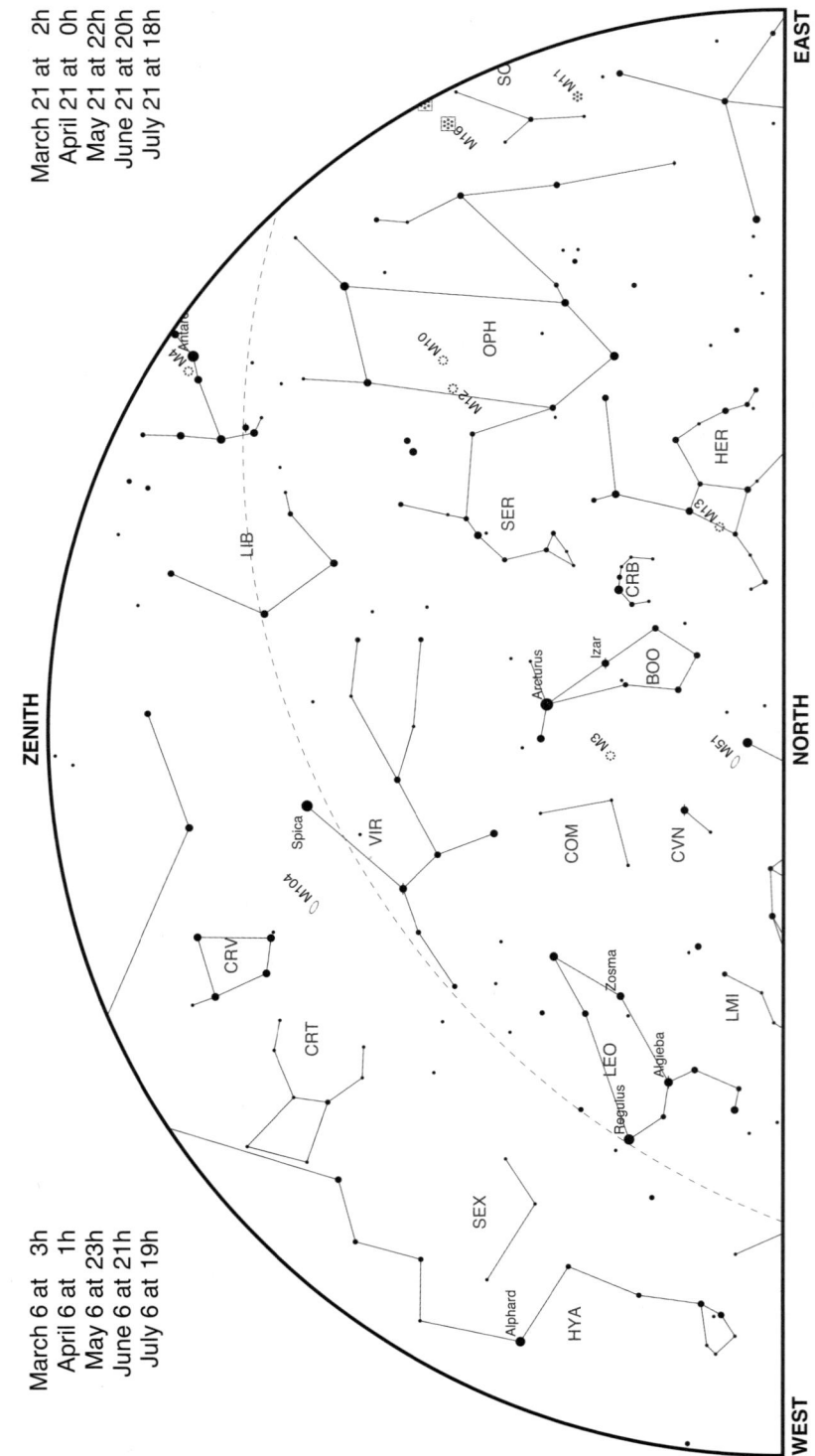

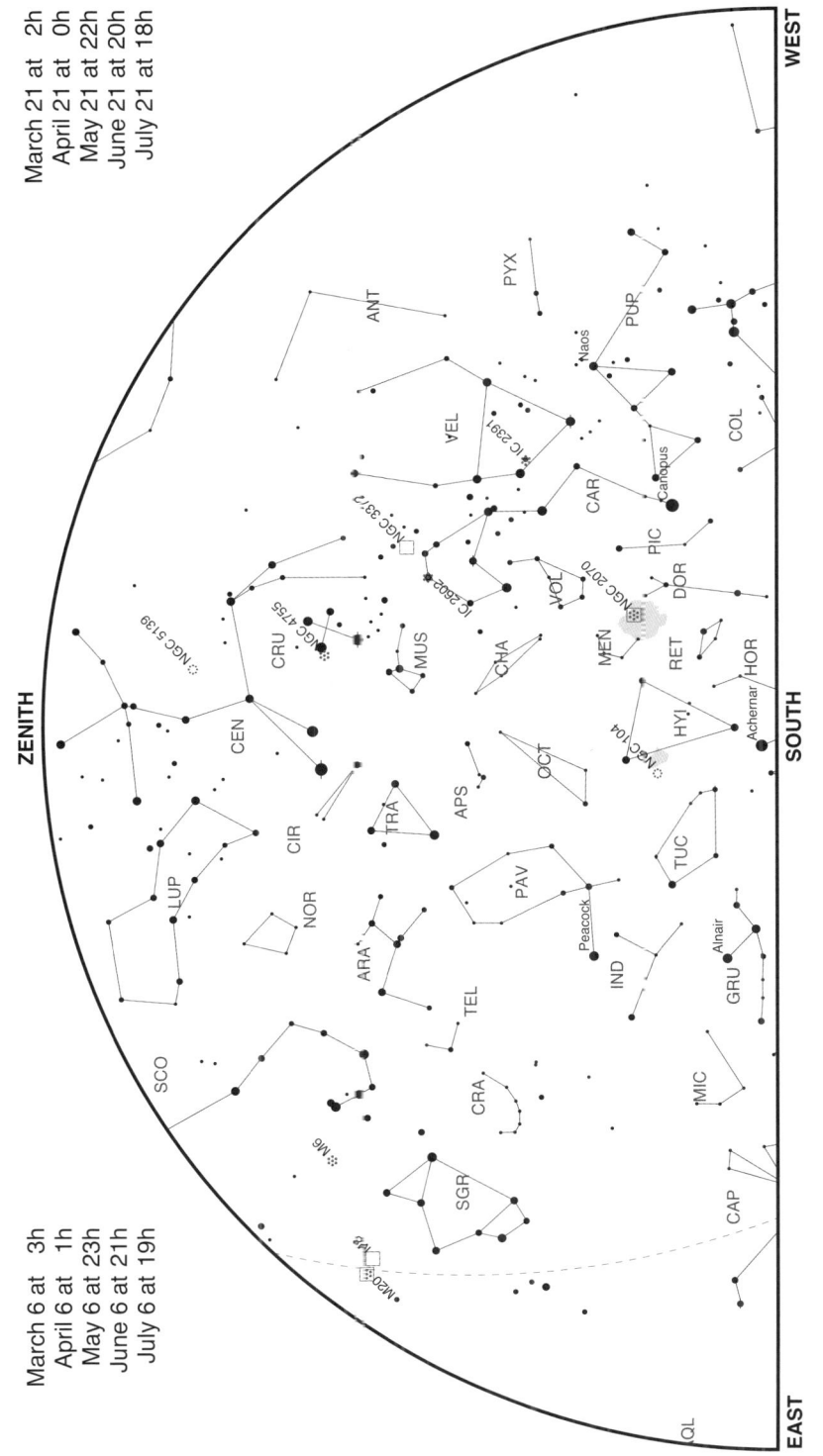

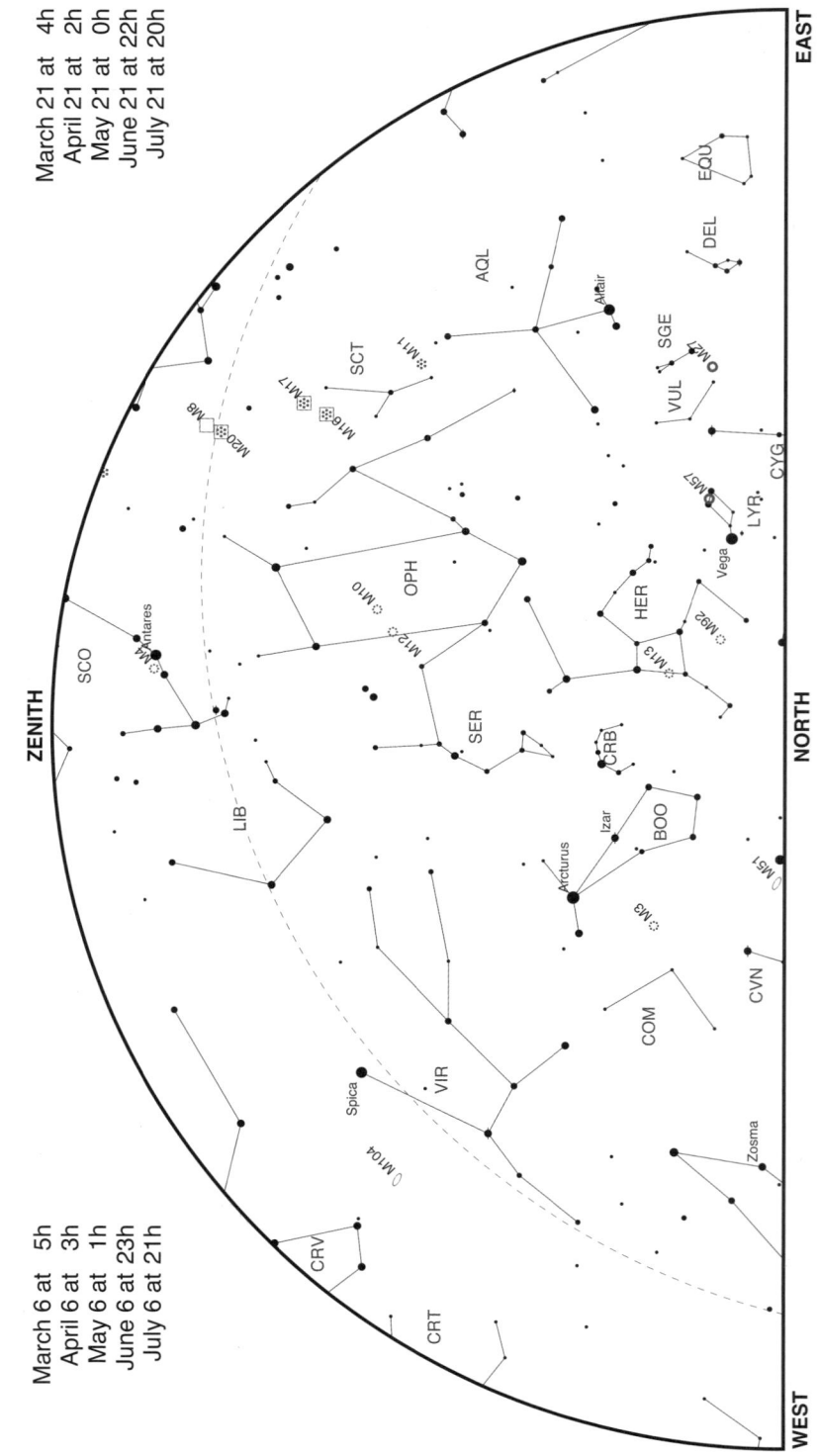

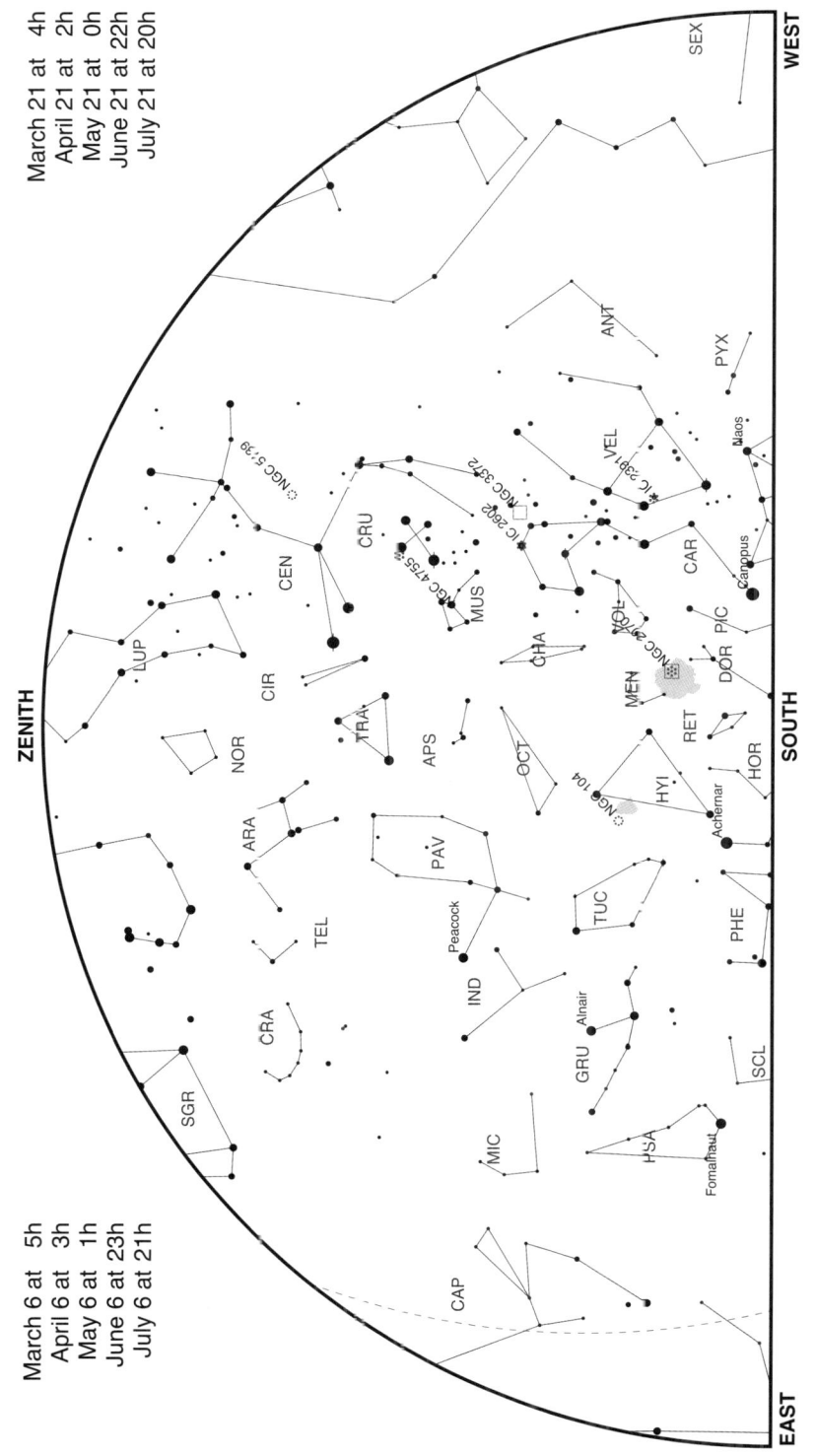

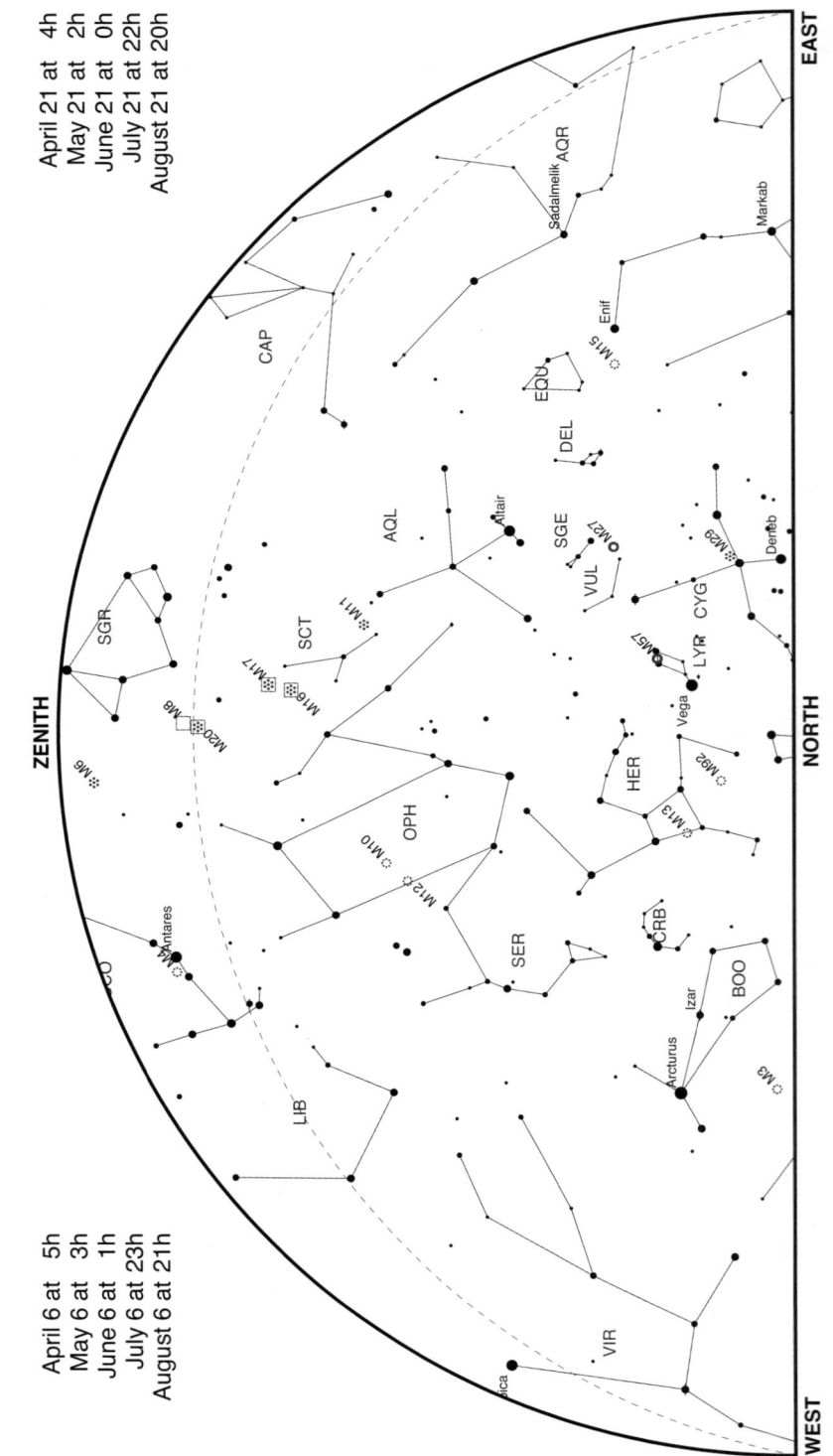

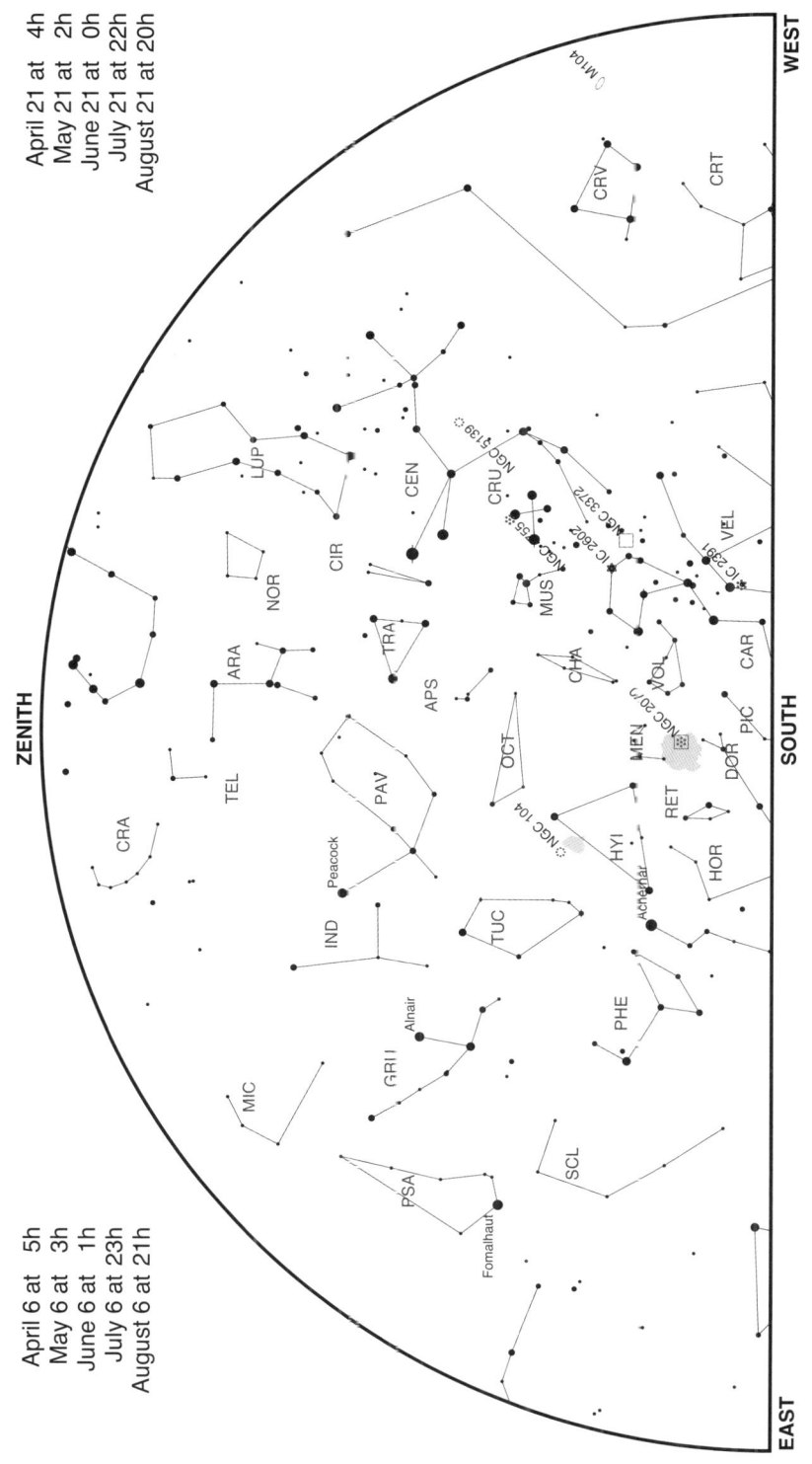

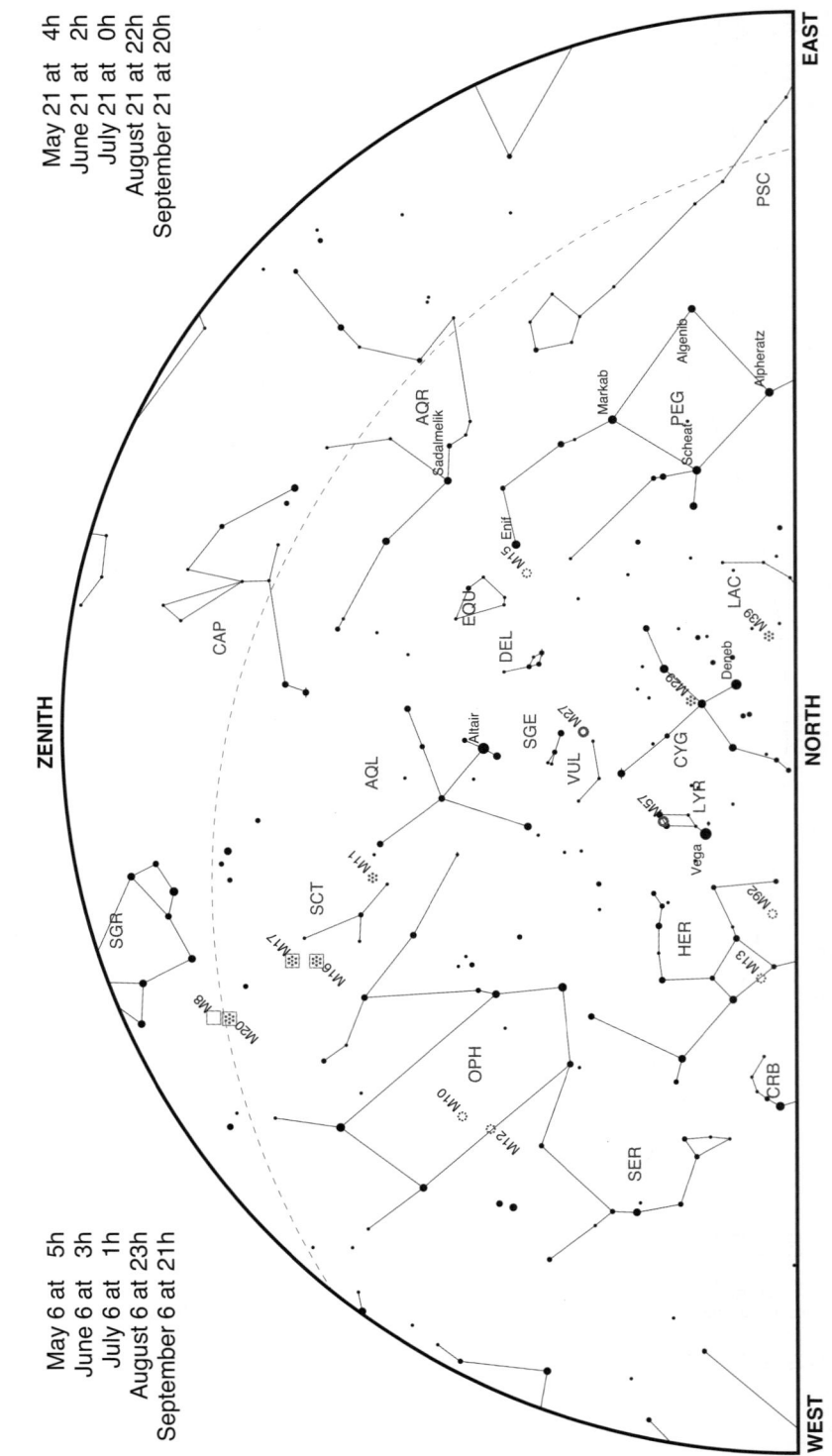

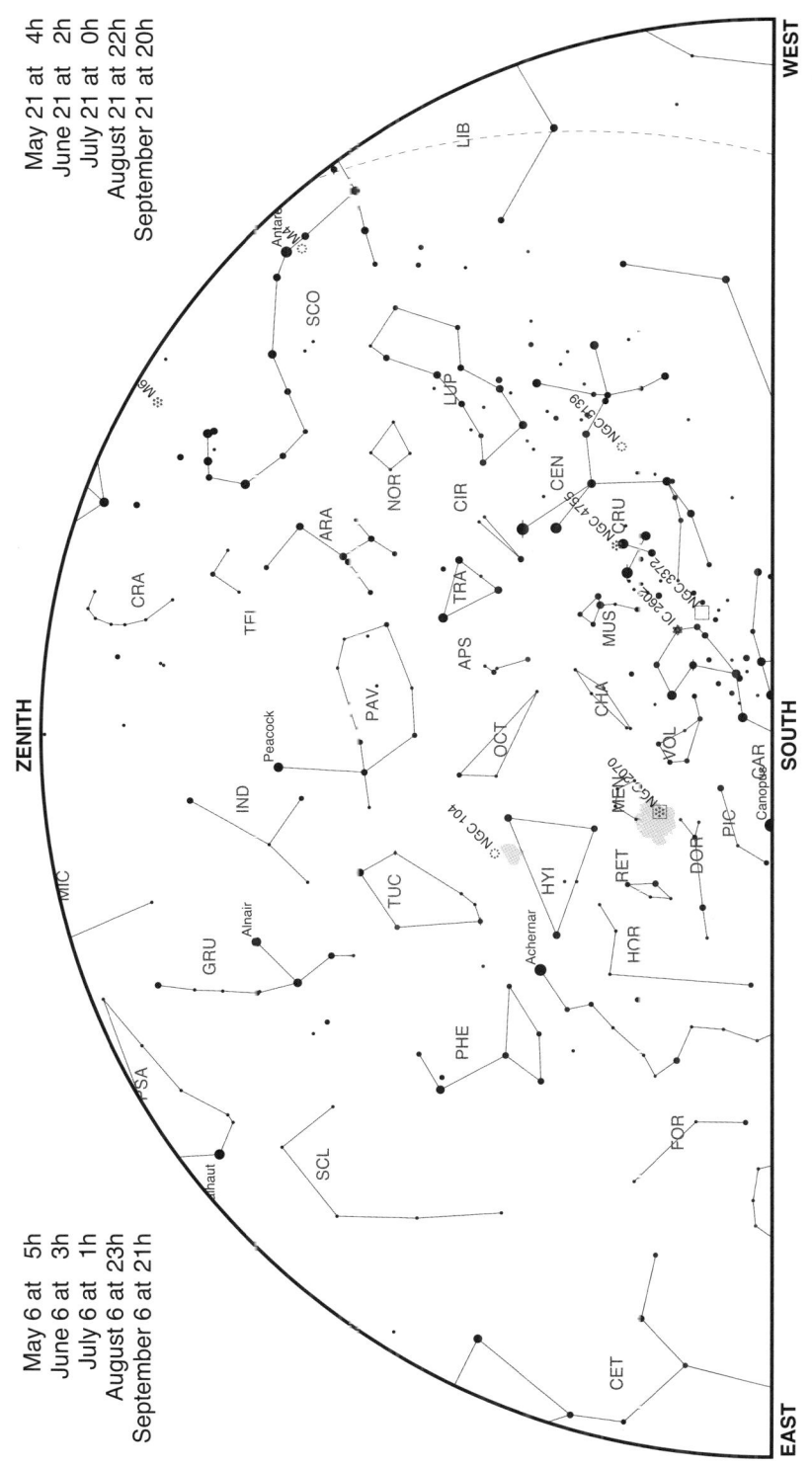

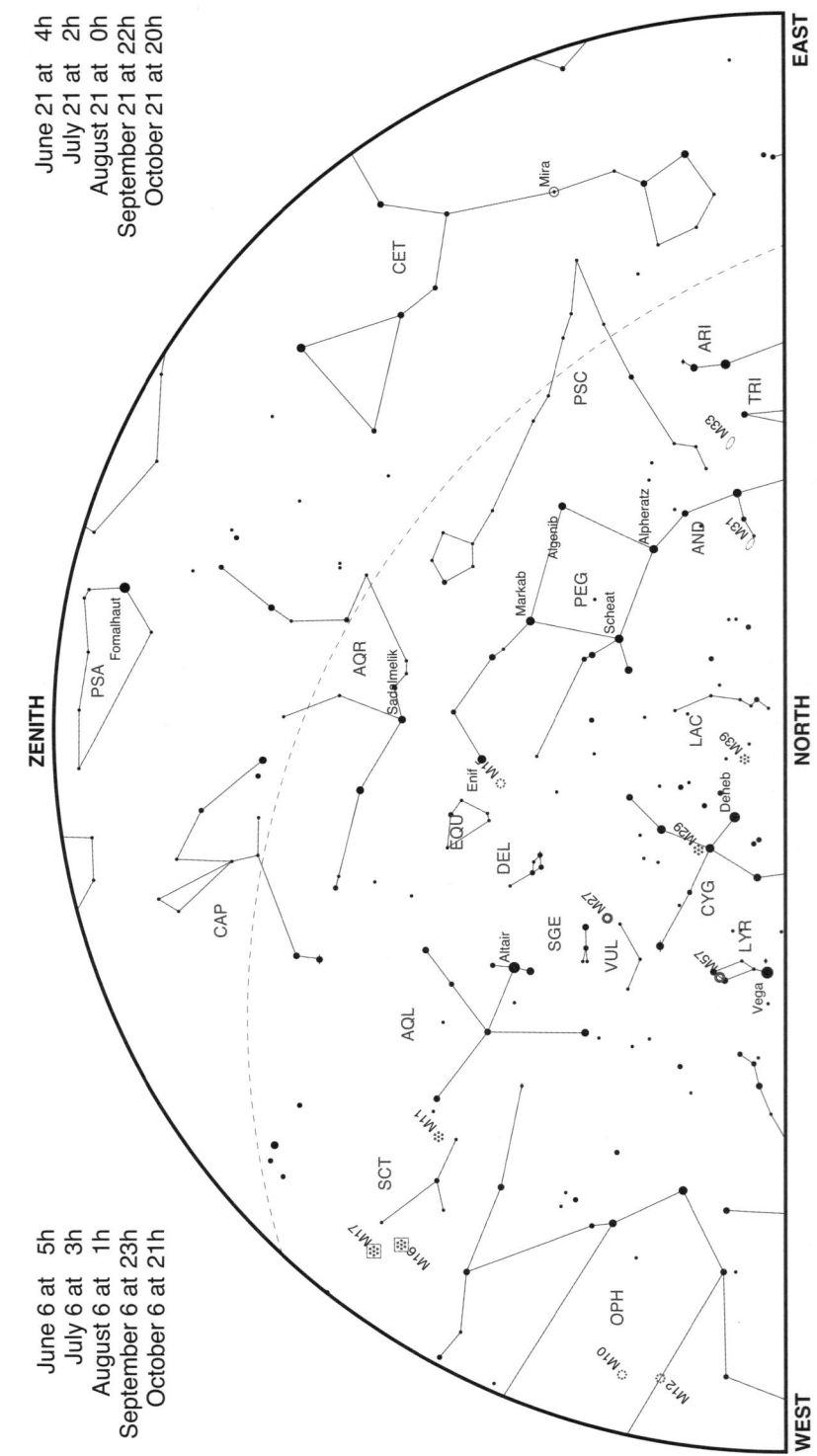

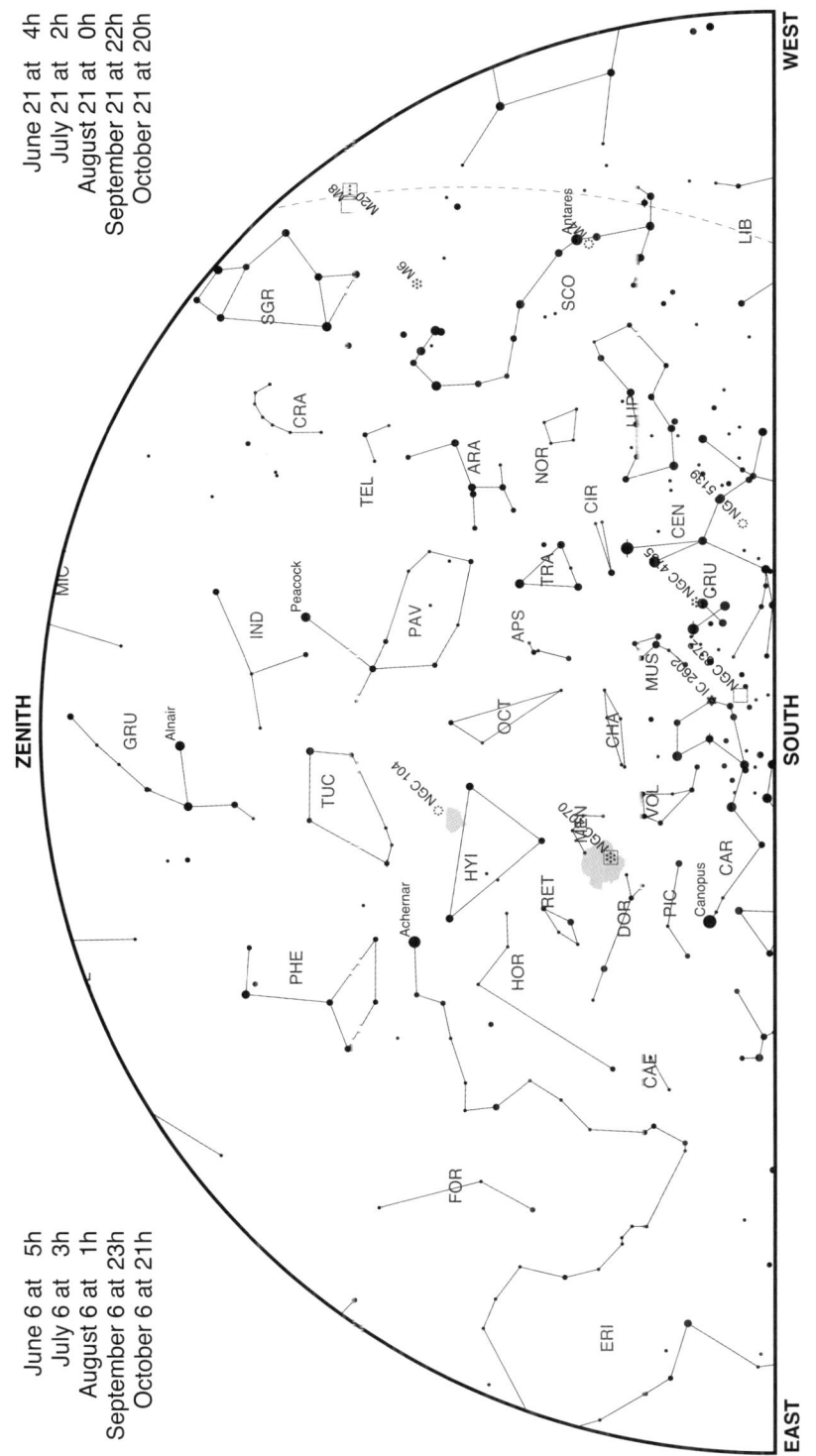

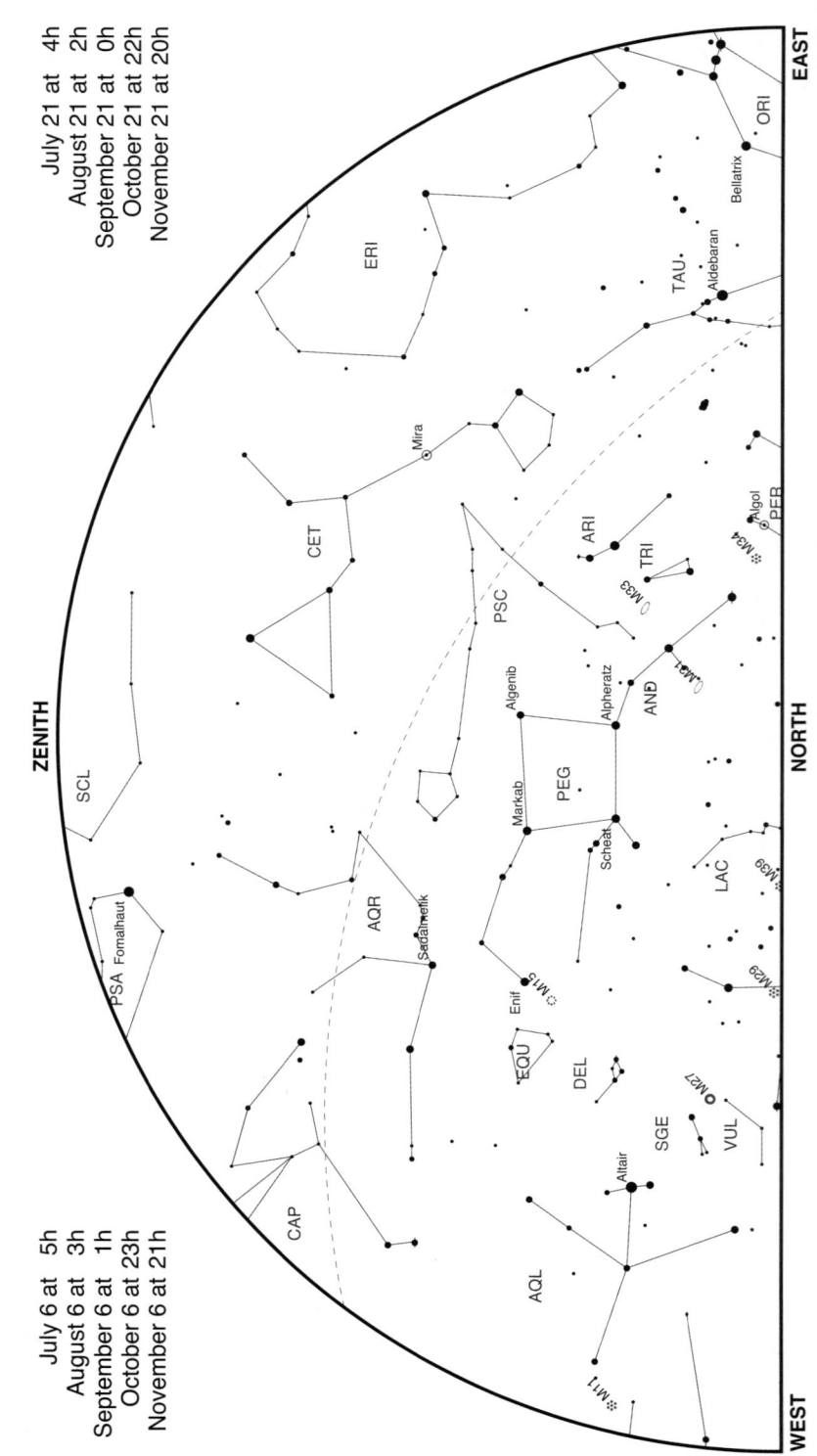

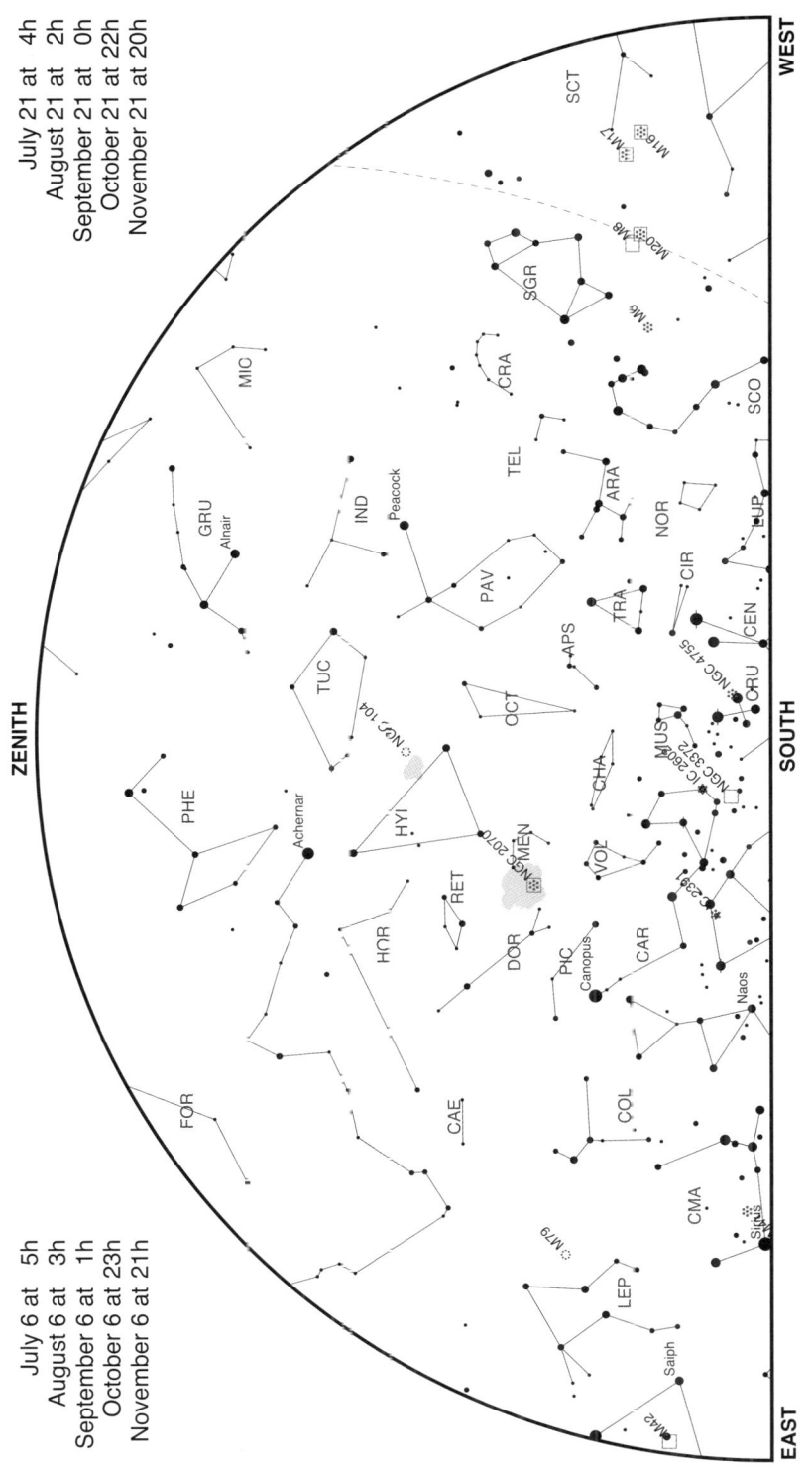

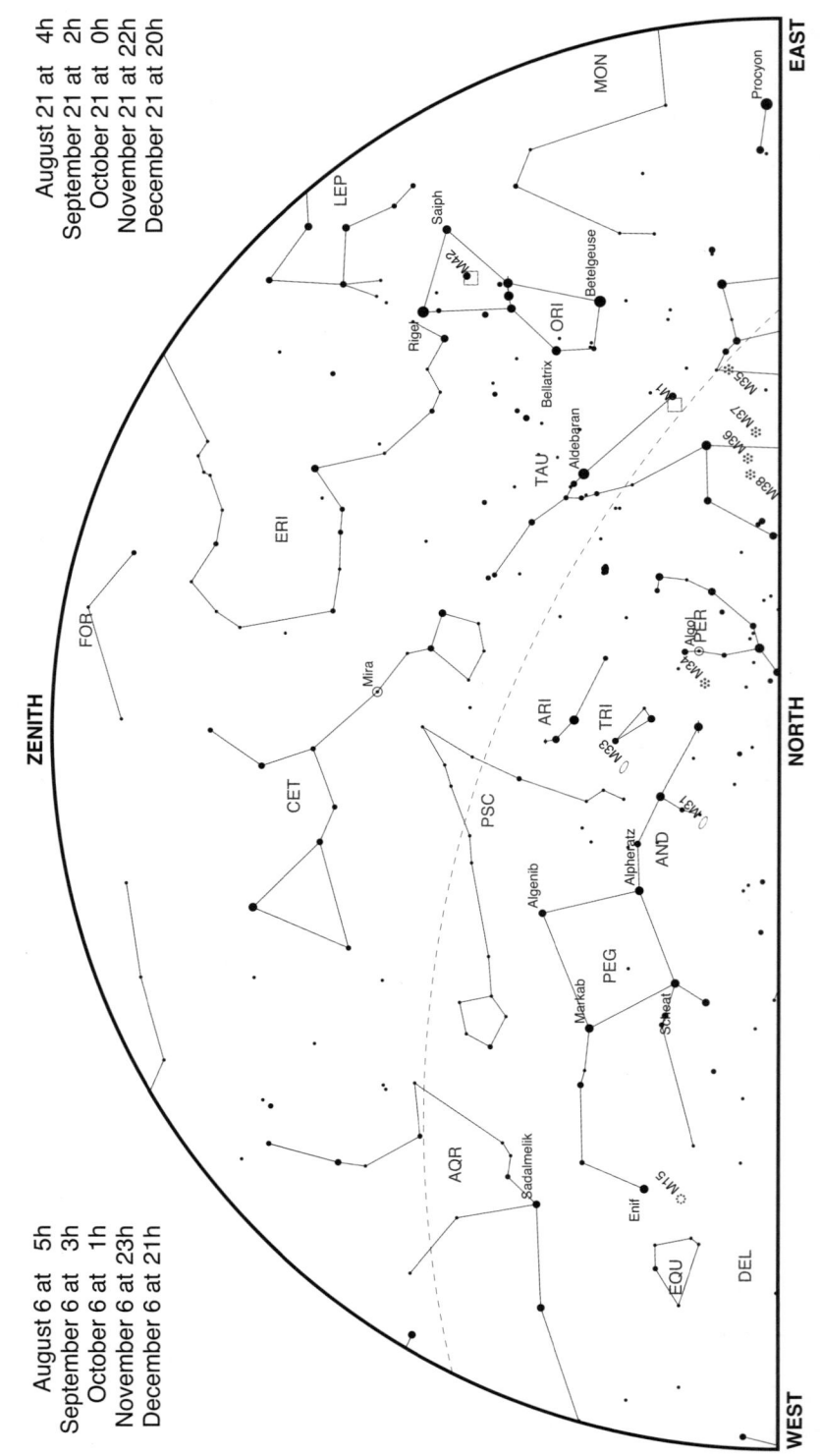

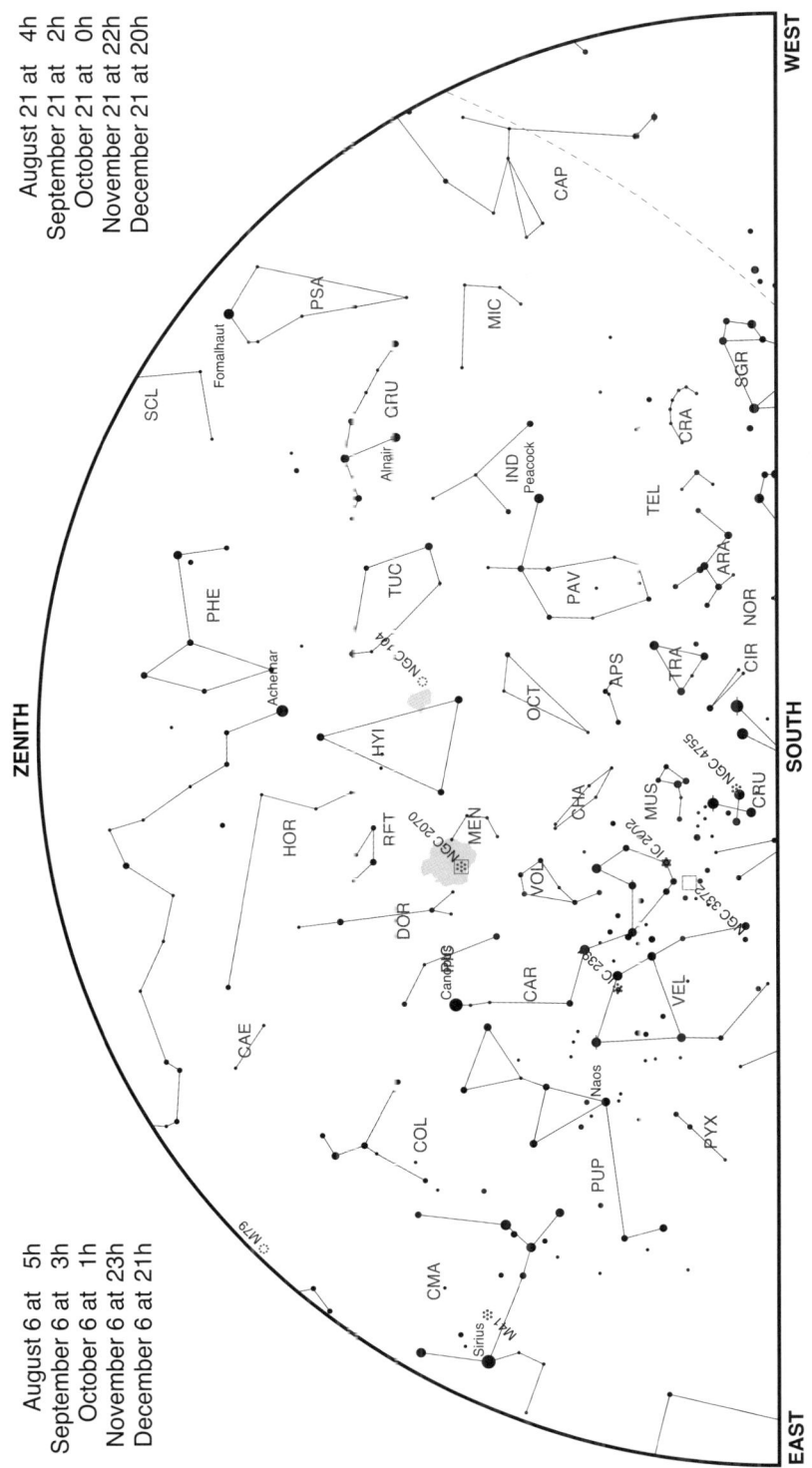

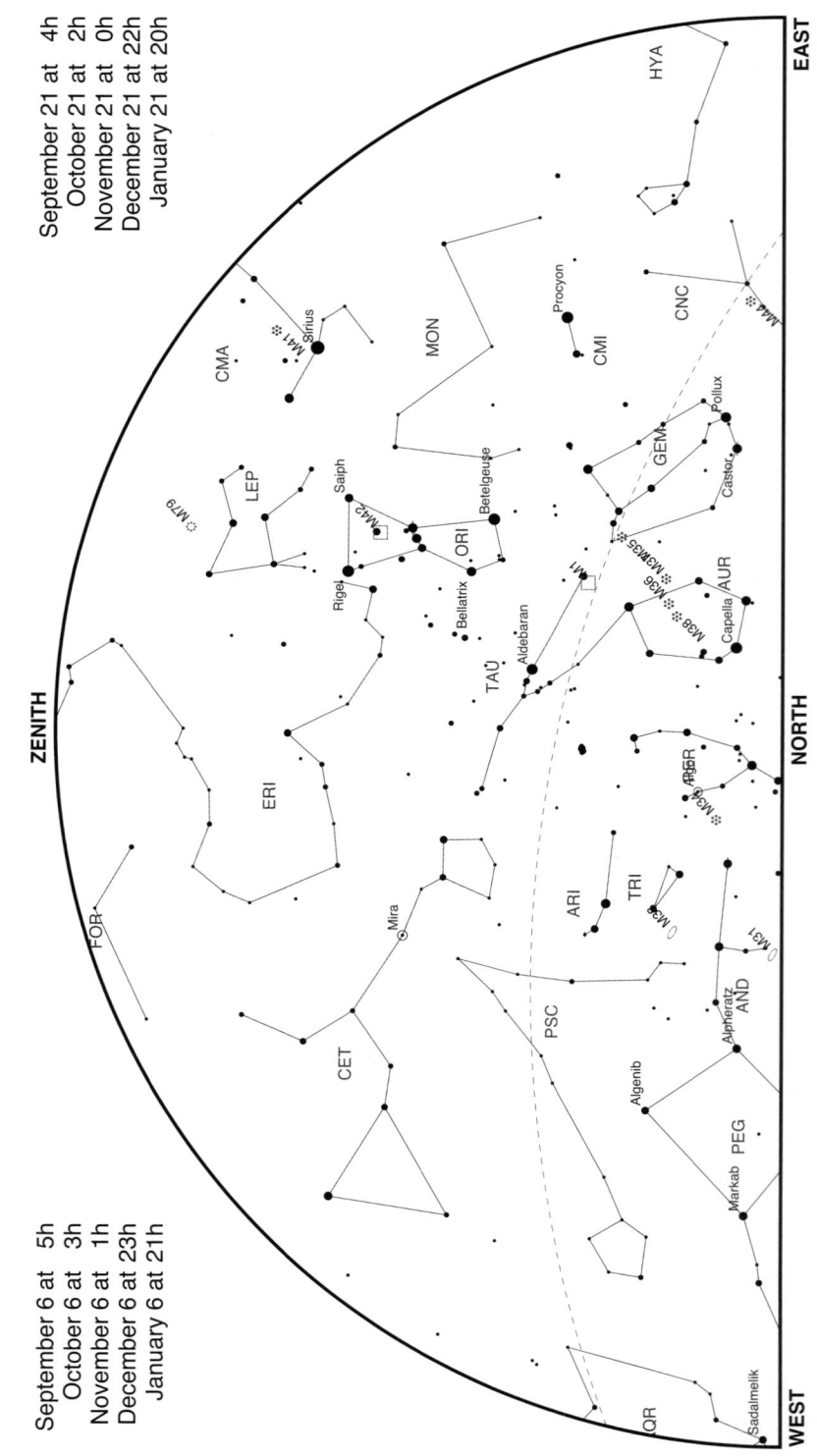

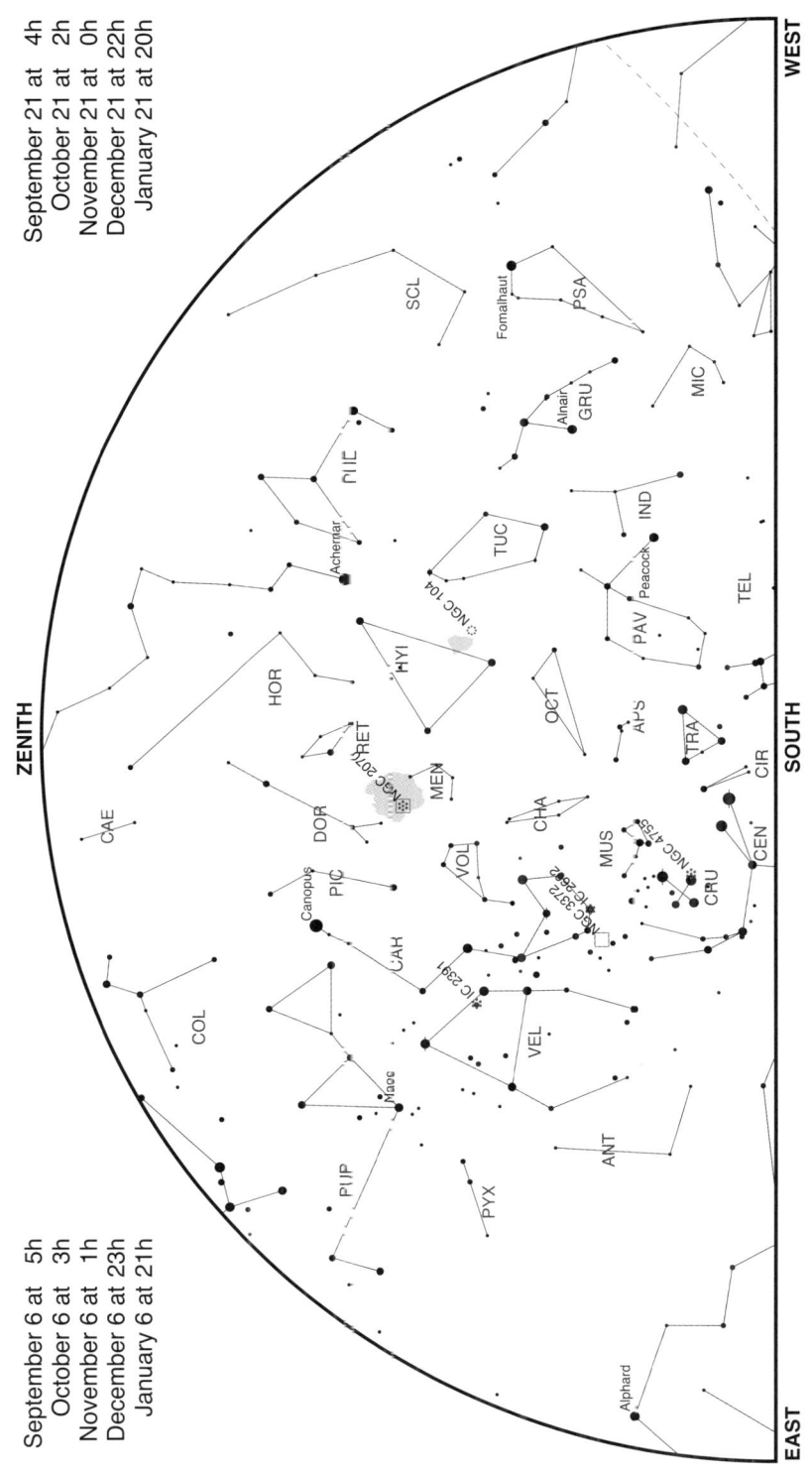

# The Planets in 2026

## Lynne Marie Stockman

### Mercury

Early risers will have to be quick in order to catch **Mercury** as it heads back toward the eastern horizon at the beginning of the year. The first full apparition, running from late January to early March, is the best evening appearance of this planet for observers in northern temperate latitudes, although the May–July apparition is nearly as good. For those in the southern hemisphere, their best evening opportunities to see Mercury occur between late August and early November with the May–July apparition also very fine. If you wish to see Mercury in the morning, then travel to the southern hemisphere for the March–May apparition. The morning appearances in the northern hemisphere are not as good, with the best ones occurring during July–August as well as the final apparition of the year. Mercury's brightness varies wildly throughout the year, with its minimum values (around sixth magnitude) occurring at inferior conjunction and its maximum values (often a negative number) at superior conjunction. Thus, the best opportunities to spot Mercury at its brightest occur early in an evening apparition and late in a morning apparition. The largest greatest elongation east (evening sky) is 25.2° in October and the largest greatest elongation west (morning sky) is 27.8° in April.

| Best Apparitions | 52° N Latitude | 35° S Latitude |
|---|---|---|
| Morning | 4 November–1 January [2027] | 7 March–14 May |
| Evening | 21 January–7 March | 27 August–4 November |

Mercury races past each of the planets this year but since it is the closest planet to the Sun, all of these encounters occur at small elongations from our star. Mercury and Venus make a pair in late January and late February, although October presents the best opportunity of seeing the two planets together. However, the two inferior planets are over 5° apart at that time. Mercury and Mars are found together in late January but both are very close to conjunction with the Sun at that point. Mercury passes Mars twice more, in March and April, with April being the best time to spot the two bodies near one another. Mercury is near Saturn and Neptune in mid-April but twilight is never a good time to look for these fainter outer planets. Uranus is days away from conjunction with the Sun when Mercury comes to call in May, and Jupiter is only 12° from Sun in August when Mercury moves past. The waxing crescent Moon occults Mercury once this year, in February

Apparition diagrams showing the position of Mercury above the western horizon at sunset and eastern horizon at sunrise may be found throughout the *Monthly Sky Notes*.

## Venus

Conspicuous by its absence as 2026 opens, **Venus** soon appears in the west as the evening star, a title it will hold until late October. This is an excellent evening apparition for southern hemisphere observers, with Venus soaring high above the western horizon and peaking in August around the time of greatest elongation east. Those looking for the evening star from equatorial regions will also enjoy good views of the bright planet. However, the astronomical community inhabiting northern temperate latitudes will have a mediocre showing of the evening star, with the planet gaining altitude only until around June before heading back toward the horizon. The morning star takes over in November, with those near the Earth's equator seeing a wonderful display as Venus vaults high above the eastern horizon. Views from the southern hemisphere are also quite good but once again, the northern temperate latitudes get a poorer showing. Venus and Mars are together in January but both are too close to conjunction with the Sun to be visible. The two approaches of Mercury in late January and February also take place perilously close to our star. Venus passes by both Saturn and Neptune in March but these events occur only 15° from the Sun. Venus overtakes Uranus in late April, when they are at 25° solar elongation, but faint Uranus may be lost in evening twilight. Jupiter comes to call in June, with the two brightest planets in the sky 1.6° apart. Mercury makes one final distant pass in October, when Venus is just weeks away from inferior conjunction. The Moon occults Venus three times this year, in June, September, and November, but the third event may be difficult to observe due to the proximity of the Sun. Venus is at a maximum magnitude of −4.7 during late November and early December and at its dimmest (−3.9) in February and early March.

Apparition diagrams showing the position of Venus above the western horizon at sunset and eastern horizon at sunrise may be found in the January and November *Sky Notes* respectively.

## Mars

The red planet is in direct motion for the entire year. It is missing from the sky at the beginning of 2026 as conjunction occurs in early January, but it finally appears low in the east at dawn in February. For those in northern temperate latitudes, **Mars** remains mired in morning twilight for the first half of the year, not returning to evening skies until September. Southern hemisphere observers have better views of the planet from January to July but for them, Mars will be strictly a morning sky object for most of the rest of 2026. A first-magnitude object in January, it brightens to −0.1 by the end of the year. The planet will present a distinctly gibbous appearance at west quadrature in November, with 89% of the disk illuminated. Opposition takes place in February next year. Mars meets both Mercury and Venus around its time of conjunction with the Sun so these close approaches will be unobservable. Mercury swings by again in March and April, with further visitations by Saturn and Neptune in the latter month, but all of these events take place less than 25° elongation from the Sun. Astronomers should have better luck seeing the close approach to Uranus in July. Bright Jupiter is the last planetary visitor in November. Mars also passes near a number of interesting deep sky objects, including the open cluster M35 in Gemini in August, the open cluster M44 (Praesepe or Beehive Cluster) in Cancer in October, and several galaxies (M95, M96, and M105) in Leo in December. The Moon occults Mars three times this year, in February, October, and November, but only the latter two occultations are easily observable. Beginning the year in the constellation Sagittarius, Mars traverses the zodiac – Capricornus, Aquarius, Pisces (with excursions into nearby Cetus), Aries, Taurus, Gemini – before ending in Leo.

An apparition diagram showing the position of Mars above the eastern horizon at sunrise for the first eight months of 2026 may be found in the March *Sky Notes*. A finder chart showing the position of Mars throughout the last eight months of 2026 may be found at the end of this article.

## Jupiter

The largest planet in the solar system rules the night skies at the outset of 2026, with opposition taking place in January. **Jupiter** is visible during evening hours until June or July, depending on the observer's latitude, when it finally succumbs to twilight. After conjunction at the end of July, the gas giant reappears in the morning sky and is rising by midnight for all sky watchers before the end of the year. Jupiter is in retrograde in the constellation of Gemini at the beginning of the year, returning to direct motion in March, and moving into the constellation of Cancer in June. It passes less than a degree south of the open cluster M44 in early August, enters the constellation of Leo in September, and goes back into retrograde motion in December. Jupiter teams up with Venus above the western horizon in early June but its encounter with Mercury in mid-August takes place too close to the Sun to be visible. Jupiter and Mars make a bright pair in mid-November. The Moon occults Jupiter four times this year, once in September, once in October, and twice in November.

A finder chart showing the position of Jupiter throughout 2026 may be found at the end of this article.

## Saturn

The ringed planet moves from the constellation Aquarius into its neighbour Pisces in January. It then cuts through a corner of Cetus from early April to early June, after which it returns to Pisces. Finally, in September, **Saturn** moves back into Cetus for the remainder of the year. It is an evening sky object at the beginning of 2026, gradually moving toward the setting Sun and getting lost in evening twilight by March. It is best seen from the southern hemisphere after its late March conjunction but moves back into the evening sky for all observers by the middle of the year. Opposition takes place in October. The rings of Saturn are its glory and they are opening up after their ring plane crossing last year. The southern side of the rings are on view and are open only to an angle of 1° as the year commences but they reach a maximum opening angle of over 9° in July before closing up again. A shallow minimum opening angle of 6° is reached in December. Last year, Saturn and Neptune began a triple conjunction in right ascension; the last leg of that meeting takes place in February along with a single conjunction in ecliptic longitude. Venus and Saturn meet in early March but the two objects are just 15° away from the Sun at the time. The morning conjunctions with Mercury and Mars in April take place a little further away from the Sun but those planets are much fainter than Venus so these encounters may be difficult to see. Retrograde motion begins in late July and continues until mid-December.

A finder chart showing the position of Saturn throughout 2026 may be found at the end of this article.

## Uranus

Found in the constellation of Taurus, **Uranus** zigzags between the bright star Aldebaran and the open star cluster known as the Pleiades for the entirety of 2026. It ceases retrograde motion in early February and remains in direct motion until mid-September. The planet is well-placed for observing in the evening sky as the year gets underway, not vanishing into the evening twilight until late April or early May. Conjunction occurs in late May and opposition takes place near the end of November. Mercury and Venus come to call around the time of conjunction but these encounters will be unobservable. However, the conjunction of Uranus with Mars in July takes place nearly 40° from the Sun; the waning gibbous Moon may provide some light interference. The sixth-magnitude planet creeps past two sixth-magnitude stars, 13 Tauri and 14 Tauri, in March, and visits fifth-magnitude 37 Tauri and sixth-magnitude $\omega^1$ Tauri (43 Tauri) both in July and November.

A finder chart showing the position of Uranus throughout 2026 may be found at the end of this article.

## Neptune

Found all year in the constellation of Pisces, **Neptune** opens 2026 already aloft as darkness falls but it sets before midnight. It is getting lost in the glow of twilight by the end of February and vanishes well before conjunction in March. As a morning sky object, it is best viewed from the southern hemisphere through the middle part of the year as the lingering morning twilight of northern temperate latitudes interferes with observations of the eighth-magnitude planet. From July, the planet begins to appear during the evening hours, reaching opposition in September. It completes the third leg of a triple conjunction in right ascension with Saturn in February and also undergoes a single conjunction in ecliptic longitude with the ringed planet in the same month. Its close encounters with the rocky planets take place close to conjunction and are unobservable. In April, Neptune crosses into the northern celestial hemisphere but returns to negative declinations in September. Retrograde motion begins in July and ends in December.

A finder chart showing the position of Neptune throughout 2026 may be found at the end of this article.

## Mars
### May to December 2026

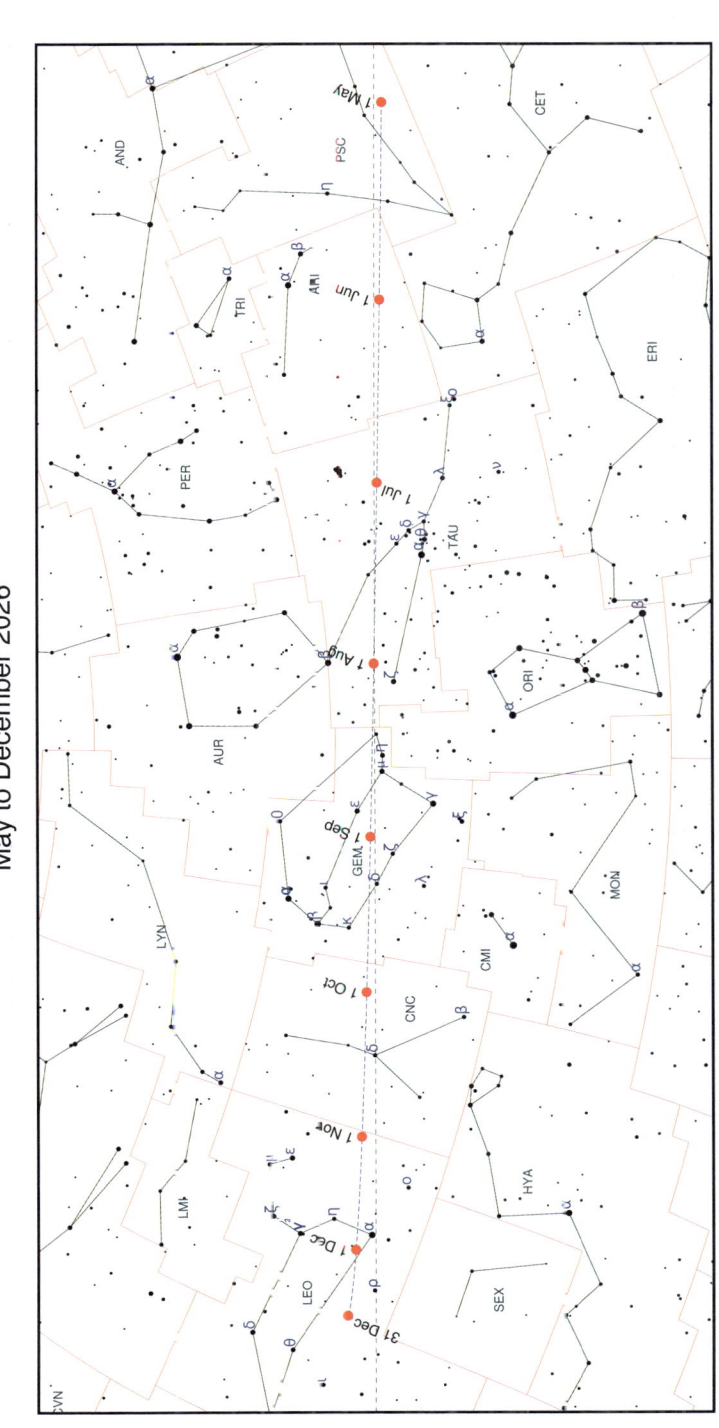

Background stars are shown to magnitude +5.5. (David Harper)

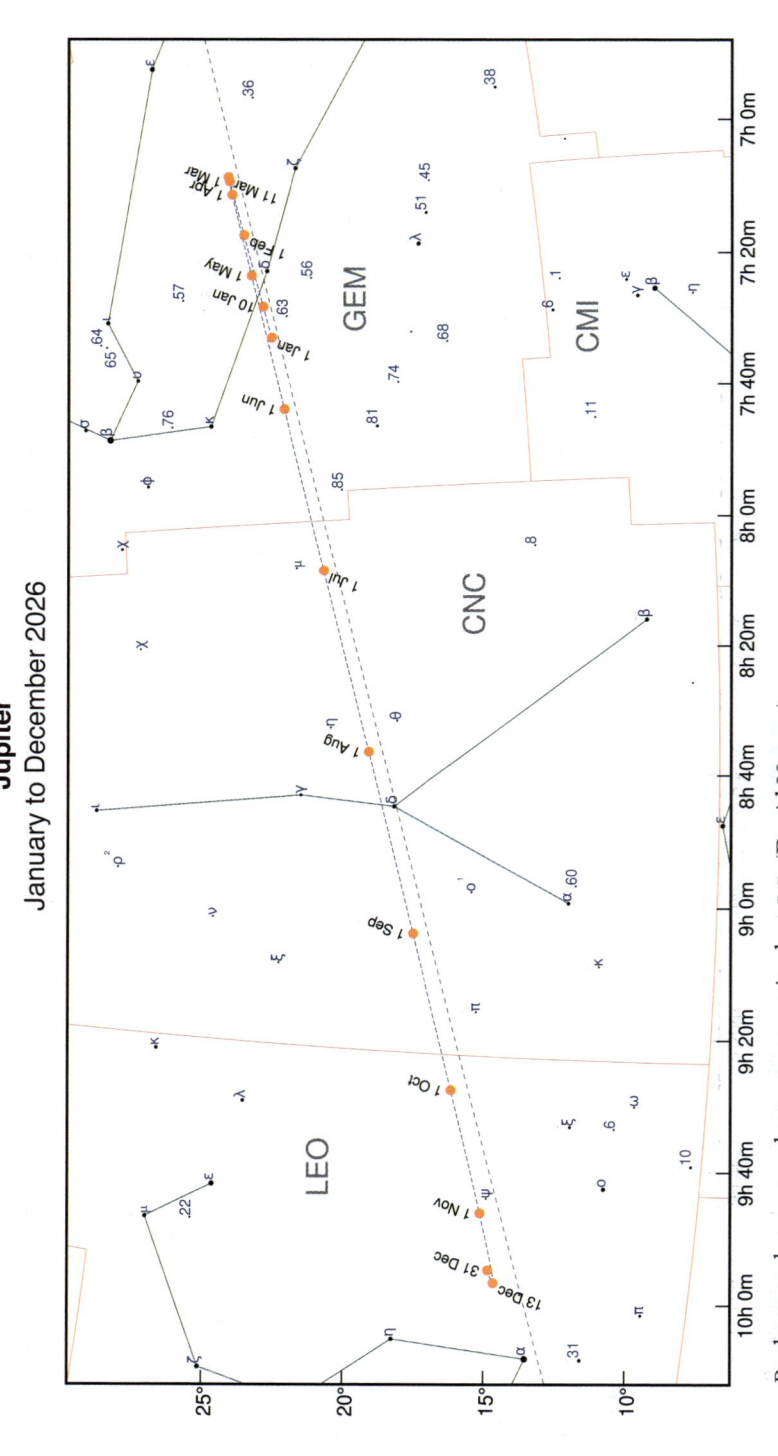

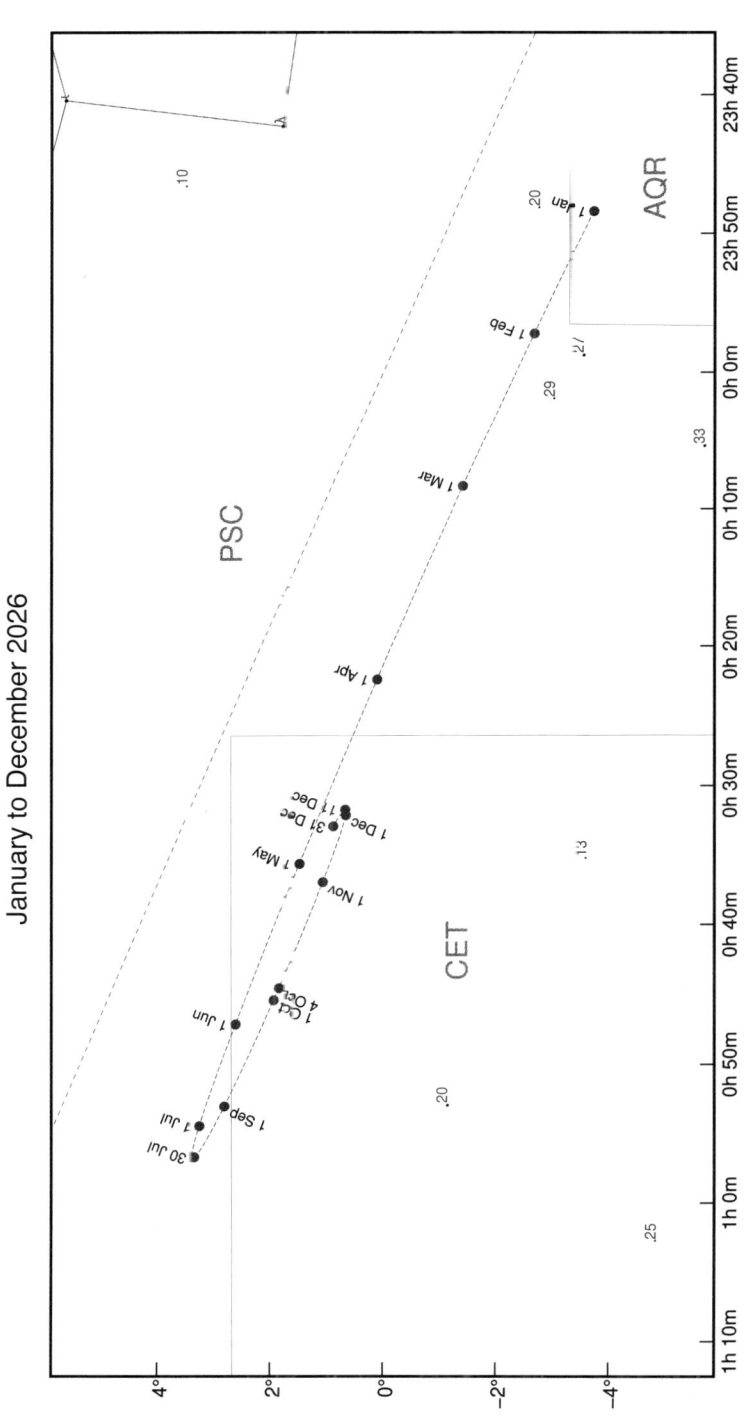

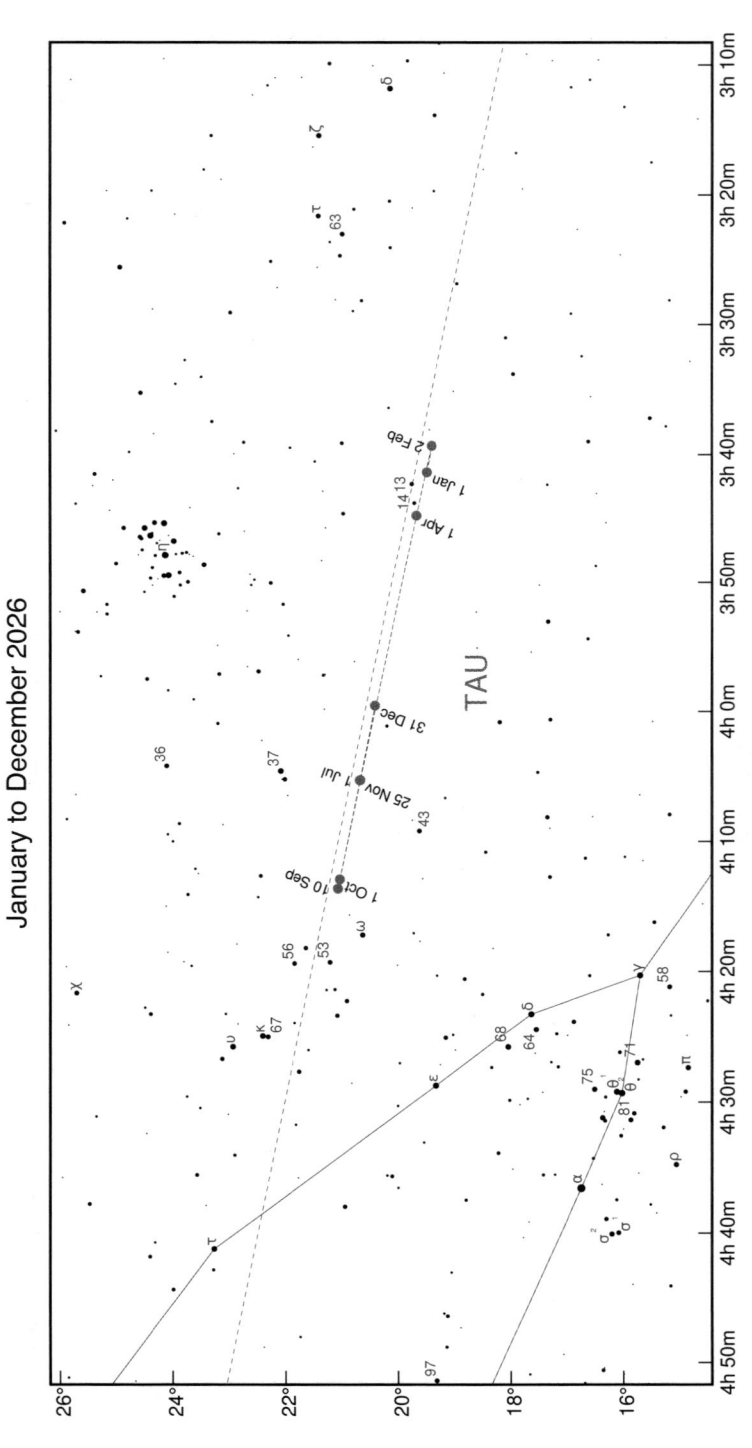

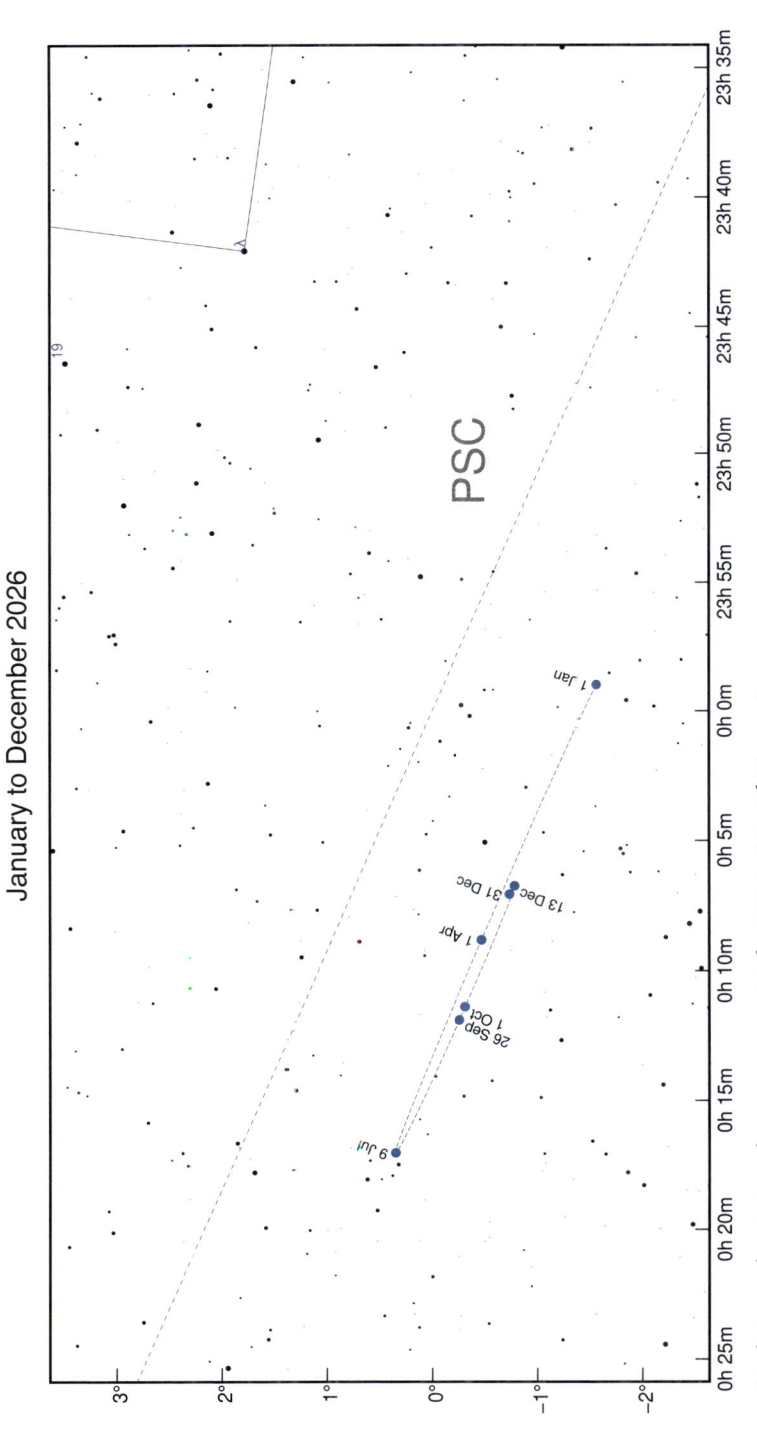

# Lunar Phenomena and Eclipses in 2026

The dates and circumstances of all lunar phenomena are calculated from the JPL DE406 solar system ephemeris. All dates are given in UT. Positions are geocentric apparent places, referred to the equator and equinox of date.

## Phases of the Moon

The apparent shape of the Moon is ever changing; this is called its *phase*. A *lunation* or lunar month typically begins with the unlit New phase. The Moon's illuminated portion becomes larger, progressing through the waxing crescent, first quarter, and waxing gibbous phases before becoming fully lit. After Full Moon, the phase decreases, going through waning gibbous, last quarter, and waning crescent, before returning to New and the beginning of the next lunation. This cycle of phases is called a *synodic month* and its average length is 29.5 days.

| BLN | New | First Quarter | Full | | Last Quarter |
|---|---|---|---|---|---|
| 1274 | | | 3 January | 1979″ | 10 January |
| 1275 | 18 January | 26 January | 1 February | 1931″ | 9 February |
| 1276 | 17 February | 24 February | 3 March | 1874″ | 11 March |
| 1277 | 19 March | 25 March | 2 April | 1822″ | 10 April |
| 1278 | 17 April | 24 April | 1 May | 1784″ | 9 May |
| 1279 | 16 May | 23 May | **31 May** | 1765″ | 8 June |
| 1280 | 15 June | 21 June | 29 June | 1769″ | 7 July |
| 1281 | 14 July | 21 July | 29 July | 1794″ | 6 August |
| 1282 | 12 August | 20 August | 28 August | 1836″ | 4 September |
| 1283 | 11 September | 18 September | 26 September | 1890″ | 3 October |
| 1284 | 10 October | 18 October | 26 October | 1944″ | 1 November |
| 1285 | 9 November | 17 November | 24 November | 1987″ | 1 December |
| 1286 | 9 December | 17 December | **24 December** | 2010″ | 30 December |

The first column is the Brown Lunation Number.[1] The fifth column gives the apparent diameter of the Full Moon in arc-seconds. Dates of the largest and smallest Full Moons are given in bold.

---

1. Brown, E. W. (1933). The Motion of the Moon, 1923–1931. *Monthly Notices of the Royal Astronomical Society*, **93** (8), 603–619. **doi.org/10.1093/mnras/93.8.603**

# Apsides

An *apsis* (plural: apsides) is either the nearest or the farthest point in an elliptical orbit of one body about another. *Periapsis* is the nearest point and *apoapsis* is the farthest point. In the case of Earth, the term *perigee* is used to describe the point where a body (in this case, the Moon) is closest to Earth and *apogee* is where a body is most distant from Earth. The difference between lunar perigee and apogee distances amounts to several tens of thousands of kilometres. The line of apsides of the Moon is not fixed in space but moves approximately 41° eastwards each year, making one full circuit around the Earth in 8.85 years. In contrast, the apsidal precession of Earth around the Sun takes about 112,000 years. The time it takes the Moon to return to the same apsis is just over 27.5 days and is called an *anomalistic month*.

| Perigee | Distance (km) | Apogee | Distance (km) |
|---|---|---|---|
| 1 January | 360,348 | 13 January | 405,438 |
| 29 January | 365,871 | 10 February | 404,576 |
| **24 February** | 370,135 | 10 March | 404,384 |
| 22 March | 366,857 | 7 April | 404,970 |
| 19 April | 361,630 | 4 May | 405,839 |
| 17 May | 358,075 | 1 June | 406,366 |
| 14 June | 357,197 | 28 June | 406,265 |
| 13 July | 359,112 | 25 July | 405,548 |
| 10 August | 363,284 | 22 August | 404,642 |
| 6 September | 368,259 | **19 September** | 404,221 |
| 1 October | 369,334 | 16 October | 404,643 |
| 28 October | 364,407 | 13 November | 405,619 |
| 25 November | 359,347 | **11 December** | 406,419 |
| **24 December** | 356,650 | | |

Distances are measured from Earth's centre to Moon's centre. Dates of annual maxima/minima are marked in **bold**.

# Nodes

The orbit of the Moon is inclined with respect to the ecliptic by just over 5° and a lunar *node* is a point in the Moon's orbit where it intersects the ecliptic plane. The *ascending node* is where the Moon crosses the ecliptic south to north and the *descending node* is where the Moon crosses the ecliptic north to south. The time it takes for the Moon to pass through the same node twice in succession is called a *draconic month*; this is just over 27 days long. These nodes are not fixed in space but precess westward by around 19° each year, making one full circuit every 18.6 years. Solar and lunar eclipses only take place when a New or Full Moon occurs very near to a node crossing. For details on this year's eclipses, see the article *Eclipses* following *Lunar Occultations*.

| Ascending | | Descending | |
|---|---|---|---|
| | | 7 January | |
| 22 January | | 3 February | |
| **18 February** | New Moon occurs 18 hours before node crossing: **annular solar eclipse on 17 February** | **3 March** | Full Moon occurs 7 hours after node crossing: **total lunar eclipse on 3 March** |
| 17 March | | 30 March | |
| 13 April | | 26 April | |
| 11 May | | 23 May | |
| 7 June | | 19 June | |
| 4 July | | 17 July | |
| 31 July | | **13 August** | New Moon occurs 16 hours before node crossing: **total solar eclipse on 12 August** |
| **27 August** | Full Moon occurs 10 hours after node crossing: **partial lunar eclipse on 28 August** | 9 September | |
| 24 September | | 7 October | |
| 21 October | | 3 November | |
| 17 November | | 30 November | |
| 14 December | | 27 December | |

Dates of nodal crossings resulting in eclipses are marked in **bold**.

# Lunar Occultations

A *lunar occultation* occurs when the Moon passes in front of a more distant celestial body as seen from Earth, partially or totally obscuring this object and blocking its light. Like total solar eclipses, occultations are seen only at particular times and from particular places, but they have been noticed for centuries; the first known observation dates to 357 BCE when Aristotle spotted Mars disappearing behind the Moon's disk. Occultation data can help astronomers measure the heights and depths of lunar features and improve knowledge about the lunar orbit; such information can be used to detect close companions in multiple star systems and improve the precision of stellar positions. Lunar occultation data was even employed to pinpoint the location of 3C 273, the first quasar to be identified by astronomers.

## Table of Lunar Occultations in 2026

|  | Jan | Feb | Mar | Apr | May | Jun | Jul | Aug | Sep | Oct | Nov | Dec |
|---|---|---|---|---|---|---|---|---|---|---|---|---|
| α Scorpii (Antares) | 14 | 11 | 10 | 6 | 4, 31 | 27 | 24 | 21 | 17 | 14 | 11 | 8 |
| α Leonis (Regulus) | 6 | 3 | 2, 29 | 26 | 23 | 19 | 17 | 13 | 9 | 7 | 3, 30 | 27 |
| M44 (Praesepe/Beehive) |  |  |  |  |  | 18 | 15 | 11 | 8 | 5 | 1, 29 | 26 |
| M45 (Pleiades) | 27 | 24 | 23 | 19 | 17 | 13 | 10 | 7 | 3, 30 | 28 | 24 | 21 |
| Mercury |  | 18 |  |  |  |  |  |  |  |  |  |  |
| Venus |  |  |  |  |  | 17 |  |  | 14 |  | 7 |  |
| Mars |  | 16 |  |  |  |  |  |  |  | 5 | 2 |  |
| Jupiter |  |  |  |  |  |  |  |  | 8 | 6 | 2, 30 |  |

The dates given in the above table are based on Universal Time (UT). Occultations of M44 (Praesepe/Beehive) refer to sixth-magnitude star ε Cancri (Meleph). Occultations of M45 (Pleiades) refer to third-magnitude star η Tauri (Alcyone).

This year, the Moon regularly occults two first-magnitude stars – Antares and Regulus – as well as the brightest stars in the Beehive and Pleiades open clusters. Mercury is near greatest elongation east when it is blotted out by the Moon in February, both Venus and Mars are claimed three times over the course of the year, and Jupiter vanishes behind the disk of the Moon four times. Details of individual occultations listed in the above table can be found in the corresponding sections of the *Monthly Sky Notes*.

Visit the website of the International Occultation Timing Association (IOTA) at **occultations.org** for more information on lunar and planetary occultations, including observing predictions for the current year.

# Eclipses in 2026

Solar and lunar eclipses do not happen every month but take place only when a New or Full Moon occurs very near to a node crossing (see the *Nodes* section earlier in this article). The New Moon's shadow covers only a small portion of the Earth's surface so a solar eclipse is visible only from a limited region of our planet. A lunar eclipse, on the other hand, is potentially visible from all locations on the night side of Earth. Depending on the exact location of the observer, the eclipse may already be in progress as the Moon rises (or sets), or the entire eclipse sequence may be visible. There are a minimum of four eclipses in any one calendar year, comprising two solar eclipses and two lunar eclipses. In fact, most years have only four although it is possible to have five (2028), six (2029), or even seven (2038) eclipses during the course of a year. This year there will be just four: an annular solar eclipse, a total lunar eclipse, a total solar eclipse, and a partial lunar eclipse.

All dates and times listed below are given in Universal Time (UT). It is important to note that the times quoted for each event refer to the start, maximum, and ending of the eclipse on a global scale rather than with reference to any specific location.

## Annular Solar Eclipse (Saros 121): 17 February

The first eclipse of the year is the annular solar eclipse of 17 February, the annular phase of which will only be visible from parts of Antarctica and the southern Indian Ocean. Regions seeing a partial eclipse include the southern Pacific Ocean, the southern tip of South America, South Africa, Zimbabwe, Mozambique, Madagascar and Mauritius and the southern Indian Ocean.

| Time (UT) | Event |
|---|---|
| 09:56 | Partial eclipse begins |
| 11:43 | Annular eclipse begins |
| 12:01 | New Moon |
| 12:12 | Maximum eclipse |
| 12:41 | Annular eclipse ends |
| 14:28 | Partial eclipse ends |
| 06:18 (+1 day) | Moon at ascending node |

## Total Lunar Eclipse (Saros 133): 3 March

The whole or parts of the total lunar eclipse of 3 March will be visible from China, Japan and most of eastern Asia, eastern Australia, New Zealand, the Pacific Ocean, the westernmost regions of South America, Alaska, much of Canada and the United States of America.

| Time (UT) | Event |
|---|---|
| 04:35 | Moon at descending node |
| 08:44 | Penumbral eclipse begins |
| 09:50 | Partial eclipse begins |
| 11:04 | Total eclipse begins |
| 11:34 | Maximum eclipse |
| 11:38 | Full Moon |
| 12:03 | Total eclipse ends |
| 13:17 | Partial eclipse ends |
| 14:23 | Penumbral eclipse ends |

## Total Solar Eclipse (Saros 126): 12 August

On 12 August there will be a total solar eclipse, the path of totality of which will commence in the Arctic Ocean near the northern coast of Russia before travelling across Greenland and Iceland and the northern reaches of the Atlantic Ocean before ending in Spain and the Balearic Islands. Locations from which a partial eclipse will be seen include Alaska, Canada, the northern United States, parts of northern Europe and much of western Africa.

| Time (UT) | Event |
|---|---|
| 15:34 | Partial eclipse begins |
| 16:58 | Total eclipse begins |
| 17:37 | New Moon |
| 17:46 | Maximum eclipse |
| 18:34 | Total eclipse ends |
| 19:58 | Partial eclipse ends |
| 09:56 (+1 day) | Moon at descending node |

## Partial Lunar Eclipse (Saros 138): 28 August

There will be a partial lunar eclipse on 28 August which will be visible from Antarctica, Africa and Europe, the Atlantic Ocean, Central and South America and most of North America and the Pacific Ocean.

| Time (UT) | Event |
| --- | --- |
| 18:47 (−1 day) | Moon at ascending node |
| 01:23 | Penumbral eclipse begins |
| 02:33 | Partial eclipse begins |
| 04:13 | Maximum eclipse |
| 04:18 | Full Moon |
| 05:52 | Partial eclipse ends |
| 07:02 | Penumbral eclipse ends |

The *saros* is the period of 6585.32 days (18 years 10.32 or 11.32 days), after which solar or lunar eclipses recur in almost identical circumstances, due to the Earth, Moon and Sun returning to almost the same relative positions. A *saros series* is a series of eclipses which recur at intervals of this period; there are many interleaved *saros series* running concurrently.

A fuller description/definition of *saros* and other astronomical terms used in this volume can be found in the downloadable Glossary available on the *Yearbook of Astronomy* website at **yearbookofastronomy.com**

Eclipse circumstances, with maps, are available from a variety of online sources, including **EclipseWise.com** (Fred Espenak), **ssp.imcce.fr/forms** (Institut de mécanique céleste et de calcul des éphémérides), and **aa.usno.navy.mil** (United States Naval Observatory).

All eclipse circumstances are taken from **EclipseWise.com**, Fred Espenak's web site dedicated to solar and lunar eclipse predictions and information.

# Monthly Sky Notes and Articles

**Evening Apparition of Venus**
6 January 2026 to 24 October 2026

52° North
35° South

# January

| 1 | Saturn | Minimum ring opening (−1.0°) | 76° |
|---|---|---|---|
| 3 | Moon | Full Moon | 176° |
| 3 | Earth | Perihelion | |
| 3/4 | Earth | Quadrantids (ZHR 120) | |
| 6 | Venus | Superior conjunction: morning → evening | 1° |
| 6 | Moon, α Leonis (Regulus) | 0.4° apart: occultation | 136° |
| 9 | Mars | Conjunction | 1° |
| 10 | Jupiter | Opposition | 180° |
| 10 | Moon | Last Quarter Moon | 90° |
| 14 | Moon, α Scorpii (Antares) | 0.6° apart: occultation | 45° |
| 18 | Moon | New Moon | 3° |
| 19 | Jupiter, δ Geminorum (Wasat) | 0.5° apart | 170° |
| 21 | Mercury | Superior conjunction: morning → evening | 2° |
| 26 | Moon | First Quarter Moon | 90° |
| 27 | Moon, M45 (Pleiades) | 1.1° apart: occultation | 112° |

Dates are based on Universal Time (UT). Occultations of M45 (Pleiades) refer to third-magnitude star η Tauri (Alcyone). Peak activity dates and ZHRs for meteor showers are estimates. The last column gives the approximate elongation from the Sun of the event.

**Mercury** inhabits the morning sky at the outset of 2026 but is losing altitude quickly. A close encounter with Mars on 18 January takes place too close to superior conjunction to be visible. Afterwards, Mercury begins its first evening apparition of the year. It is best seen from the southern hemisphere this month but ultimately, this evening appearance favours northern latitudes.

**Venus** is missing! Although technically the morning star at the outset of 2026, Venus is only a degree away from the Sun and impossible to see. Superior conjunction takes place on 6 January after which Venus moves into the evening sky. However, it will remain extremely low in the west immediately after sunset for the rest of the month. Venus approaches Mars on 8 January and Mercury drops by near the end of the month but all three planets are less than 6° from the Sun, rendering these encounters unobservable.

**Earth** reaches perihelion, when it is closest to the Sun, on 3 January. We are approximately 0.9833 au (147,100,000 kilometres) from our star at 17:30 UT on this date. (Compare these numbers to the aphelion figures in early July.) The Full Moon of this date ruins observations of the Quadrantid meteor shower which peaks around this time. See *Meteor Showers in 2026* for more information.

The **Moon** occults two first-magnitude stars this month, Regulus on 6 January and Antares eight days later. The waxing gibbous Moon also passes through the Pleiades open star cluster on 27 January.

**Comet 24P/Schaumasse** reaches perihelion on 8 January, with its closest approach to Earth taking place a few days earlier. This short-period Jupiter-family comet will probably get no brighter than ninth magnitude as it passes through Virgo this month. See *Comets in 2026* for more information on this icy visitor.

**Mars** is at conjunction this month. The close passages of Venus and Mercury take place too close to the Sun to observe, as does the Martian encounter with the globular cluster M75 on 22 January. Mars begins the year in the constellation of Sagittarius but moves to Capricornus on 23 January.

**Jupiter** begins the year in reverse, moving in retrograde through the constellation of Gemini. It rises at sunset and on the second day of the month, is just over a degree north of the planetary nebula C39 (NGC 2372). Discovered by William Herschel in 1787, this is a tenth-magnitude nebula, and long-exposure photography is necessary to bring out the details in the nebulosity. When opposition takes place on 10 January, Jupiter is 4.2 au distant and shining at magnitude −2.7. In a telescope, it appears as a disk approximately 46.5″ in diameter. Nine days after opposition, Jupiter passes 0.5° north of fourth-magnitude Wasat (δ Geminorum), an F-type main sequence star and spectroscopic binary.

**Saturn** enters 2026 with its rings open a mere 1° relative to Earth. However, this tilt will increase over the next six months, affording a much better view later in the year. Saturn starts in the constellation of Aquarius but moves into Pisces mid-month. It sets before midnight so look for the first-magnitude planet as soon as the skies get dark.

**Uranus** is located in Taurus, currently in retrograde between the Hyades and the Pleiades open star clusters. The green ice giant is well-placed for viewing in the evening hours, not setting until well after midnight for observers in northern temperate latitudes. At sixth magnitude, this planet is a difficult naked-eye object so choose a dark night to seek it out.

**Neptune** is an evening sky object although it sets before midnight for all observers. Found in the constellation of Pisces, this eighth-magnitude object is best viewed with a telescope on a moonless night when the sky is at its darkest.

# Gone But Not Forgotten
## Argo Navis

### Lynne Marie Stockman

Today the professional astronomical community recognises 88 constellations but where did they originate? Nearly half of the figures are relatively modern, with Dutch astronomer-cartographer Petrus Plancius (1552–1622) responsible for 16, Polish astronomer Johannes Hevelius (1611–1687) for seven, and French astronomer Nicolas-Louis de Lacaille (1713–1762) for 17. Except for Coma Berenices, elevated from asterism to constellation in the early sixteenth century, the remaining 47 constellations hark back to the *Almagest*, a second-century treatise on astronomy written by Greek mathematician and astronomer Claudius Ptolemy (c.100–c.170). But the *Almagest* outlined 48 constellations. What happened to Argo Navis?

'Argo Navis' is Latin for 'the ship Argo' and is named after the famous vessel from Greek myth. The Argo was the oared galley in which Jason and his crew, the Argonauts, sailed on their quest to acquire the Golden Fleece. The constellation covers a huge area, spanning five hours in right ascension, and 60° or more in declination. It also contains many bright stars, including Canopus (the second brightest star in the night sky), eight second-magnitude stars, and over a dozen at third magnitude.

In antiquity, the constellation was visible as far north as the Mediterranean. Ptolemy assigned 45 stars to the figure, missing out many of the stars near the prow of the ship, possibly because they did not rise high enough to be easily observed. Arabic astronomers knew Canopus as *Suhayl* (this name, spelled differently, is now assigned to λ Velorum), with nearby τ Puppis called *Bulqayn*, a proper name of unknown origin. The trio of second- and third-magnitude stars, $γ^2$ Velorum, δ Velorum, and τ Velorum, were also known by variations on the name of 'Suhayl', and were part of a larger complex encompassing the modern constellations of Vela and most of Carina regarded as 'the journeying camels' (al-aʿbār) although some authorities give them the name 'the wild asses' (al-aʿyār) or 'the cows' (al-baqar).

In India, the star Canopus was identified with Agastya, a Hindu sage. In the far east, the bright star was often known as an old man. Traditional Chinese astronomy divided the sky into Three Enclosures encompassing the northern heavens surrounding the pole, with the rest of the stars residing in one of Four Symbols.

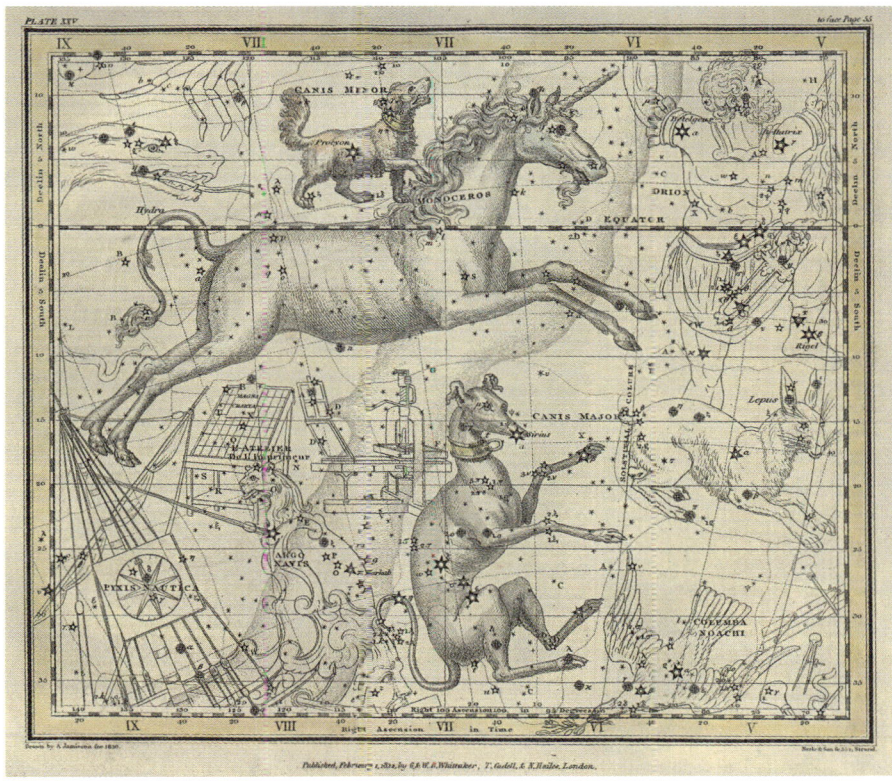

Argo Navis sails off the lower left side of Plate XXV from Alexander Jamieson's 1822 *Celestial Atlas*. (Wikimedia Commons / Alexander Jamieson / United States Naval Observatory Library)

The parts of Argo Navis that were visible from China resided in the Jǐng (well) and Guǐ (ghost) mansions in the Symbol of the Vermillion Bird of the South. Canopus itself was variously known as Lǎorénxīng (the star of the old man), Shòu xīng (the god of longevity), or Nánjílǎorén (the old man of the south pole), amongst other titles. Not only was the star revered by numerous southern hemisphere societies, Polynesian sailors also used it as an important navigational aid. Canopus still serves a similar purpose today; because of its brightness and position far from the ecliptic, spacecraft use both it and the Sun to navigate between the planets.

Lacaille examined the southern skies from the Cape of Good Hope between 1750 and 1754. The catalogue that resulted from his observations, *Coelum Australe Stelliferum*, was published in 1763, the year after his death. In it, Lacaille introduced a number of new constellations, many of which are still in use today. For the stars in Argo Navis brighter than third magnitude, he re-used (subject to occasional alterations) the original Greek lettering scheme introduced by Johann Bayer (*Uranometria*, 1603) and labelled them simply as 'Argûs', the Latin genitive of 'Argo'. For many of the fainter stars, Lacaille subdivided the large constellation into more manageable chunks. These stars were assigned lower- and upper-case Latin letters and appended with 'Argûs in carina', 'Argûs in puppi' or 'Argûs in velis', depending on their position on the ship (the keel, the poop deck, or the sails). The faintest unlettered stars were also simply labelled 'Argûs'.

Despite being sliced up by Lacaille, Argo Navis remained afloat; some astronomers embraced Lacaille's scheme whilst others kept Argo intact. For example, French astronomer Joseph Jérôme Le Français de Lalande used 'Navire' (the French equivalent of 'Navis') in his 1801 star catalogue whilst Italian astronomer Guiseppe Piazzi used Lacaille's notation two years later in his Palermo star catalogue. Star maps tended to label the constellation simply as 'Argo' or as 'Navis', not its constituent parts. The celebrated Harvard Observatory computer Antonia Maury used the 'Argo' nomenclature right up to the end of the nineteenth century, but when the observatory released the *Henry Draper Catalogue* between 1918 and 1924, the stars were labelled 'Carina', 'Puppis' and 'Vela'.

When the International Astronomical Union (IAU) met in 1922, one of their first tasks was to decide which constellations were canonical and which were not. They began by assigning three-letter abbreviations to the various constellations: Andromeda (And), …, Carina (Car), …, Puppis (Pup), …, Vela (Vel), …, Vulpecula (Vul). The IAU ended up defining 88 principal constellations, but Argo Navis was so embedded in the astronomical mindset that it too was assigned an abbreviation, Arg. Belgian astronomer Eugène Joseph Delporte finally sank the ship in the 1930s when he drew the modern boundaries of the IAU-sanctioned constellations. Carina, Puppis and Vela were here to stay.

And yet the Argo sails on. The Lacaille-inspired Greek and Latin lettering system for the figure – Greek letters for the brightest stars of the entire ship, Latin letters for the fainter stars in each of the three parts of the ship – was retained by the IAU. For this reason, only Carina has an alpha and beta star, gamma and delta reside in Vela, the highest-ranking letter in Puppis is zeta, and so forth. The Greek letters preserve the fossilised remains of Jason's legendary ship.

The principal stars of Argo Navis are detailed in the following table.

| Star | Name | Notes |
|---|---|---|
| α Car | Canopus | The name is derived from the Greek proper name Κάνωβος. At magnitude −0.74, Canopus is second in brightness only to Sirius (α Canis Majoris). |
| β Car | Miaplacidus | The name is thought to be derived from the Arabic *miyāh* (waters) and the Latin *placidus* (calm). |
| γ Vel | | The radiant of the December Puppid-Velid meteor shower is in the vicinity of this second-magnitude star system. It will be the southern pole star, with a minimum separation from the true pole of 3.1°, for seven centuries beginning around 10,850 CE. |
| δ Vel | Alsephina | The name is derived from the Arabic *al-safinah* (ship). This second-magnitude star system will be the southern pole star, with a minimum separation from the true pole of 0.2°, for seven centuries beginning around 8750 CE. |
| ε Car | Avior | The name was invented by H.M. Nautical Almanac Office in the early twentieth century at the behest of the Royal Air Force. |
| ζ Pup | Naos | The name is derived from the Greek ναύς (ship). |
| η Car | | Recorded as fourth magnitude in the late sixteenth century, it erupted to become the second brightest star in the sky in 1843 before fading. Now the star is enclosed by the Homunculus Nebula, a small cloud of gas and dust thought to have formed in the mid-nineteenth century during the Great Eruption event |
| θ Car | | This third-magnitude star is the brightest member of the open star cluster IC 2502 which is also known as the 'Southern Pleiades'. It was last the southern pole star in 45,000 BCE and will be again in 31,200 CE. |
| ι Car | Aspidiske | The name is derived from the Greek Ἀσπιδίσκη (little shield). This second-magnitude star will be the southern pole star, with a minimum separation from the true pole of 0.2°, for five centuries beginning around 7700 CE. |
| κ Vel | Markeb | The name is derived from the Arabic *markab* (something to ride). This second-magnitude spectroscopic binary star was last the southern pole star in 17,400 BCE and will be again in 33,900 CE. |
| λ Vel | Suhail | The name is derived from the Arabic proper name *Suhayl*. |
| μ Vel | | A huge and unexpected X-ray flare from the primary star in this visual binary was detected in 1998. |
| ν Pup | Pipit | The name is derived from the word for 'sparrow' by the Dayak Kanayatn people of West Kalimantan province, Borneo, Indonesia. This third-magnitude, rapidly-rotating, oblate B-type giant star will be the southern pole star, with a minimum separation from the true pole of 2.2°, for 14 centuries beginning around 12,800 CE. |
| ξ Pup | Azmidi | The name is derived from the Greek Ἀσπιδίσκη (little shield). |

| Star | Name | Notes |
|---|---|---|
| o Vel | | This fourth-magnitude variable star is the brightest member of the open star cluster IC 2391. Lacaille labelled this star as (omicron) o Argûs and another fainter star in Puppis as (oh) o Argûs in Puppi. It will be the southern pole star, with a minimum separation from the true pole of 1.0°, for seven centuries beginning around 9500 CE. |
| π Pup | | The radiant of the April Pi (π) Puppids meteor shower is in the vicinity of this star. |
| ρ Pup | Tureis | The name is derived from the Arabic *turays* (little shield). |
| σ Pup | | This third-magnitude star system will be the southern pole star, with a minimum separation from the true pole of 4.5°, for ten centuries beginning around 11,600 CE. |
| τ Pup | | This third-magnitude star was discovered to be a long-period spectroscopic binary in 1908 when only six such stars were known. It will be the southern pole star, with a minimum separation from the true pole of 5.0°, for two centuries beginning around 12,600 CE. |
| υ Car | | This third-magnitude visual binary will be the southern pole star, with a minimum separation from the true pole of 0.8°, for seven centuries beginning around 6300 CE. |
| φ Vel | | This brilliant blue supergiant was last the southern pole star in 43,300 BCE. |
| χ Car | | Along with the Sun, this third-magnitude variable star lies within the Local Bubble, a region of relatively low gas density thought to be the result of a supernova explosion some millions of years ago. It will be the southern pole star, with a minimum separation from the true pole of 2.9°, for six centuries beginning around 10,200 CE. |
| ψ Vel | | This visual binary is made up of two F-type subgiants. |
| ω Car | | This third-magnitude star will be the southern pole star, with a minimum separation from the true pole of 0.8°, for nine centuries beginning around 5400 CE. |

Pole star calculations are taken from *Mathematical Astronomy Morsels V* by Jean Meeus (2009: Willmann-Bell, Inc.), pp. 353–363, and include the effects of both spatial proper motion (proper motion, radial motion, and distance of the star) and long-term precession.

## Further Reading

Allen, Richard Hinkley. (1963). *Star Names: Their Lore and Meaning*. Dover Publications, Inc.

Barentine, John C. (2016). *The Lost Constellations: A History of Obsolete, Extinct, or Forgotten Star Lore*. Springer Praxis Books.

Ridpath, Ian. (2018). *Star Tales: Revised and Expanded Edition*. The Lutterworth Press.

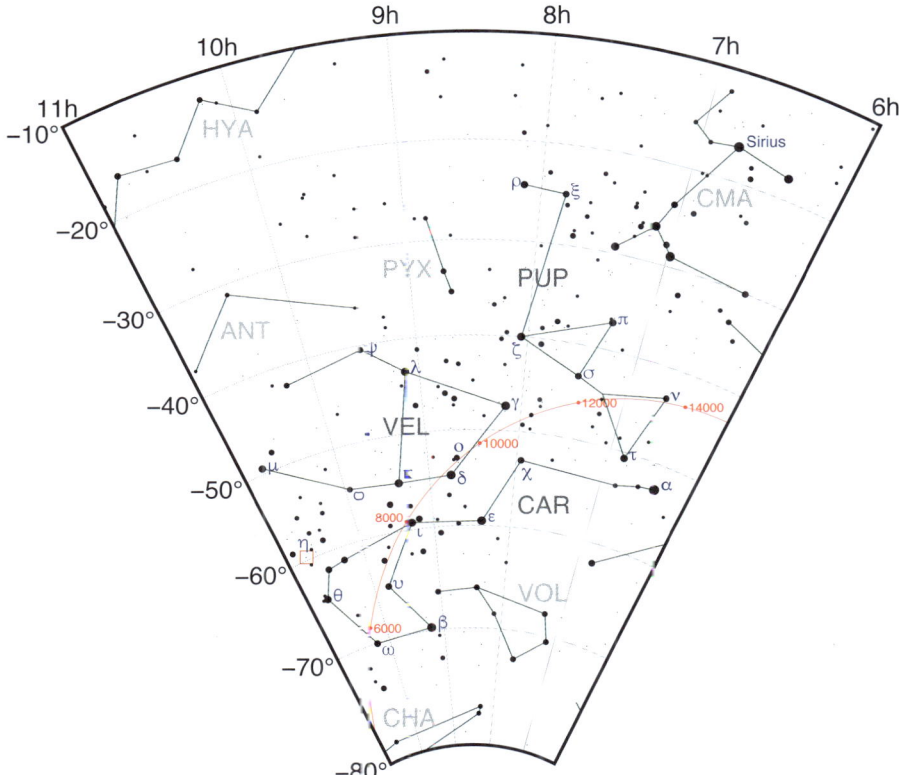

**The stars of Argo Navis.** Precession will bring a number of the bright stars of Argo Navis (Carina, Puppis, Vela) into contention as the southern pole star over the coming millennia. The red curve is the path of the south celestial pole from 4500 CE to 14,500 CE. Stars are shown to magnitude +5.5. (David Harper)

# February

| | | | |
|---|---|---|---|
| 1 | Moon | Full Moon | 178° |
| 3 | Moon, α Leonis (Regulus) | 0.3° apart: occultation | 164° |
| 9 | Moon | Last Quarter Moon | 90° |
| 11 | Moon, α Scorpii (Antares) | 0.7° apart: occultation | 72° |
| 16 | Saturn, Neptune | 0.9° apart: conjunction in right ascension | 33° |
| 16 | Uranus | East quadrature | 90° |
| 16 | Moon, Mars | 0.7° apart: occultation | 9° |
| 17 | Moon | New Moon: annular solar eclipse | 1° |
| 18 | Moon, Mercury | 0.1° apart: occultation | 18° |
| 19 | Mercury | Greatest elongation east (evening) | 18° |
| 20 | Saturn, Neptune | 0.8° apart: conjunction in ecliptic longitude | 29° |
| 24 | Moon, M45 (Pleiades) | 1.2° apart: occultation | 85° |
| 24 | Moon | First Quarter Moon | 90° |

Dates are based on Universal Time (UT). Occultations of M45 (Pleiades) refer to third-magnitude star η Tauri (Alcyone). The last column gives the approximate elongation from the Sun of the event.

**Mercury** appears in the west after sunset and is best seen from northern temperate latitudes. On 18 February, the waxing crescent Moon passes 0.1° south of the tiny planet, leading to an occultation for some lucky observers. A greatest elongation east of 18.1° takes place the following day. Mercury enters into retrograde motion late in the month and makes a distant pass by Venus on the last day of February. As is always the case with evening apparitions, Mercury begins bright but dims throughout its western appearance as the planet races toward Earth following superior conjunction. The reason is simple: although the planet's apparent diameter is getting larger (5.0″ to 9.4″), its phase is getting smaller (97% to 10%) and ultimately, it is the diminishing phase that wins out. Accordingly, Mercury shines a magnitude −1.2 at the beginning of February but is at +2.1 by the end of it.

**Venus** is the evening star but is extremely low in the west after sunset and barely 11° from the Sun by the end of the month. The new crescent Moon, less than a day old, passes 1.5° north of Venus on 18 February when the two objects are just 10° from the Sun. Mercury comes to call on the last day of the month but the two planets are separated by 4.5°. Venus is at its minimum magnitude of −3.9 for most of the month.

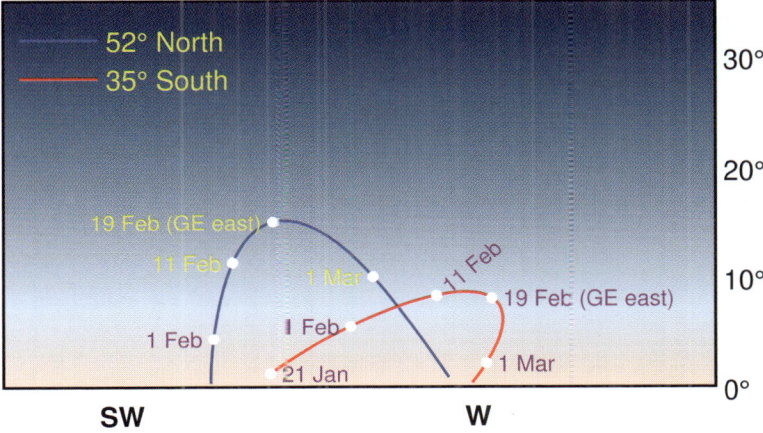

## Evening Apparition of Mercury
### 21 January to 7 March

The **Moon** continues its monthly obscurations of the first-magnitude stars Regulus (3 February) and Antares (11 February) as well as an occultation of the Pleiades (24 February). However, a planet joins this mix on 16 February when the waning crescent Moon occults Mars. Unfortunately, this disappearance of the red planet behind the disk of the Moon takes place less than 10° from the Sun. On 18 February, another planet, Mercury, is occulted by the waxing crescent Moon in an event that may be visible from sunset. In this instance, both Mercury and the Moon will be 18° from the Sun. Splitting these two occultations is an annular solar eclipse on 17 February when the New Moon casts its shadow on Earth. More information on observing circumstances is available in the *Eclipses* article preceding the *Monthly Sky Notes*. Finally, the most distant perigee of the year occurs on 24 February when the Moon comes to within 370,135 kilometres of Earth.

**Mars** is extremely low in the east in dawn skies this month. Conjunction took place in January and Mars is only reluctantly parting ways with the Sun. The red planet passes close by fourth-magnitude stars θ Capricorni, ι Capricorni, and Nashira (γ Capricorni) this month but these events all take place less than 10° from the Sun. The waning crescent Moon occults Mars on 16 February but this occurs less than a day before the Moon reaches its New phase. The brightest star in Capricornus is Deneb Algedi (δ Capricorni), a third-magnitude eclipsing binary, and Mars is found 1.5° north of this star on 23 February, but morning twilight should render the Deneb Algedi invisible. Mars leaves Capricornus behind just before the end of the month when it enters Aquarius.

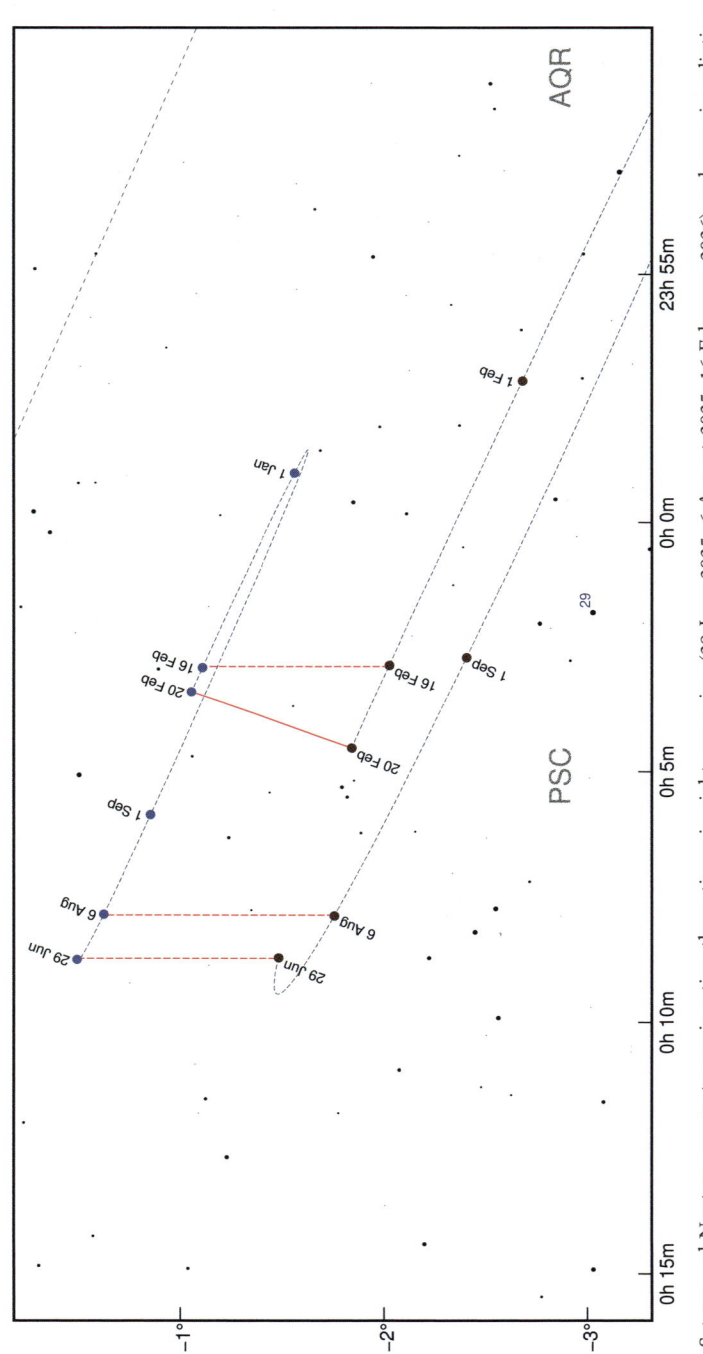

Saturn and Neptune come to conjunction three times in right ascension (29 June 2025, 6 August 2025, 16 February 2026) and once in ecliptic longitude (20 February 2026). The two planets were last in conjunction in 1989 and will not meet again until 2061. (David Harper)

**Jupiter** was at opposition last month and is already above the horizon when the Sun sets. Shining at magnitude −2.6, the largest planet in the solar system is retrograding through the constellation of Gemini.

**Saturn** and Neptune are less than a degree apart in Pisces. They have been participating in a triple conjunction in right ascension since last year; the final leg takes place on 16 February. But that is not the end of the story. The two planets also have a single conjunction in ecliptic longitude which occurs on 20 February. Saturn's ring tilt increases to $3.6°$ by the end of the month but the planet is getting more difficult to see in the west as it heads toward conjunction next month.

**Uranus** ends its retrogression in the opening days of February; it will remain in direct motion until September. The distant world reaches east quadrature mid-month and is best viewed from the northern hemisphere where it sets after midnight. Located in Taurus, Uranus shines at magnitude +5.7 so keen eyes and dark skies are required if you wish to see this planet without telescopic aid.

**Neptune** sets mid- to early evening so seek it out in Pisces as soon as skies darken. It undergoes its third and final conjunction in right ascension with Saturn on 16 February and then four days later, aligns with the ringed planet for a single conjunction in ecliptic longitude. Interestingly, Saturn and Neptune were in the middle of triple conjunctions in both right ascension and ecliptic longitude when Neptune was discovered on 23 September 1846. A telescope will be necessary to spot this eighth-magnitude planet.

# The Cosmic Microwave Background:
## Black Body Radiation

### David M. Harland

Owing largely to a serendipitous discovery made by astronomers at Bell Laboratories in 1965, it is now widely accepted that the 'Big Bang' created the universe in an infinitely dense and incredibly hot state. But there was a time when this possibility seemed nonsensical.

In presenting the cosmological implications of his Theory of General Relativity in 1917, Albert Einstein noted that his equations indicated the universe must be

After serving as a ground station for communications satellites in the early 1960s the microwave antenna built by Bell Laboratories at Holmdel, New Jersey, was used for astronomical research. (NASA/Wikipedia/Bammesk)

'dynamic', but astronomers believed it to be 'static' in the sense of having always existed, pretty much as we see it now, so he introduced a mathematical fix to stave off gravitational collapse.

However in 1931 Georges Lemaître at the University of Louvain in Belgium, familiar with relativity and aware of Edwin Hubble's redshift evidence for the universe being in a state of expansion, reasoned that 'winding the clock back' implied that there was a 'time zero' when all of the matter in the universe emerged from a single superdense entity. He called this the 'primeval atom' and proposed that it disintegrated in an analogous manner to spontaneous radioactive decay. Since then, the debris has been scattering.

George Gamow, a Russian-born physicist who moved to the United States in 1933 and gained a position at George Washington University in Washington, D.C., argued in 1935 that when Lemaître's superdense entity disintegrated it had issued only neutrons, and since free neutrons are unstable they would have gradually decayed into pairs of protons and electrons. He further argued that atomic nuclei were created by the late-decaying neutrons joining with protons.

This 'primordial nucleosynthesis' process was studied by his student Ralph Alpher, who found the neutron capture process to be so efficient that in order for it not to have turned all the hydrogen into helium there must have been an intense radiation field that acted to break up helium, with the outcome being a balance between all these processes.

Alpher teamed with postdoc Robert Herman, and after calculating a ratio of one billion photons of energy per nucleon (the general name for protons and neutrons in the nucleus) they realised that at time-zero this radiation field would have had the characteristics of a 'black body' with a given temperature, initially in the billions of degrees with its emission peak in the gamma-ray range.

In 1948 Gamow announced this 'hot fireball' theory. The next year, Alpher and Herman said that because the universe was expanding, the energy from the primordial radiation field would have been redshifted down to a temperature of roughly 5K (the actual value would depend on the age of the universe, which was uncertain). The peak intensity of the distribution would now lie in the microwave portion of the radio spectrum. Gamow reckoned this ought to be detectable as an isotropic field (meaning its intensity would be the same in all directions) but he did not know of any sufficiently sensitive detector.

In 1963 Robert Dicke at Princeton, was intrigued by the possibility of the universe eventually collapsing, bouncing, and expanding again. He set postgraduate Jim Peebles the task of calculating conditions in the final stage of such a collapse. In doing this, Peebles independently trod the same ground that Gamow's team had explored, and he realised that the radiation (from either a singular origin event or the start of the current cycle of the 'oscillating universe', as appropriate) ought now to have a black body temperature of about 10K. As Dicke's team was preparing an instrument to seek this field, they heard that a microwave antenna at Bell Laboratories at Holmdel, New Jersey had serendipitously found an isotropic field with a temperature of about 3K that had them baffled. In 1965 the teams published separate papers, with the Bell Laboratories team reporting the detection of an isotropic radiation field without committing themselves to an explanation, and the Princeton team explaining why an isotropic field ought to exist.

John C. Mather, with a picture on the wall showing the Cosmic Background Explorer (COBE) satellite that was launched in 1989 to measure what is often referred to as the 'afterglow' of the Big Bang. (NASA)

But did this energy field have the characteristics of a black body. The only way to find out was to measure its intensity across a broad range of wavelengths. In addition to operating detectors on the ground, they were sent aloft on stratospheric balloons and on sounding rockets, with some results being consistent with expectation and others not.

To settle the matter, in November 1989 NASA launched the Cosmic Microwave Background Explorer (COBE) satellite equipped with three instruments, one of which was designed to measure the intensity of the isotropic field at 67 wavelengths that were chosen to span the predicted peak of the distribution for a black body at about 3K. The data from the first ten minutes were sufficient to settle the issue.

When John C. Mather, the instrument's principal investigator at NASA Goddard Space Flight Center, reported the initial data to the American Astronomical Society in January 1990, he presented a graph for the rise and fall of the predicted black body, on which were superimposed the observed data. The error bars on the individual

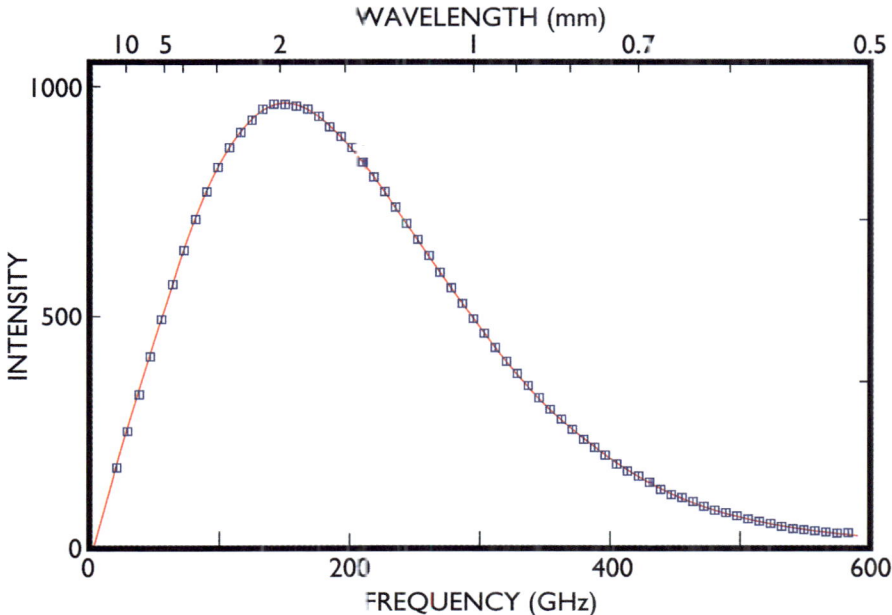

The energy spectrum of the cosmic microwave background, as measured by the Far Infrared Absolute Spectrophotometer (FIRAS) instrument of the COBE satellite was a precise match for a black body radiation field having a temperature of 2.72K, plus or minus one-hundredth of a degree. (Adapted by W. D. Woods from Figure 9 of Mather's memoir *The Very First Light*)

measurements were so small that the points precisely matched the curve. The audience of two thousand scientists, alerted to expect something special, rewarded him a standing ovation; a rare honour for such a professional gathering.

In 2006 Mather shared a Nobel Prize in Physics in recognition of his studies of the cosmic microwave background, seeking proof of the Big Bang origin of the universe.

This story is a magnificent demonstration of our ability to calculate, with accuracy, something as strange as to appear implausible, and then to build instruments capable of detecting the evidence needed to prove that it really happened.

## Further Reading
*The Very First Light: The True Inside Story of the Scientific Journey Back to the Dawn of the Universe*, John C. Mather and John Boslough, Basic Books, 1996.

# March

| | | | |
|---|---|---|---|
| 1 | Mars, ι Aquarii | 1.0° apart | 12° |
| 2 | Moon, α Leonis (Regulus) | 0.3° apart: occultation | 168° |
| 3 | Moon | Full Moon: total lunar eclipse | 180° |
| 7 | Mercury | Inferior conjunction: evening → morning | 4° |
| 7 | Venus, Neptune | 0.1° apart | 14° |
| 8 | Venus, Saturn | 0.9° apart | 15° |
| 10 | Moon, α Scorpii (Antares) | 0.7° apart: occultation | 100° |
| 11 | Moon | Last Quarter Moon | 90° |
| 17 | Mars, λ Aquarii | 0.7° apart | 15° |
| 18 | Uranus, 13 Tauri | 0.2° apart | 61° |
| 19 | Moon | New Moon | 2° |
| 20 | Earth | Equinox | |
| 22 | Neptune | Conjunction | 1° |
| 23 | Moon, M45 (Pleiades) | 1.1° apart: occultation | 58° |
| 24 | Mars, φ Aquarii | 47″ apart | 17° |
| 25 | Saturn | Conjunction | 2° |
| 25 | Moon | First Quarter Moon | 90° |
| 26 | Uranus, 14 Tauri | 0.1° apart | 52° |
| 29 | Moon, α Leonis (Regulus) | 0.3° apart: occultation | 141° |

Dates are based on Universal Time (UT). Occultations of M45 (Pleiades) refer to third-magnitude star η Tauri (Alcyone). The last column gives the approximate elongation from the Sun of the event.

**Mercury** plummets back toward the Sun, undergoing inferior conjunction on 7 March and re-entering the morning sky. It scoots past Mars on 15 March but the two planets keep their distance, not coming closer than 3.4° in dawn skies. Retrograde motion ceases 19–20 March. This first complete morning sky apparition favours southern latitudes and indeed, is the best one of the year for the southern hemisphere. The tiny planet will not get very high for observers in northern temperate latitudes however. Mercury is sixth-magnitude around the time of inferior conjunction but brightens steadily throughout the morning apparition. The rapid increase in phase more than compensates for the decreasing apparent size of the disk as Mercury moves away from Earth.

## Morning Apparition of Mercury
### 7 March to 14 May

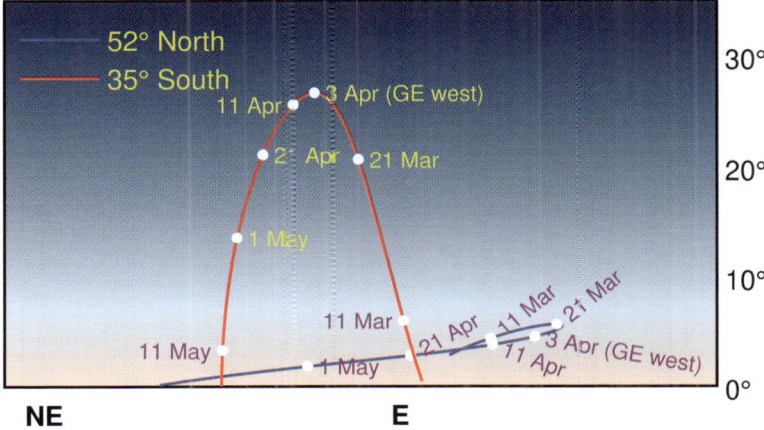

**Venus** is finally beginning to distance itself from the Sun and is getting easier to see in the west after the Sun goes down. It passes a scant 0.1° north of Neptune on 7 March and a more distant 0.9° north of Saturn the following day but these events take place during evening twilight and will not be easily visible due to the relative faintness of the superior planets. The view of the evening star is slightly better from northern temperate latitudes than the southern hemisphere this month.

The waxing gibbous **Moon** occults Regulus twice this month, on 2 and 29 March. Antares follows on 10 March and the open star cluster known as the Pleiades vanishes on 23 March. The Full Moon of 3 March takes place just hours after passing through its descending node, resulting in a total lunar eclipse. See *Eclipses* elsewhere in this volume for more information. On 20 March, **Earth** reaches an equinox. This ushers in astronomical spring to the northern hemisphere and autumn to the south. This equinox is known as the *Vernal Equinox* in astronomy and marks the instant when the Sun moves south to north across the celestial equator. The name derives from the Latin word for 'spring', but this is the autumnal equinox for anyone living south of the equator.

**Comet 88P/Howell** reaches perihelion mid-month although its closest approach to Earth does not take place until October. As it is unlikely to get any brighter than tenth magnitude, a telescope will be necessary to see this short-period Jupiter-family comet. See *Comets in 2026* for more information.

## Morning Apparition of Mars
### 1 January 2026 to 1 September 2026

— 52° North
— 35° South

**Mars** continues its tour of the zodiac, spending the entire month in Aquarius. Found low in the east before sunrise, the first-magnitude planet is best sought from southern hemisphere vantage points. The red planet encounters Mercury on 15 March but it is a distant meeting with the two planets 3.4° apart and barely 15° away from the Sun. Mars also passes within a degree of several Bayer-lettered stars this month: fourth-magnitude ι Aquarii on 1 March; fourth-magnitude λ Aquarii on 17 March; and fourth-magnitude φ Aquarii on 24 March. This last fly-past is a particularly close one, with the planet and the star closing to within 47″ of one another. The star φ Aquarii is an M-type giant which makes it a similar hue to Mars. Mars is at perihelion on 26 March when it is 1.3813 au from the Sun.

**Jupiter** returns to direct motion on 11 March, reversing course through Gemini. The largest planet in the solar system is an evening sky object, not setting until after midnight for favoured northern hemisphere observers. It diminishes slightly in brightness as it heads away from opposition, beginning the month at magnitude −2.5 and ending at −2.2.

**Saturn** is at conjunction with the Sun late this month and is largely hidden in twilight. Venus glides past on 8 March but this event takes place less than 15° from the Sun. Saturn also crosses the celestial equator on 26 March, moving south to north.

**Uranus** sets before midnight for all observers but is best viewed from the northern hemisphere. It passes by two naked-eye stars with Flamsteed designations this month. On 18 March, the sixth-magnitude planet passes 0.2° south of 13 Tauri, a B-type (blue) star whose visual apparent magnitude almost exactly matches that of Uranus. Eight days later, on 26 March, Uranus is only 0.1° south of 14 Tauri. This is a G-type giant star of magnitude +6.1. Long-exposure photographs might show the colour contrast between green Uranus and yellow-orange 14 Tauri.

**Neptune** is at conjunction with the Sun this month and lost to view in the glare of the Sun. Unfortunately, the close approach by Venus on 7 March takes place only 14° from our parent star.

# Managing Healthcare in Space with an On Demand Pharmacy

## Martin Braddock

### Introduction
The potential for space missions to become longer in duration means it will be impossible to transport from Earth and replenish a cargo full of every medication needed and in sufficient volume to take from Earth. Resupplying any mission in transit is cost limiting and the prospect of restocking Lunar and Martian bases with medicines will pose huge cost and logistical challenges. This means that we will need to find ways to produce drugs in space.

### The Problem Statement
Human beings are, in essence highly efficient and effective biological machines. Like all machines they require servicing, will almost certainly need repairing periodically and possibly require parts replacing at some point during their lifetime. On Earth, where access to medical facilities is the norm, and dependent upon age, it is routine for many people to have annual health checks where adjustments are sometimes made to diet and exercise regimes. Where surgical procedures are carried out, patients will typically visit a hospital and their local doctor to ensure a satisfactory outcome and that there are no complications such as infection. How would medical situations be managed in space?

### The Need for Medicines in Space
The effects of the space environment in low Earth orbit (LEO) and from missions to the Moon on human physiology and psychology have been very well documented since the first manned spaceflight in 1961, (e.g. Arone et al 2021, Hart 2023, Tomsia et al 2024). The pharmacy that has been taken into space has evolved over the years and today on the International Space Station (ISS), approximately 190 medicines[1] are available and the crew is trained to deal with emergencies which may arise

---

1. 'Evolution of NASA Medical Kits: From Mercury to ISS', *Space Safety Magazine*, spacesafetymagazine.com/spaceflight/space-medicine/evolution-medical-kits-mercury-iss

I'm in space and I need medical help! (Pixabay)

during a mission typically of six months duration. The designated medical officer is trained for administering first-aid, stitching wounds and administering injections and all astronauts are trained for emergency resuscitation should a crew member suffer a heart attack.[2,3]

## No Space for Being Left on the Shelf

Imagine a future when astronauts are on a journey for many years or are living in a space colony on another world. What challenges does this pose to mission planners (Blue et al 2019a)? The 'use by date' of the medicine has expired and every month that passes makes the medicine less effective. Moreover, it is very unlikely we will be able to provide medication 'in date' as it may take too long to get to

---

2. 'Medical Procedures for Astronauts on the International Space Station (ISS)' newspaceeconomy.ca/2024/02/22/medical-procedures-for-astronauts-on-the-international-space-station-iss
3. 'How Do Astronauts Handle Medical Emergencies in Space?' (Northrop Grumman) now.northropgrumman.com/how-do-astronauts-handle-medical-emergencies-in-space

you as a patient. This is a very stark reality and made the more so given that space radiation may play an additional role in inducing active pharmaceutical ingredient (API) instability during long-duration spaceflight (Blue et al 2019b). How the body 'sees' and responds to medicines may differ in space (Dello Russo et al 2022) and the way medicines are packaged may need to be changed to adapt for use in space (Daniels and Williams 2023, Reichard et al 2023).

## Keeping the Sell by Date Current and Drugs Safe

With the prospect of medications becoming less effective as the duration of missions become extended possibly for years, the obvious way to circumvent this obstacle is to manufacture them in space (Seoane-Viaño et al 2022, Tran et al 2022). To achieve this goal a number of initiatives have been implemented and two of many are now described. NASA has defined a multi-phase programme to deliver a flexible, personalised – including DNA-sequence informed – on-demand astropharmacy.[4] Concentrating efforts in the first instance on protein-based drugs, phase one of the programme has successfully genetically engineered pre-launch bacteria to produce filgrastim, a growth factor that can restore the bone marrow after radiation damage and teriparatide, a peptide drug which may prevent bone demineralization. The next phase will address whether the design system can manufacture drugs of acceptable quantity and purity for astronaut use. The second initiative described by the UK's space agency[5] is by VITA (Visualising In-Space TxTl Astropharmaceuticals), an astropharmacy team at the University of Nottingham. VITA is aiming to create a light, room-temperature storage and system that will allow genetically engineered bacteria to manufacture protein-based medicines to be transported in space in an inert form. The technology involves defining the right conditions for adhering medicine components to cellulose freeze-drying the composites and reconstituting with water when needed. Importantly this also has applications for medical intervention in extreme environments on Earth.

Just as in the discovery and development of drugs on Earth, rigorous safety assessment of any drug made or reconstituted in space is mandatory to ensure that the benefits outweigh any risks associated with taking the medicine. The field

---

4. 'A Flexible, Personalized, On-Demand Astropharmacy' (NASA)
   nasa.gov/general/a-flexible-personalized-on-demand-astropharmacy
5. 'Space medicine for the future of space exploration' (UK Space Agency)
   space.blog.gov.uk/2024/02/14/space-medicine-for-the-future-of-space-exploration

of astrotoxicology,[6] which includes understanding of the extremes of the space environment in addition to how it may influence the behaviour of drugs is very much in its infancy. It will be essential to determine whether the safety profile of any 'made in space drug' remains acceptable, how to dose the drug and capitalise on the knowledge and skill of pharmacists (Aziz et al 2022) and evaluate the drug's benefit/risk profile in clinical trials in space.[4] These steps will be needed in the years to come, in the planning for and especially when lunar or Martian settlements are established.

## Discovering Drugs in Space, for Space and for Earth

Our ability to provide medication for use in space for long term missions does not rely solely on research applicable simply to the space environment. Indeed, one of the best and, therefore, fund-worthy uses of the space environment in LEO is to use the microgravity on the ISS to help discovery and develop new drugs for use on earth (Ryder and Braddock 2020). Many multi-national pharmaceutical companies[7,8] are investing in technologies to protect and extend the API of drugs, develop systems to manufacture drugs on demand including 3D printing and 3D bioprinting, to assess the health status of the astronaut and as is becoming practice on Earth, starting to plan for the development of personalised medicines for use in space (Pavez Lorie et al 2021). All of these innovations have direct relevance for increasing both the efficiency and speed to manufacture while reducing drug production costs on Earth (Wani et al 2024).

The future promises many exciting developments in medical operations that are by necessity independent from Earth. The design of a medical facility – a 'sick bay' is a balance between essential capabilities and technologies and the restrictions of available room and mass of material taken from Earth. As is becoming more widely available on Earth for many of us who invest and can invest in our health, point-of-care, that is in-situ digital diagnostics are critical for supporting astronaut health and performance. Challenges remain especially in understanding how

---

6. 'Quite Detached—Astropharmacy and Astrotoxicology in Space Flight', *ChemistryViews* **chemistryviews.org/quite-detached-astropharmacy-and-astrotoxicology-in-space-flight**
7. 'The future of medicine in space: drug development, manufacturing, and beyond' **clinicaltrialsarena.com/features/medicine-in-space/?cf-view**
8. 'Space – the next frontier in drug development', *PharmaVoice* **pharmavoice.com/news/pharma-space-research-merck-bristol-myers-NASA/640079**

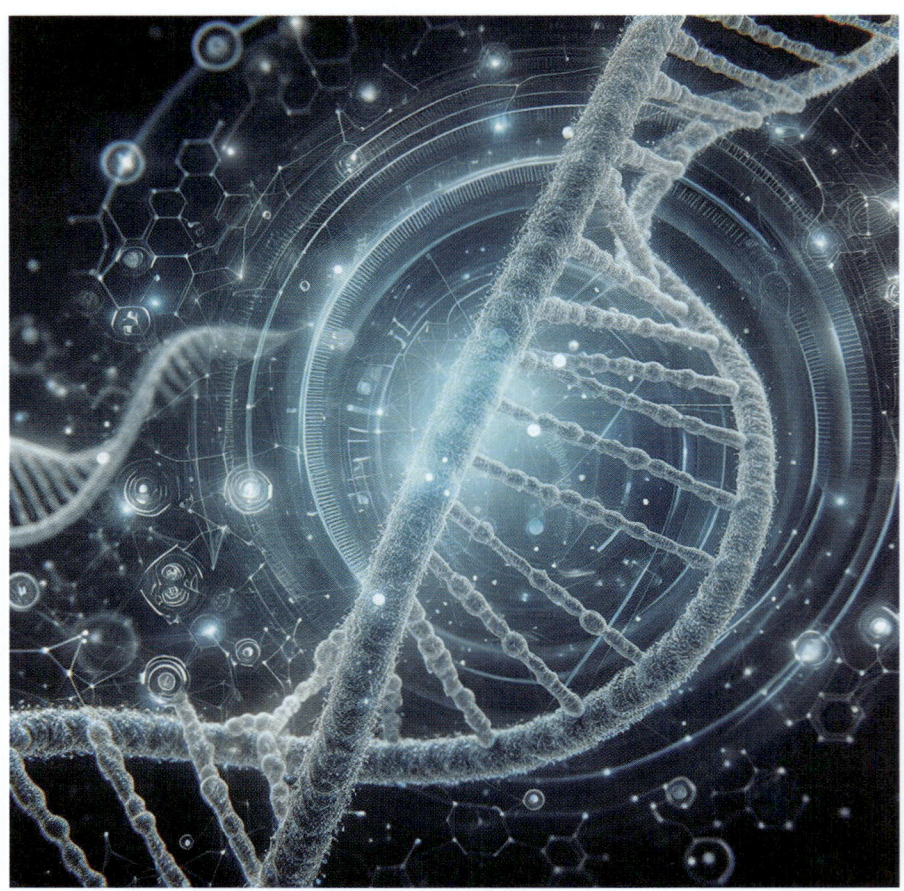

Designing medicines to enable future living in space. (Pixabay)

humans function in space over long periods of time and the field of space medicine is poised to make an essential contribution not only to space travel but to making helping medicines become more quickly available to Earth dwellers.

As quoted by Dr Thais Russomano,[7] a professor of space physiology and CEO of the think tank InnovaSpace:

> "Once humans enter space, every single cell and every single cellular process change. This could really alter the way we process medicines."

## References

Arone, A., Ivaldi, T., Loganovsky, K., Palermo, S., Parra, E., Flamini, W., Marazziti, D. 2021, 'The Burden of Space Exploration on the Mental Health of Astronauts: A Narrative Review'. *Clinical Neuropsychiatry*. 18(5):237–246.
doi:10.36131/cnfioriieditore20210502

Aziz, S., Raza, M.A., Noreen, M., Iqbal, M.Z., Raza. 2022, 'Astropharmacy: Roles of Pharmacist in Space', *Innovations in Pharmacy* 13(3):10.24926
pubmed.ncbi.nlm.nih.gov/36627911

Blue, R.S., Bayuse, T.M., Daniels V.R., Wotring, V.E., Suresh, R., Mulcahy, R.A., Antonsen, E.L. 2019a 'Supplying a pharmacy for NASA exploration spaceflight: challenges and current understanding'. *npj Microgravity* 5, 14 (2019).
nature.com/articles/s41526-019-0075-2

Blue, R.S., Chancellor, J C., Antonsen, E.L., Bayuse, T.M., Daniels, V.R., Wotring, V.E. 2019b, 'Limitations in predicting radiation-induced pharmaceutical instability during long-duration spaceflight'. *npj Microgravity* 5, 15.

Daniels V.R., Williams E.S., 2023. 'Exploring the complexities of drug formulation selection, storage, and shelf-life for exploration spaceflight', *British Journal of Clinical Pharmacology* doi:10.1111/bcp.15957. Epub ahead of print. PMID: 37940128.

Dello Russo, C., Bandiera, T., Monici, M., Surdo, L., Yip V.L.M., Wotring, V., Morbidelli, L. 2022, 'Physiological adaptations affecting drug pharmacokinetics in space: what do we really know? A critical review of the literature'. *British Journal of Pharmacology* 179(11) 2538–2557.

Hart, D.A. 2023, '*Homo sapiens*—A species not designed for space flight: health risks in low Earth orbit and beyond, including potential risks when travelling beyond the geomagnetic field of Earth'. *Life*, 13, 757.

Pavez Lorié, E., Baatout, S., Choukér, A., Buchheim, J.I., Baselet, B., Dello Russo, C., Wotring, V., Monici, M., Morbidelli, L., Gagliardi, D., Stingl, J.C., Surdo, L., Yip, V.L.M. 2021, 'The Future of Personalized Medicine in Space: From Observations to Countermeasures'. *Frontiers in Bioengineering and Biotechnology* 9
doi.org/10.3389/fbioe.2021.739747

Reichard, J.F., Phelps, S.E., Lehnhardt, K.R. et al. 'The effect of long-term spaceflight on drug potency and the risk of medication failure'. *npj Microgravity* 9, 35 (2023).

Ryder, P., Braddock, M. (2022), 'Harnessing the Space Environment for the Discovery and Development of New Medicines' in *Handbook of Space Pharmaceuticals*, edited by Pathak, Yashwant V.; Araújo dos Santos, Marlise; Zea, Luis. ISBN: 978-3-030-05526-4. Cham: Springer International Publishing, 2022, 823–857.

Seoane-Viaño, I., Ong, J.J., Basit, A.W., Goyanes, A. 2022, 'To infinity and beyond: Strategies for fabricating medicines in outer space'. *International Journal of Pharmaceutics: X* 4, 100121.

Tomsia, M., Cieśla, J., Śmieszek, J., Florek, S., Macionga, A., Michalczyk, K. and Stygar, D. (2024), 'Long-term space missions' effects on the human organism: what we do know and what requires further research'. *Frontiers in Physiology.* **15**:1284644. doi: **10.3389/fphys.2024.1284644**

Tran, Q.D., Tran, V., Toh, L.S., Williams, P.M., Tran, N.N., Hessel, V. 2022, 'Space medicines for space health', *ACS Medicinal Chemistry Letters* 13 (8), 1231–1247.

Wani, A., Prabhakar, B., Shende, P. 2024, 'Strategic Aspects of Space Medicine: A Journey from Conventional to Futuristic Requisites'. *Space Science and Technology.* 4:0123.

# April

| 2 | Moon | Full Moon | 177° |
|---|---|---|---|
| 3 | Mercury | Greatest elongation west (morning) | 28° |
| 5 | Jupiter | East quadrature | 90° |
| 6 | Moon, α Scorpii (Antares) | 0.6° apart: occultation | 127° |
| 10 | Moon | Last Quarter Moon | 90° |
| 13 | Mars, Neptune | 0.3° apart | 21° |
| 17 | Moon | New Moon | 4° |
| 19 | Moon, M45 (Pleiades) | 1.0° apart: occultation | 31° |
| 20 | Mercury, Saturn | 0.5° apart | 23° |
| 22/23 | Earth | Lyrids (ZHR 18) | |
| 23 | 136108 Haumea | Opposition | 151° |
| 23/24 | Earth | Pi Puppids (ZHR varies) | |
| 24 | Moon | First Quarter Moon | 90° |
| 24 | Venus, Uranus | 0.8° apart | 26° |
| 26 | Moon, α Leonis (Regulus) | 0.2° apart: occultation | 114° |
| 30 | Jupiter, δ Geminorum (Wasat) | 0.6° apart | 68° |

Dates are based on Universal Time (UT) Occultations of M45 (Pleiades) refer to third-magnitude star η Tauri (Alcyone). Peak activity dates and ZHRs for meteor showers are estimates. The last column gives the approximate elongation from the Sun of the event.

**Mercury** arrives at greatest elongation west, 27.8°, on the third day of the month. This is the best morning apparition of the year for astronomers in the southern hemisphere but the brightening planet (magnitude +0.4 to −0.8) heads back toward the eastern horizon early in the month. This is a poor morning apparition for observers further north, with the tiny planet scarcely clearing the horizon. Mercury slips past Neptune on 17 April, and Saturn and Mars three days later, but these encounters will be difficult to see in dawn skies.

**Venus** is the evening star, adorning the western skies after sunset. It approaches to within 0.8° of Uranus on 24 April but the two planets are aloft during evening twilight, making sixth-magnitude Uranus an improbable find.

Earth should enjoy an excellent renewal of the Lyrid meteor shower which peaks in the latter half of the month when the Moon in its waxing crescent phase. The Pi Puppids reach maximum activity at around the same time and favour southern observers. See *Meteor Showers in 2026* for more information.

The **Moon** continues with its occultation series of Antares (6 April), the Pleiades (19 April), and Regulus (26 April), but the occultation of the open cluster is getting increasingly difficult to observe as the Sun approaches that position.

**Mars** moves into Pisces on the second day of the month. Whilst in this fishy constellation, the red planet encounters Neptune on 13 April. Mars then ducks into the constellation of Cetus where it meets Saturn on 19 April. Saturn is slightly brighter (magnitude +0.9) than Mars (magnitude +1.2). Mercury swings past the following day, with the tiny planet coming to within 2° of Mars. Mars returns to Pisces on 21 April and reaches a solstice, with summer beginning in the southern hemisphere and winter in the north, on 24 April. Look for Mars in the morning sky, not long before sunrise. It is best viewed from the southern hemisphere.

**Jupiter** arrives at east quadrature on 5 April so it is exactly 90° away from the Sun and visible in the evening sky after sunset. The gas giant is located in Gemini and is best viewed from the northern hemisphere where it sets after midnight. (Southern hemisphere observers lose the bright planet in late evening.) Jupiter encounters the fourth-magnitude spectroscopic binary Wasat ($\delta$ Geminorum) for the second time this year, passing 0.6° north of the star on the last day of the month.

**Saturn** is joined in the morning sky first by Mars (1.2° apart on 19 April) and then by Mercury (0.5° apart on 20 April). Both of these conjunctions take place less than 25° from the Sun so the best place to attempt observations is from the southern hemisphere where Saturn rises ahead of the dawn. Saturn is in Pisces at the beginning of the month but soon leaves it for the constellation of Cetus where the planet will remain until June.

**Uranus** is approaching conjunction with the Sun and is now setting mid-evening. Venus makes a close approach on 24 April when the two planets appear 0.8° apart, but the two worlds are just 26° from the Sun at the time. Look for Uranus in Taurus as soon as darkness falls.

**Neptune**, located in Pisces, should become visible to observers in the southern hemisphere by the end of the month but the eighth-magnitude gas giant lingers in dawn twilight for planet watchers in northern temperate latitudes. Mars passes 0.3° north of the faint planet on 13 April, with the two planets appearing 21° away from the Sun. Mercury is in attendance four days later, with the two bodies 1.3° apart with an elongation of 24° from the Sun. On 24 April, the planet passes through its ascending node with respect to the equatorial plane; it will have a positive declination until September.

**136108 Haumea** comes to opposition on 23 April. Situated nearly 49 au away from Earth on this day, the dwarf planet shines at a meagre seventeenth magnitude in the constellation of Boötes. Opposition in terms of right ascension occurs early next month. See *Minor Planets in 2026* for more information on this distant world.

# Freedom Will Not Be Shackled

## Jonathan Powell

As the USA and Soviet Union jostled for position before and after the 'Space Race' era that was to dominate the 1950s, 1960s, and a fair proportion of the 1970s, talk was to quickly turn, after the triumphs of both Yuri Gagarin and Neil Armstrong, to the next big challenge in space.

With suggestions of a base on the Moon, or even the idea of a voyage to Mars, crosshairs were more realistically being aligned with something much closer to home, in the form of a crewed orbiting space station. Granted, not a new idea, but one that could now be seen as being not only achievable but, more importantly, sustainable. However, it remained questionable as to whether or not such a station

Artist Alan Chinchar's 1991 conception of the completed space station *Freedom* in orbit. (NASA)

would remain a singular proposition for one country alone. Yet it was a vision that was to quickly gain momentum, as so many people who had been involved in getting humans into space, and then to the Moon, now turned their focus towards this next as-yet-unwritten chapter.

The space station idea became even more pressing following the end of the *Apollo* missions which had sapped both government funding and public interest alike. Furthermore, the Americans and Soviets became aware that their own 'superpower' status may have a shorter shelf life than anticipated, with European countries galvanising and enhancing their own technological capabilities, and even the Chinese looking on with a growing urge to pave their own path in space.

With NASA's *Skylab*, the first American space station, cutting its teeth in 1973, and two years previously the Soviet Union's experimental space station *Salyut* gaining much attention, both countries had already made inroads into the potential for an orbiting platform that would ultimately provide years of service and address more than just basic scientific research.

On 25 January 1984, in President Ronald Reagan's State of the Union Address, NASA was directed to build an international space station within the decade that followed, with the project outline duly christened *Space Station Freedom*. In the years leading up to the White House's vision, James Montgomery Beggs, who had been appointed Administrator of NASA by the Reagan administration in 1981, had stated that *Freedom* had been "the next logical step" in space following the Apollo missions, something Reagan was now keen to build upon and see to fruition.

With an eye on the shifting political landscape, NASA was keen to make sure it was not too many moves behind in this new chess game that was afoot, with the Soviet Union's opening gambit of the *Mir* space station making its debut in 1986. During the years that followed the Soviet Union would fall, but *Mir* was to serve the country that replaced it, the space station lasting until 2001 before having its funding withdrawn.

Prior to *Mir*, *Apollo* had also had its funding cut, leading to the subsequent cancellation of missions 18, 19, and 20, all of which were destined for the lunar surface. However, and more worryingly, *Freedom* was still on the drawing board and facing similar challenges. The logistical side of placing the necessary hardware into orbit to construct the space station was already in hand, courtesy of the revolutionary reusable Space Shuttle, but even with that in place, *Freedom's* progress was to be painfully slow, with the gradient of the road ahead becoming increasingly challenging.

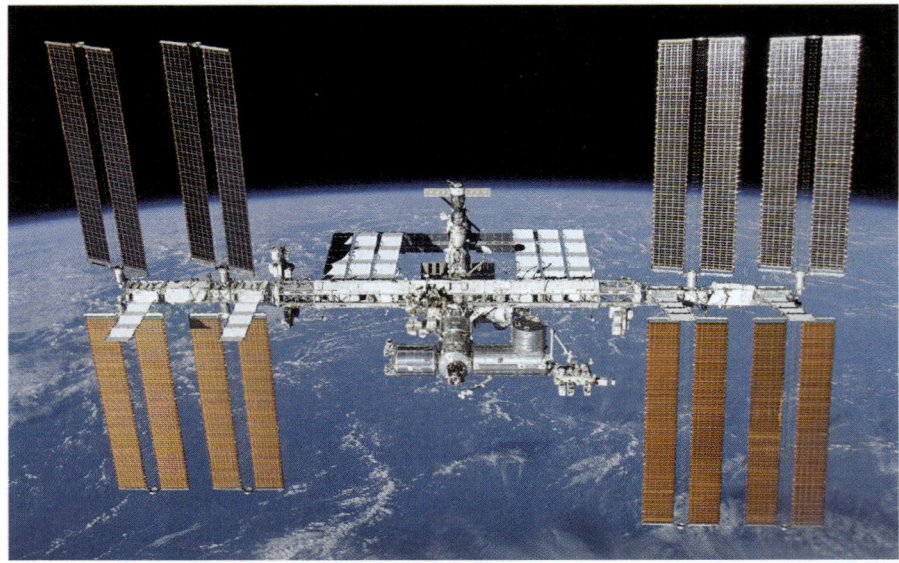

The International Space Station is seen here in this image captured on 23 May 2010 by an STS-132 crew member on board the Space Shuttle *Atlantis*. Construction of the ISS involved several space agencies, these being NASA; Roscosmos, ESA, the Japan Aerospace Exploration Agency (JAXA) and the Canadian Space Agency. (NASA/Crew of STS-132)

Despite NASA signing contracts in 1988 to finally develop *Freedom*, and move to the actual fabrication of the hardware stage, there followed a succession of repeated requests from US Congress to redesign the proposed space station and to reduce costs. The unrelenting pressure was to back NASA into a corner with the only possible option left being to relinquish sole ownership of the *Freedom* project and make it an international collaboration, with other countries helping to shoulder the increasing burden. This is turn meant that the identity of *Freedom* as an American dream alone would be lost, and whereas the grip was already becoming tenuous with more than just a passing Russian hand involved in the project, everything was to grind to a halt during the administration changes from Ronald Reagan to George H. W. Bush.

However, the project was to be reborn under the next US president, Bill Clinton, who in 1993 set the wheels in motion once more, but with a reduced size of space station and, initially, a reduced crew capacity, with the overall functionality of the station also compromised. This transformation phase also saw the name of *Freedom* replaced with the *International Space Station*.

Whilst the end goal of placing a vastly superior space station into orbit had been achieved, a lot of *Freedom's* former character had been stripped away, gutted almost, with just the kernel of the original idea left in place. It had all come at a great cost, not just financially, but to those who had displayed unwavering dedication and commitment to the project.

Suffice to say that America's space station was never in doubt. Their own interests in furthering the country's involvement in space could not be seen to falter. However, *Freedom's* eventual existence as a coalition project was to subsequently make for uneasy and suspicious bedfellows in the years that followed, especially with the ever-present tensions between nations back on Earth.

# May

| 1 | Moon | Full Moon | 175° |
|---|---|---|---|
| 4 | Moon, α Scorpii (Antares) | 0.4° apart: occultation | 153° |
| 6/7 | Earth | Eta Aquariids (ZHR 50) | |
| 9 | Moon | Last Quarter Moon | 90° |
| 14 | Mercury | Superior conjunction: morning → evening | 0° |
| 16 | Mars, o Piscium (Torcular) | 0.9° apart | 28° |
| 16 | Moon | New Moon | 5° |
| 17 | Moon, M45 (Pleiades) | 0.9° apart: occultation | 7° |
| 22 | Uranus | Conjunction | 0° |
| 23 | Moon, α Leonis (Regulus) | 0.1° apart: occultation | 88° |
| 23 | Moon | First Quarter Moon | 89° |
| 31 | Moon | Full Moon | 175° |
| 31 | Moon, α Scorpii (Antares) | 0.4° apart: occultation | 175° |

Dates are based on Universal Time (UT). Occultations of M45 (Pleiades) refer to third-magnitude star η Tauri (Alcyone). Peak activity dates and ZHRs for meteor showers are estimates. The last column gives the approximate elongation from the Sun of the event.

**Mercury** is too low in the east to see from northern temperate latitudes and is soon lost to view from the southern hemisphere as it approaches the Sun and then reaches superior conjunction on 14 May. The nearest planet to the Sun moves past Uranus on 18 May but Uranus is also at conjunction with the Sun this month so this event will not be observable. Mercury returns to the west after superior conjunction and ends the month at a bright magnitude −0.6.

**Venus** continues to climb above the western horizon after sunset but that upwards progress ends for observers in northern temperate latitudes. Planet watchers in the southern hemisphere will have the best views of the evening star for the remainder of this apparition, with the bright planet continuing to climb in altitude until greatest elongation east in August.

**Earth** runs into comet 1P/Halley this month, or rather, parts of it, when the Eta Aquariid meteor shower peaks during the first week. Unfortunately, the waning gibbous Moon will provide considerable light interference. See *Meteor Showers*

## Evening Apparition of Mercury
## 14 May to 13 July

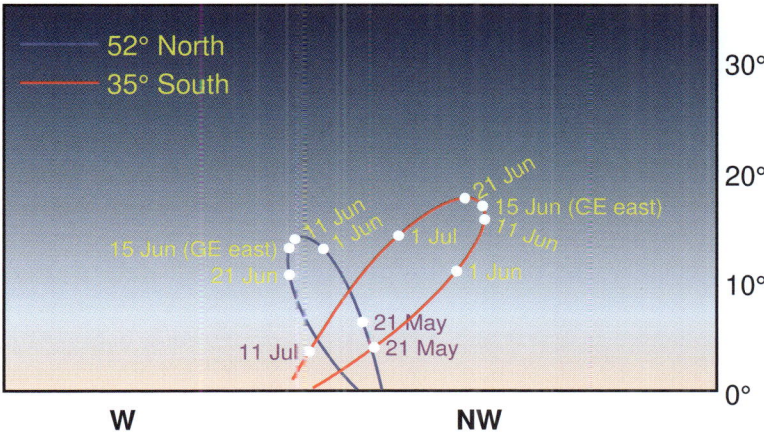

in 2026 for information about this event. Lunar occultations of stellar and deep-sky objects continue, with Antares disappearing on both 4 May and 31 May; the Pleiades on 17 May; and Regulus on 23 May. However, the occultation of Pleiades will not be visible as the Sun is just 7° distant.

The **Moon** reaches its fully illuminated phase twice this month; the second occurrence of a Full Moon in a calendar month is commonly referred to as a 'Blue Moon'. With apogee taking place less than 24 hours later, this Blue Moon is also the 'smallest' of the Full Moons in terms of apparent size, with a diameter of 1765″. Compare this with the 'largest' Full Moon which occurs in December.

**Mars** remains an extremely difficult morning sky object for observers in northern temperate latitudes, being mired in morning twilight. However, it is rising in dark skies when viewed from the southern hemisphere. It begins the month in Pisces, passing by Torcular (o Piscium), a fourth-magnitude star whose official name is derived from the Latin for 'wine press', on 16 May. Mars leaves Pisces for Aries two days later.

**Jupiter** is an evening sky object, shining at magnitude −2.0. It is now setting around or before midnight for all observers, but is best seen from the northern hemisphere. On 15 May, Jupiter again passes north of the famous planetary nebula

C39 (NGC 2392). (See the January *Sky Notes* for more information about this notable deep sky object.)

**Saturn** rises after midnight in the non-zodiacal constellation of Cetus, the whale or sea monster. The lengthy morning twilight of northern temperate latitudes means that ideal viewing conditions of this first-magnitude planet lie in the southern hemisphere. The rings continue to open up relative to Earth, beginning the month at $7.0°$ and ending at $8.3°$.

**Uranus** is at conjunction with the Sun this month and lost to view. Mercury's close approach on 18 May takes place too close to the Sun (just $4°$ away) to observe.

**Neptune** is located in Pisces and rises in the early morning hours. At eighth-magnitude, it is not a naked-eye object; observation requires both optical aids and dark skies.

# Craters of Eternal Darkness

## Katrin Raynor

If you listen to heavy metal, then you may think the following is about a band named Craters of Eternal Darkness. Whilst an excellent name for a metal group, the term craters of eternal darkness also refers to permanently shadowed craters on the Moon that lie close to the lunar poles.

The majority of the Moon's surface receives sunlight, both sides of the Moon seeing the light of day in almost equal amounts. Due to the absence of an atmosphere, the sunlight that reaches the lunar surface creates huge contrasts in temperature, varying from 120°C during the day to −130°C at night. Neither are

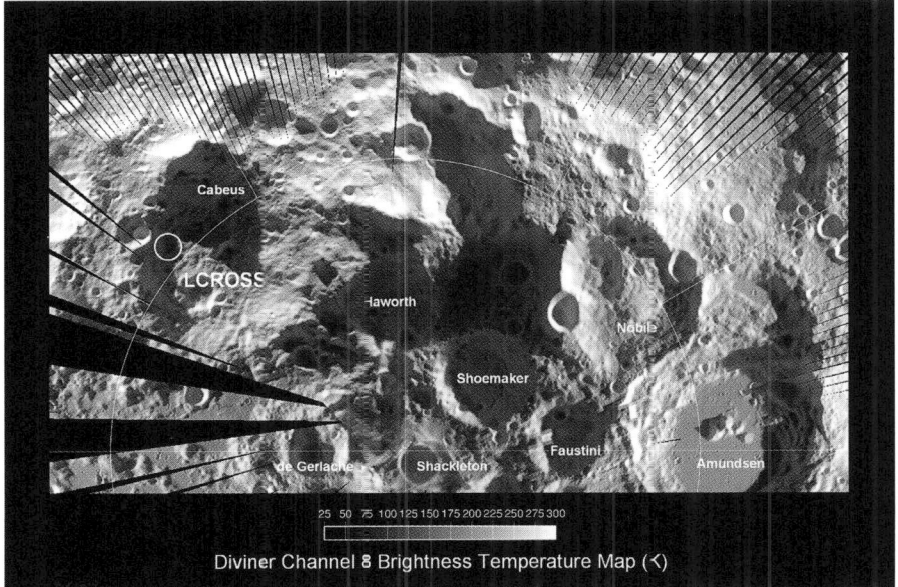

This image, taken by the *Diviner Lunar Radiation Experiment* infrared radiometer aboard the *Lunar Reconnaissance Orbiter*, displays the temperature variations across numerous craters located at the lunar south pole, and which range in temperature between −247°C and 26°C. Some of the featured craters are locations on the Moon that lie in permanent shadow, with Shackleton, de Gerlache and Shoemaker lying closest to the south pole. (Wikimedia Commons/NASA)

there any lunar seasonal variations in temperature because of the angle at which the Moon rotates.

The Moon and Earth do not spin upright on their axes, the imaginary line running through their poles around which they rotate. The Earth is instead tilted on its axis at an angle of 23.5°. Imagine the Earth as a spinning top but one which is slightly lopsided. As we orbit the Sun and the year progresses, we are either tilting towards the Sun or away from it causing variations in the amount of direct sunlight hitting the surface. The rotation axis of the Moon is tilted at a much shallower angle (1.53°) compared to Earth, limiting the amount of sunlight that reaches its surface, the effect of which is that the floors of impact craters located at latitudes close to one of the lunar poles often lie in permanent shadow.

During the early history of the solar system – around 3.9 to 4.4 billion years ago – the Moon was bombarded by asteroids, the impacting objects during the period generally being much larger than anytime since. This resulted in huge impact basins and craters ranging in size from hundreds of kilometres right down to craters microscopic in size. The Moon is particularly vulnerable to surface impacts because, unlike Earth, it has no atmosphere and therefore no protection from asteroids or other cosmic rocks hurtling towards its surface. There is also very little erosion on the Moon – there being no water, weather or active tectonic plates to continually change the surface – so the craters we see are very well preserved. The Moon hosts over a million craters and, to date, 9,137 of these are recognised by the International Astronomical Union, with over 324 of them never being exposed to sunlight.

One of the best known craters of eternal darkness is 21 kilometre Shackleton. Located at the lunar south pole, the peaks along Shackleton's rim are the only areas exposed to sunlight, whilst its dark 4.2 kilometre deep interior receives no warmth from the Sun, plunging to a chilly −183°C. As with Shackleton, the floor of the nearby 50.9 kilometre diameter crater Shoemaker receives no light or warmth from the Sun, remaining hidden in darkness and maintaining a temperature below −173°C.

At the lunar north pole, 9.9 kilometre diameter Erlanger crater experiences the same conditions as Shackleton, where only its rim is lit by the Sun. Images taken by NASA's *Lunar Reconnaissance Orbiter (LRO)* Camera show the crater with a bright rim and an ominous, black-as-ink interior where sunlight does not penetrate.

The absence of sunlight and low temperatures recorded in craters at the lunar poles lead scientists to believe that there is a great abundance of water ice at these locations mixed into the lunar soil or lying exposed within the craters. But the idea of the presence of water on the Moon is nothing new. Since 1645, when Dutch

This highly detailed combined photograph of Shackleton crater was created by NASA's *Lunar Reconnaissance Orbiter* and *ShadowCam* teams. Located at the lunar south pole, Shackleton lies in permanent shadow and has been the subject of study for years. Despite the interior lying in shadow, parts of the crater rim are continually exposed to sunlight, thereby making Shackleton an excellent destination for future lunar landings that could utilise the sunlight to convert to electricity. (Lunar Reconnaissance Orbiter and ShadowCam teams with images provided by NASA/KARI/ASU)

astronomer and cartographer Michael van Langren (1598–1675) published the first known map of the Moon, he theorised that the darker areas on the lunar surface were seas or *maria*.

The possibility of water being present on the Moon was never fully resolved until 1994, when data from NASA's *Clementine* mission suggested that ice may be present in permanently shadowed craters. In 1998, the *Lunar Prospector* mission concentrated on analysing these frigid craters and, although results indicated that water ice may be present, nothing was conclusive.

The best evidence of water at the lunar south pole came following the launch of NASA's *Lunar Crater Observation and Sensing Satellite (LCROSS)* with the *Lunar Reconnaissance Orbiter (LRO)* on 18 June 2009. The *LRO* separated from the *Centaur* rocket shortly after launch and just four days later entered into orbit around the

Hermite crater was recorded as the coldest place in the solar system by the Lunar Reconnaissance Orbiter (LRO) Diviner instrument in 2009 – a teeth chattering temperature of −247 °C. Located within a few degrees of the lunar north pole, Hermite's interior lies mostly in permanent shadow; its satellite crater Hermite A – seen here adjoining the main crater near top right of image – is thought to contain water ice along its walls and floor. (James Stuby based on NASA image)

Moon, kicking off its mission to map the lunar surface, determine impact hazards and locate potential new landing sites for future missions.

The *Centaur* upper stage rocket and *LCROSS* continued to orbit the Moon until 9 October at which point the two separated, the *Centaur* being deliberately crashed into the shadowed 98 kilometre diameter Cabeus crater, located 100 kilometres from the lunar south pole. *LCROSS* flew through the plume of debris to collect sample data, crashing onto the Moon's surface shortly after. The results of the plume collected by *LCROSS's* spectrometer discovered the presence of water, mainly in the form of pure ice crystals as well as metals including silver, magnesium and mercury.

Given that the Moon has no atmosphere or weather system, you may be wondering how or why water is present on the lunar surface. This has been the subject of much debate with theories as to its origin including either that water ice and other volatiles are a result of volcanic activity or are due to surface chemistry caused by solar wind particles. However, it is now widely accepted that these deposits were left over from impacts by comets and asteroids, the water ice from these rocky, frozen bodies scattering over the lunar surface upon impact depositing water ice in the shadowed craters.

Craters of eternal darkness are of great interest to space agencies and private companies. The discovery of water, minerals and metals on the Moon means that they can potentially be utilised for commercial purposes, including converting the ice to drinking water and oxygen and using hydrogen for propellant, all of which are important resources for future habitation and exploration of the Solar System.

# June

| 8 | Moon | Last Quarter Moon | 90° |
|---|---|---|---|
| 9 | Earth | Tau Herculids (ZHR low) | |
| 9 | Venus, Jupiter | 1.6° apart | 37° |
| 13 | Moon, M45 (Pleiades) | 1.0° apart: occultation | 22° |
| 15 | Moon | New Moon | 5° |
| 15 | Mercury | Greatest elongation east (evening) | 25° |
| 17 | Moon, Venus | 0.3° apart: occultation | 39° |
| 18 | Moon, M44 (Praesepe) | 0.9° apart: occultation | 41° |
| 19 | Moon, α Leonis (Regulus) | 0.3° apart: occultation | 62° |
| 20 | Venus, M44 (Praesepe) | 0.8° apart | 39° |
| 21 | Earth | Solstice | |
| 21 | Moon | First Quarter Moon | 90° |
| 25 | Neptune | West quadrature | 90° |
| 27 | Moon, α Scorpii (Antares) | 0.4° apart: occultation | 154° |
| 29 | Moon | Full Moon | 176° |

Dates are based on Universal Time (UT). Occultations of M44 (Praesepe/Beehive) refer to sixth-magnitude star ε Cancri (Meleph). Occultations of M45 (Pleiades) refer to third-magnitude star η Tauri (Alcyone). Peak activity dates and ZHRs for meteor showers are estimates. The last column gives the approximate elongation from the Sun of the event.

**Mercury** continues its ascent above the western horizon, reaching a greatest elongation east of 24.5° on 15 June, after which the tiny planet heads back toward the Sun. It begins the month at a bright magnitude −0.6 but is fading throughout the apparition, down to +0.6 at greatest elongation and then second magnitude before the end of the month. Mercury enters into retrograde motion on the penultimate day of June.

**Venus** makes a date with mighty Jupiter on 9 June when the two planets are less than 2° apart in the western sky. Although Jupiter is a much larger planet than Venus, the brilliant clouds and proximity of the evening star more than makes up for this inequality, and Venus outshines Jupiter by over two magnitudes. On 17 June, Venus is occulted by the waxing crescent Moon, and three days later, is found only 0.8° north of M44, the faint Beehive Cluster. Venus is rising higher every evening for southern hemisphere viewers but those in northern temperate latitudes can only watch the planet decline in altitude over the month.

**Earth** intersects the Tau Herculid meteor stream early in the month. Hopefully the waning crescent Moon will not prove too troublesome for observations of this poorly-understood shower. The article *Meteor Showers in 2026* has further information. Our planet reaches its first solstice this year on 21 June when the Sun reaches it northernmost declination in 2026. This marks the beginning of astronomical summer in the northern hemisphere and winter in the south.

Lunar occultations continue, with the waning crescent **Moon** passing through the Pleiades on 13 June, and the waxing crescent Moon blotting out Venus (17 June), sixth-magnitude M44 (18 June), and first-magnitude Regulus (19 June). Bright Antares is also a target, passing behind the disk of the waxing gibbous Moon on 27 June. Calendars in the United Kingdom, Republic of Ireland, Portugal, Madeira, the Canary Islands, and the Faroe Islands may disagree with the date of the Full Moon which appears in the events table. This is because the Moon becomes Full at 23:56 UT on 29 June which is 00:56 BST/IST/WEST on 30 June.

**Mars** is best seen from the darker winter skies of the southern hemisphere. The red planet is a morning sky object but rises during morning twilight as seen from northern temperate latitudes. Mars is in Aries at the beginning of June but passes into Taurus on the twentieth. Its phase is decreasing slightly over the course of the month, with the disk illumination going from 97% to 95%. This causes the planet to dim ever so slightly, from magnitude +1.2 to +1.3. Telescopically, the planet will continue to look more gibbous as the year progresses.

**Jupiter** and Venus, the two brightest planets in our sky, are 1.6° apart on 9 June. The proximity of an inferior planet is a sure sign that the Sun is not far away and indeed, the two planets' solar elongations are just 37°. Still located in Gemini at the beginning of the month, Jupiter is in the west as skies begin to darken and sets during twilight for observers in northern temperate latitudes. The gas giant moves into the constellation of Cancer on 22 June where it will remain for the next three months.

**Saturn** returns to Pisces from Cetus early in the month. Saturn is a morning sky object, rising around or just after midnight, and shining at magnitude +0.9.

**Uranus** is found in Taurus which is a morning sky constellation at this time of year. Because the planet was at conjunction late last month, sixth-magnitude Uranus will be difficult to spot at dawn for most of June.

**Neptune** attains west quadrature on 25 June, appearing 90° from the Sun in the sky. The faint planet is found in Pisces and rises before midnight by the end of the month. Look for it (with at least a small telescope) on a moonless night.

# Allan Rex Sandage

## David M. Harland

Allan Rex Sandage was born in Iowa City, Iowa, on 18 June 1926 as the only child of Charles Harold Sandage, a professor of advertising at Miami University in Oxford, Ohio, and Dorothy Briggs Sandage.

Being raised during the hardships of the Depression years taught him to seek academic excellence, something which his parents encouraged, and it was presumed that he would attend college and push on to postgraduate school.

His introduction to astronomy came at age nine when, on a trip to Washington D.C., he was invited to look through a backyard telescope. Upon returning home to Ohio his father bought him a telescope. During the day he monitored sunspots and at night he explored the wonders of the heavens. At school his focus was on mathematics and the sciences, and in his spare time he taught himself the essentials of astronomy, such as celestial mechanics.

As a teenager he read *The Realm of the Nebulae*, published in 1936 by Edwin Hubble, the astronomer at the Mount Wilson Observatory in the San Gabriel mountains north of Los Angeles who had used the 100-inch Hooker telescope, then the largest such instrument in the world, to reveal (1) that, contrary to what had been presumed, the system of stars that we call the Milky Way is just one of a great many similar systems, and (2) that, as indicated by a relationship between the redshift of the 'lines' in the spectrum of such a system and its distance from us, the universe is in a state of expansion. The young man was captivated by the sheer audacity of attempting to infer the evolution of the universe.

After majoring in physics and minoring in philosophy for two years at Miami University (in Ohio) he was drafted into the navy, becoming

A sketch of Allan Sandage at the peak of his career in the 1970s. (John McCue)

an electronics technician. By the time of his release, his father had relocated to the University of Illinois at Urbana, so the young Sandage opted to pick up his physics education there. On finding that the university had an observatory, he signed on as a volunteer and learned (1) how to expose and analyse photographic plates of the sky, and (2) that observatory domes are cold during a Midwest winter. When visiting his father on sabbatical at Berkeley in California, Allan took a trip to Mount Wilson and realised that this peak was his destiny.

The astronomers of Mount Wilson were based at the California Institute of Technology in Pasadena. Although it had no undergraduate astronomy course, the institute opened a graduate school in 1948, and Sandage, with his newly minted degree in physics, joined the first intake.

The 200-inch Hale telescope on Mount Palomar, in the mountains well away from the light pollution of Los Angeles, was just entering service and Hubble was keen to extend his redshift-distance relationship far beyond the technical limits imposed by the 100-inch, but after suffering a heart attack in 1949 he began to send Sandage up to the telescope to expose plates. As a result, on Hubble's death in 1953, a year after Sandage graduated with his doctorate, the young astronomer was the obvious person to continue Hubble's work.

Thus Sandage's career came to be dominated by his seeking (1) to refine the slope of the redshift-distance relationship, known as the 'Hubble constant', which measures the rate at which the universe is expanding, and (2) to try to measure the rate at which that expansion should be slowing by self-gravitation in order to determine whether the universe is 'open' and will expand forever or is 'closed' and will eventually collapse back on itself.

In part because for many years he was the only person working in the field, his body of work became an obstacle for others to surmount in order to compete, so most left him to it, but some accepted the challenge and by the mid-1990s there were two competing values for the Hubble constant, each specified to within 10% but without an overlap. At that point the Hubble Space Telescope enabled the value to be pinned down at 21.5 kilometres per second per million light-years to an accuracy of 10%, with the two values derived from terrestrial observations being just outside this range, one above it and the other below. This was a satisfying convergence. Utterly astonishing, however, was the finding by others that the rate at which the universe is expanding is not decelerating, it is *accelerating*. Taking all of this evidence into account, the universe began with the Big Bang some 13.8 billion years ago and it will expand forever. As Sandage had surmised upon choosing astronomy as his career, it was indeed audacious to set out to understand the evolution of the universe, and in this endeavour he certainly made a major contribution.

In 1959 Sandage married fellow astronomer Mary Connelley. They had two sons, David and John. He spent his entire career as a Carnegie astronomer based in Pasadena, and published over five hundred scientific papers. In addition to gaining membership of the American Philosophical Society, the National Academy of Sciences and the Royal Astronomical Society, plus fellowship of the Royal Society of London, he received many awards. He died of pancreatic cancer on 13 November 2010 at his home in San Gabriel, California, aged 84 years.

# July

| 4 | Mars, Uranus | 0.1° apart | 38° |
|---|---|---|---|
| 6 | Saturn | West quadrature | 90° |
| 6 | Earth | Aphelion | |
| 7 | Moon | Last Quarter Moon | 90° |
| 7 | Mars, $\omega^2$ Tauri | 0.5° apart | 39° |
| 9 | Venus, α Leonis (Regulus) | 1.0° apart | 43° |
| 10 | Mars, $\kappa^2$ Tauri | 0.7° apart | 40° |
| 10 | Mars, $\kappa^1$ Tauri | 0.8° apart | 40° |
| 10 | Moon, M45 (Pleiades) | 1.1° apart: occultation | 49° |
| 13 | Mercury | Inferior conjunction: evening → morning | 5° |
| 14 | Moon | New Moon | 3° |
| 15 | Moon, M44 (Praesepe) | 0.8° apart: occultation | 15° |
| 16 | Mars, τ Tauri | 0.8° apart | 42° |
| 17 | Moon, α Leonis (Regulus) | 0.3° apart: occultation | 36° |
| 21 | Saturn | Maximum ring opening (−9.2°) | 104° |
| 21 | Moon | First Quarter Moon | 90° |
| 24 | Moon, α Scorpii (Antares) | 0.6° apart: occultation | 128° |
| 26 | 3 Juno | Opposition | 165° |
| 27 | 134340 Pluto | Opposition | 176° |
| 28/29 | Earth | Delta Aquariids (ZHR 25) | |
| 29 | Jupiter | Conjunction | 1° |
| 29 | Moon | Full Moon | 178° |

Dates are based on Universal Time (UT). Occultations of M44 (Praesepe/Beehive) refer to sixth-magnitude star ε Cancri (Meleph). Occultations of M45 (Pleiades) refer to third-magnitude star η Tauri (Alcyone). Peak activity dates and ZHRs for meteor showers are estimates. The last column gives the approximate elongation from the Sun of the event.

**Mercury** is very low in the west at the beginning of July and vanishes from view well before inferior conjunction on 13 July. It reappears in the morning sky, climbing quickly in the east and brightening to magnitude +0.4 by the end of the month. The closest planet to the Sun returns to direct motion ten days after conjunction.

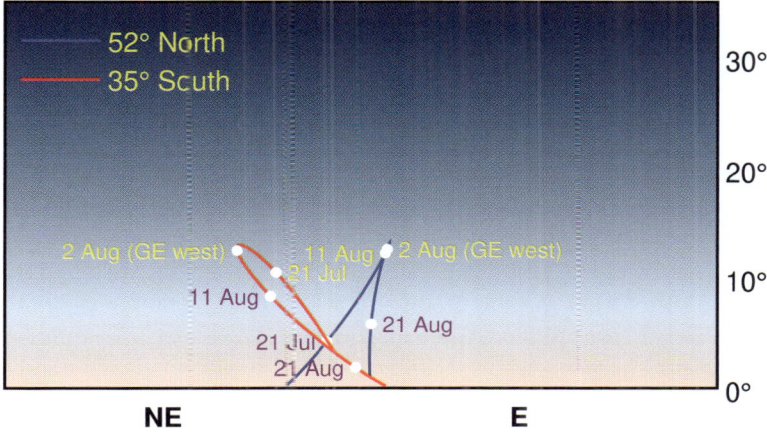

**Venus** soars high above the north western horizon for those viewing it from the southern hemisphere but observers in northern temperate latitudes will need to adjust their sights much closer to the horizon to see the evening star. Venus outshines first-magnitude Regulus on 9 July when the two objects are a degree apart in the evening sky. The waxing crescent Moon makes its closest approach to the bright planet, just under 2°, on 17 July.

**Earth** arrives at aphelion, its furthest point from the Sun on 6 July, when it is 1.017 au (152,100,000 kilometres) distant from our star. Compare these distances with those at perihelion in January. The peak of the Delta Aquariids near the end of the month is spoiled by the nearly Full Moon. Read more about this event in *Meteor Showers in 2026* elsewhere in the volume. The Moon occults M45, the Pleiades, on 10 July and M44, the Beehive Cluster, five days later. (This latter occultation will be difficult to see due to the proximity of the Sun.) Regulus is the next target, on 17 July, followed by Antares on 24 July.

**Mars** passes 0.1° south of Uranus, 1.5° south of the fourth-magnitude star 37 Tauri, and 1.1° north of sixth-magnitude $\omega^1$ Tauri (43 Tauri) on the fourth day of the month. Mars, Uranus, and the constellation of Taurus itself are all morning sky objects which are best viewed from the southern hemisphere at this time of year. The red planet comes to within a degree or so of several other stars in Taurus boasting Bayer designations: fifth-magnitude ω² Tauri on 7 July; fifth-magnitude κ² Tauri and fourth-magnitude κ¹ Tauri on 10 July; fourth-magnitude υ Tauri on 11 July; fourth-magnitude τ Tauri on 16 July; and fifth-magnitude ι Tauri on 23 July. However, it passes well north of the Hyades open star cluster. Mars moves through its ascending node on 24 July; it will remain north of the ecliptic for the rest of the year.

**3 Juno** is a member of the main asteroid belt between Mars and Jupiter, and the third minor planet to be discovered. It arrives at opposition in the non-zodiacal constellation of Aquila on 26 July but as it is only ninth-magnitude, a telescope will be necessary to view it. 3 Juno is 1.8 au from Earth at the time of opposition and has an elongation from the sun of 165°. It is opposite the Sun in right ascension three days earlier, on 23 July. For more information on this and other minor planets, see *Minor Planets in 2026*.

**Jupiter** is very low in the west at sunset and is soon lost to view. Conjunction with the Sun occurs on 29 July.

**Saturn** reaches west quadrature on 6 July and enters the evening sky for both hemispheres. It is located in Pisces and shines at magnitude +0.8. The rings have been opening all year and reach a maximum tilt of 9.2° on 21 July. Retrograde motion begins near the end of the month.

**Uranus** and Mars team up on the fourth day of the month. Located just 0.1° apart, early risers in the southern hemisphere will have the best views of this event; astronomers from northern temperate latitudes have to contend with extended periods of twilight at this time of year, making observations of sixth-magnitude Uranus difficult. Like nearby Mars, Uranus also passes by several naked-eye stars in Taurus this month, including 37 Tauri on 3 July (1.4° apart) and ω¹ Tauri (43 Tauri) on 15 July (1.2° apart).

**Neptune** now rises before midnight. Still located in Pisces, the eighth-magnitude planet begins retrograde motion early this month.

**134340 Pluto** comes to opposition on 27 July in the constellation of Capricornus. Observing the fifteenth-magnitude dwarf planet requires a large telescope and a detailed finder chart. Unfortunately, the waxing gibbous Moon is in the neighbouring constellation of Sagittarius that night and may provide considerable light interference.

# Vega

## David M. Harland

Vega is the brightest star in the constellation of Lyra, and one of the brightest in the northern sky. Along with Altair in Aquila and Deneb in Cygnus it forms the Summer Triangle. Its eminence has been remarked upon throughout the centuries, by cultures around the globe. In his 1933 poem *A Summer Night* W. H. Auden includes the line, "Vega conspicuous overhead …"

A declination of 39° places Vega far from the north celestial pole, although 14,000 years ago the effects of the 26,000 year precession of Earth's spin axis positioned the star to within several degrees of the pole.

In July 1850 William Cranch Bond of Harvard College Observatory took the first photograph of a star, namely Vega for its brightness and convenience of viewing.

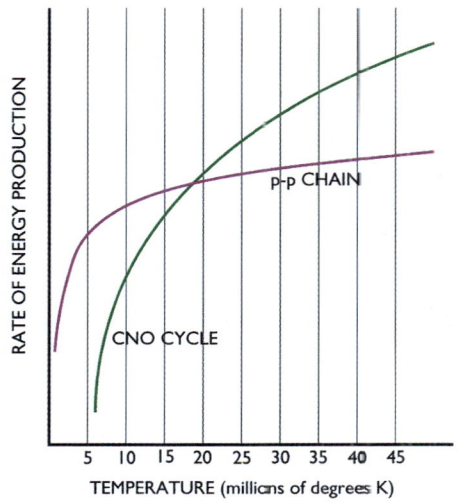

The temperature at the heart of the Sun is just short of where nuclear fusion by the carbon-nitrogen-oxygen cycle surpasses the proton-proton chain. The much hotter core of Vega makes it much more luminous. (David M. Harland)

In August 1872, Henry Draper, a wealthy amateur astronomer in New York who had a 28 inch reflector succeeded in photographing the spectrum of Vega, allowing accurate measurements of the wavelengths of its dark Fraunhofer lines.

Vega was one of the first stars to have its distance estimated by the parallax method, whereby Earth's motion around the Sun displaces a nearby star against the background of more distant stars. It is around 26 light years away, although its motion in space relative to the Sun means this distance is decreasing. The closest point of approach at 13.2 light years will occur in about 260,000 years.

Vega is a 'dwarf' star like the Sun, but it has twice the mass. With a spectral class of A0 its surface is much hotter. The dominant process of fusion within its core is not the proton-proton chain of the Sun but the carbon-nitrogen-oxygen cycle that liberates much more energy, giving it a luminosity forty times that of the Sun.

As a result of consuming its hydrogen at a faster rate, despite being only half a billion years old it is already nearing the middle-point of its time on the main sequence. As will the Sun, Vega will evolve first into a red giant and later shed its envelope and become a white dwarf.

The *Infrared Astronomical Satellite (IRAS)*, active for ten months in 1983, made the first all-sky survey at wavelengths of 12, 25, 60 and 100 micrometres in the electromagnetic spectrum.

When Vega was examined, the data was puzzling in the sense that it was brighter than predicted by models of how stars should shine. *IRAS* placed this 'infrared excess' within an angular radius of 10 seconds of arc, centred on the star, equating to a radius of 80 astronomical units. It implied the presence of dust, glowing at infrared wavelengths. Because our line of sight is only a few degrees off the spin axis of the star, we are able to 'look down' on material occupying the equatorial plane.

Although at that time astronomers suspected other stars might have planets, this had yet to be established.

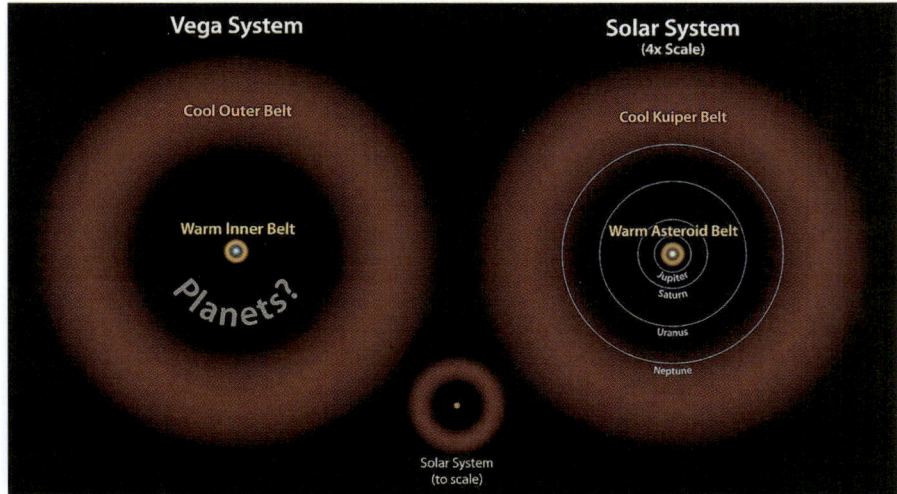

Vega has two belts of dust and rocky debris which are in similar proportions to the warm asteroid belt and cool Kuiper belt of the solar system, although in the case of Vega the overall structure is four times larger. (NASA/JPL-Caltech)

Our understanding of how the Sun formed and gained planets says that a dense interstellar cloud of gas and dust collapsed under its own gravitational attraction to form the Sun, and that by the conservation of angular momentum most of the left over material became flattened into a rapidly spinning disk within which grains clumped together to make a succession of ever-larger rocky bodies, the most massive being able to attract envelopes of hydrogen to become gas giants before the radiation pressure from the new star blew away the residual nebula.

Although the planets account for barely 0.1% of the overall mass of our solar system, they possess 98% of its angular momentum. In fact, Jupiter, larger than all the other planets combined, possesses 60% of the overall momentum.

Vega is rotating rapidly with a velocity at its equator of 236 kilometres per second, about 100 times that of the Sun. There is sufficient centrifugal force to give the star a significant equatorial bulge. By this reasoning, it might not yet have formed very massive planets orbiting sufficiently far out to capture most of the system's momentum. Nevertheless, the detection of dust around Vega was the first evidence that the processes that created planets in the solar system may indeed be underway elsewhere.

Infrared studies by the *Spitzer Space Telescope*, the *Herschel Space Observatory* and terrestrial telescopes revealed Vega to have two belts of dust and rocky debris in similar proportions to the asteroid belt and Kuiper belt of the solar system. In both systems, the gap between the inner and outer belts has a ratio of about 1 to 10, with the outer belt ten times farther away from its host star than the inner belt.

The first potential detection of a planet came in 2021 by analysing a decade of observations of the star's spectrum, where displacements in

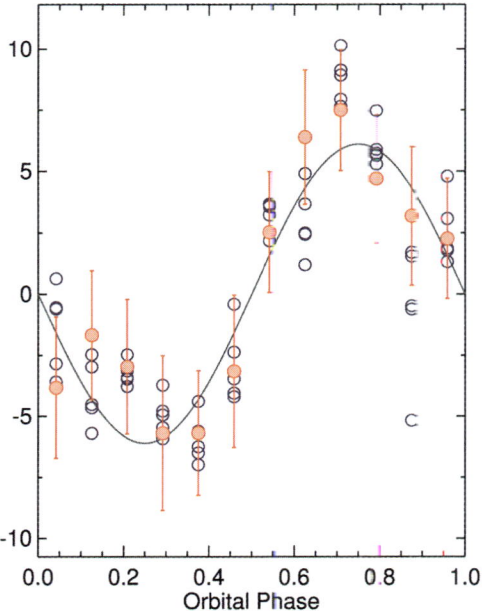

The radial velocity curve for a very massive candidate planet orbiting Vega with a period of 2.43 days. (Taken from Figure 8 of 'A Decade of Radial-velocity Monitoring of Vega and New Limits on the Presence of Planets', Spencer A. Hurt et al., *Astronomical Journal*, **161**, 157–171, 2021)

the spectral lines indicate wobbles induced by orbiting planets. In this case, the data suggested a planet with a 2.43 day orbital period. Its mass was calculated to be at least that of Neptune and possibly considerably more, depending upon the plane of its orbit relative to our line of sight. Orbiting so close to such a bright star, the planet would be exceedingly hot.

The question is whether the gap between the debris belts is occupied by planets. The near-pole-on inclination of the system makes it difficult to use the radial velocity method, and of course there can be no tell-tale transits. Hopefully the infrared instruments of the *James Webb Space Telescope* will be able to find out.

# August

| 2 | Mercury | Greatest elongation west (morning) | 20° |
|---|---|---|---|
| 6 | Moon | Last Quarter Moon | 90° |
| 7 | Moon, M45 (Pleiades) | 1.1° apart: occultation | 75° |
| 11 | Moon, M44 (Praesepe) | 0.8° apart: occultation | 12° |
| 12 | Moon | New Moon: total solar eclipse | 1° |
| 12/13 | Earth | Perseids (ZHR 100) | |
| 13 | Mars, 1 Geminorum | 0.4° apart | 49° |
| 13 | Moon, α Leonis (Regulus) | 0.5° apart: occultation | 9° |
| 13 | Mars, M35 | 0.6° apart | 50° |
| 15 | Venus | Greatest elongation east (evening) | 46° |
| 15 | Mercury, Jupiter | 0.6° apart | 13° |
| 17/18 | Earth | Kappa Cygnids (ZHR 3) | |
| 20 | Moon | First Quarter Moon | 90° |
| 21 | Moon, α Scorpii (Antares) | 0.6° apart: occultation | 102° |
| 27 | Mercury | Superior conjunction: morning → evening | 2° |
| 28 | Moon | Full Moon: partial lunar eclipse | 180° |
| 28 | Uranus | West quadrature | 90° |

Dates are based on Universal Time (UT). Occultations of M44 (Praesepe/Beehive) refer to sixth-magnitude star ε Cancri (Meleph). Occultations of M45 (Pleiades) refer to third-magnitude star η Tauri (Alcyone). Peak activity dates and ZHRs for meteor showers are estimates. The last column gives the approximate elongation from the Sun of the event.

**Mercury** attains a greatest elongation west of 19.5° on the second day of the month and heads back toward the eastern horizon. It brightens as it loses altitude and approaches superior conjunction on 27 August. Mercury appears 0.6° north of Jupiter on 15 August. Jupiter, at magnitude −1.7, is just brighter than Mercury's −1.3, but as the two planets are only 13° away from the Sun at the time, this event will be difficult to observe.

**Venus** reaches greatest elongation east (45.9°) on 15 August, with theoretical dichotomy occurring three days earlier. On 16 August, the waxing crescent Moon is just 2° from the evening star. This has been an excellent evening apparition of Venus for astronomers in the southern hemisphere but the planet reaches its peak altitude above the western horizon this month and starts to head back toward the Sun. Observers in northern latitudes have been watching Venus fall from the sky for several months; it is already getting quite low after sunset.

**Earth** has perfect viewing conditions for the annual Perseid meteor shower which peaks around 12/13 August. This is because the Moon is new the previous day and absent from the sky. However, little or no activity is expected from the episodic Kappa Cygnid shower a few days later. See *Meteor Showers in 2026* for details.

The New **Moon** of 12 August blots out the Sun in a total solar eclipse. Our satellite also occults a number of stellar objects, including the Pleiades on 7 August, Regulus on 13 August, and Antares on 21 August. The lunar occultation of the Beehive Cluster takes place the day before the eclipse and will not be visible. A partial lunar eclipse on 28 August is the last major event involving the Moon this month. See *Eclipses* elsewhere in this volume for more information on the two eclipses mentioned here.

**Comet 10P/Tempel 2** arrives at perihelion on 2 August and makes it closest approach to Earth the following day. Comets are notoriously unpredictable but 10P/Tempel 2 could reach seventh magnitude so optical aids are necessary for spotting it as it passes through the constellations of Capricornus and Piscis Austrinus this month. For more information on this Jupiter-family comet, consult *Comets in 2026*.

**Mars** remains a morning sky object, but it is getting easier to see from northern latitudes. It also begins to brighten, beginning the month at magnitude +1.3 and ending at +1.2. Mars moves from Taurus to Gemini on 12 August. The following day, the red planet passes 0.4° north of 1 Geminorum, a fourth-magnitude G-type giant star and spectroscopic binary, and 0.6° south of M35, an open cluster with an apparent magnitude of +5 and an angular diameter of nearly half a degree. On 17 August, Mars is found just over a degree north of the third-magnitude red giant star Propus (η Geminorum) and two days later, is located a similar distance north of Tejat (μ Geminorum), a third-magnitude long-period variable star of spectral type M. Mebsuta (ε Geminorum) is the next bright star visited by Mars when, on 27 August, the red planet passes 1.6° south of the third-magnitude, G-type supergiant.

**Jupiter** is just past conjunction and appears in the morning sky just ahead of the Sun. Its encounter with M44 (Praesepe or Beehive Cluster) on the fourth day of the month takes place far too close to our star to be observable. The same can be said of the planet's pass by Asellus Australis (δ Cancri), a fourth-magnitude multiple star system, on the tenth. Mercury and Jupiter come to within 0.6° on 15 August but the two planets are barely 12° away from the Sun at the time. However, the planet, now shining at magnitude −1.8, should be visible low in the east by the end of the month.

**Saturn** now rises during evening hours, shining at magnitude +0.6 in Pisces. The rings are closing again, with the tilt reducing from 9.1° to 8.5° over the course of the month. The Moon passes by the ringed planet twice this month, on the third day and the last, but never comes closer than 6°.

**Uranus** continues its slow traversal of Taurus, arriving at west quadrature near the end of the month. It is a morning sky object for observers south of the equator but rises a little before midnight for astronomers in northern temperate latitudes.

**Neptune** rises early in the evening by the end of August. Located in Pisces, the most distant planet in the solar system shines at magnitude +7.8 so a telescope is required to see it.

# The Extraordinary George Eric Deacon Alcock
## Return of the GEDA

### John McCue

Guisborough, a market town near my home in the North Riding of Yorkshire, shelters at the foot of the Cleveland Hills, and was the venue for a garden party recently. Sitting next to me was a new friend, Julie. "Have you heard of George Alcock?" she asked, "He was my teacher at primary school". As a conversational gambit, it was a winner – in fact, I nearly fell off my garden bench. The school was Southfield Primary in Peterborough.

She showed me her project exercise book, beautifully handwritten at age nine, in which appears Halley's Comet. It's well known that George discovered five comets with his large angled-eyepiece tripod-mounted

Sketch of George Alcock looking skyward. (John McCue)

Primary school drawing of Halley's Comet. (Julie Pearson)

binoculars; Julie often observed the night sky through this historic instrument. His discoveries came as a reward for his dedication to scanning the night sky ceaselessly and tirelessly, learning the relative positions and patterns of thousands of fixed stars, which eventually became an intrinsic part of his memory – he never needed to consult a star atlas! He would describe his discoveries, which also included five novae, to small groups of interested pupils at his wonderfully cluttered and old-fashioned home where, after a chilly night's search of the heavens, George's wife would serve tea in bone china cups, making sure George's many cats stayed out of the way. Julie remembers George as an inspirational and caring teacher, but he was also quite capable of keeping order. A memorable story relates how George dealt with chaotic behaviour on the top deck of the school bus which ferried pupils from the nearby villages, handily going right past George's house, which was fittingly named *Antares*. Four particular boys were incorrigible, always ignoring the bus driver's appeals. Finally, he sought George's help. "Write down their names", said George, 'and push the note through my letter box as you drive past tomorrow and I'll deal with them". The driver did just that, and so did George! The pupils' nickname for him was GEDA (Jedda), and George knew that, though to family and friends he was always Eric. His last comet discovery was in 1983, named as IRAS-Araki-Alcock.

Julie and I were told of Gill, who came to Guisborough to visit us, and delighted us with the fact that George's wife was her godmother. Her family were close to George and his wife, Mary, and spent every Christmas with them. Gill became the daughter that George and Mary never had, and she was a pupil at another of George's schools, Old Fletton Primary School.

George's house was guarded by a dyke, into which Gill remembers tumbling as she approached the house one Boxing Day; luckily there was very little water in the stream! Some years later, Gill accompanied George on his regular train journeys – a packet of biscuits always in his

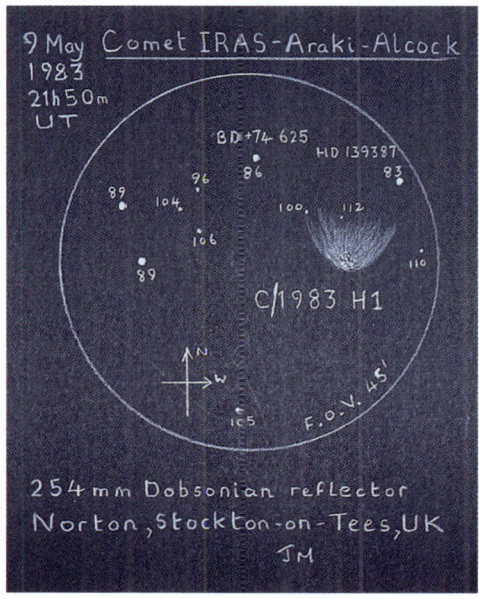

Telescopic observation of Comet IRAS-Araki-Alcock. (John McCue)

Train-spotter's railway sketch at Ely, near Peterborough. (George Alcock)

pocket – to the Royal Geographical Society in London. Peterborough was railway mad, being on the London to Edinburgh connection; indeed Julie's grandfather was a driver of the *Flying Scotsman* locomotive. As a young girl, she recalls seeing him one day in the driver's cab at the Peterborough station but hardly recognised him – his face was black with soot as the fire box glowed. He would often toot the *Flying Scotsman*'s whistle when approaching Peterborough to let his wife know, living not far from the station, when to put the dinner on!

George's early observing was telescopic. Although he sketched planetary features with extraordinary artistic skill, he did not confine himself to astronomical objects. His astonishing ability to portray details can be seen in his many railway sketches, such as the one seen here which shows the 11:00 network express from King's Lynn to London Liverpool Street as it arrives at Ely. Even more numerous are his portrayals of historical architecture which often show unforgettable minutiae, such as in his depiction of the steeple of St. Magnus the Martyr, located in Lower Thames Street near London Bridge – George was a devotee of Christopher Wren's creations.

Gill recalls, as many pupils would, George's zeal for history and geography during his secondary teaching years. He would spend time before a lesson illustrating the topics with drawings in coloured chalks on the class blackboard. One can only imagine the pupils' fascination on seeing such skill, but anyone misbehaving would be the target of flying chalks and board rubbers from George! Gill recalls with embarrassment Uncle Eric's target practice with these coloured sticks! It seems a shame as well that he had to scrub off this art work at the end of the lesson.

During World War 2 George was posted to Italy for service in the RAF Signal Corps. Recruits were always sent to training venues distant from their hometown, in order to acquaint these newcomers with absence from their families, which was often long and worrisome during the conflict. George craftily circumvented this rule by giving an address in Yorkshire, one belonging to relatives. Consequently, he was posted to Polebrook for training, which was even close enough to cycle there – and to get back home again for tea! After training, he was posted to Italy, where, in whatever spare time was available, he explored his surroundings, sketching and writing about the landscapes, buildings, nature and people he encountered, sending his drawings home. Sadly, straight after the war, George returned home to the unfortunate news that Mary had not kept his work – she was in serious trouble! What must have been marvellous artwork was lost.

Architectural sketch of St. Magnus the Martyr, a Christopher Wren masterpiece. (George Alcock)

Julie remembers George driving out to the countryside on his artistic search for birds, trees, churches, and locomotives, usually in an Austin A40 or a Morris 1000. As happens to us all, his driving career came to an end in his later years. The last straw was when a council refuse wagon reversed into George's car, which was stationary at the time! Luckily no-one was hurt.

George could be a strict and an exacting taskmaster but Gill, Julie, and many others who knew him, have fond memories of a remarkable, kind, smiling man whom they loved and admired.

# September

| 2  | Mars, ω Geminorum         | 1.0° apart              | 56°  |
|----|---------------------------|-------------------------|------|
| 3  | Moon, M45 (Pleiades)      | 1.2° apart: occultation | 101° |
| 4  | Moon                      | Last Quarter Moon       | 90°  |
| 8  | Moon, M44 (Praesepe)      | 0.8° apart: occultation | 38°  |
| 8  | Moon, Jupiter             | 0.8° apart: occultation | 31°  |
| 9  | Mars, δ Geminorum (Wasat) | 0.8° apart              | 58°  |
| 9  | Moon, α Leonis (Regulus)  | 0.5° apart: occultation | 17°  |
| 11 | Moon                      | New Moon                | 2°   |
| 14 | Moon, Venus               | 0.5° apart: occultation | 41°  |
| 17 | Moon, α Scorpii (Antares) | 0.6° apart: occultation | 76°  |
| 18 | Moon                      | First Quarter Moon      | 90°  |
| 23 | Earth                     | Equinox                 |      |
| 26 | Neptune                   | Opposition              | 179° |
| 26 | Mercury, α Virginis (Spica) | 0.9° apart            | 21°  |
| 26 | Moon                      | Full Moon               | 177° |
| 30 | Moon, M45 (Pleiades)      | 1.1° apart: occultation | 127° |

Dates are based on Universal Time (UT). Occultations of M44 (Praesepe/Beehive) refer to sixth-magnitude star ε Cancri (Meleph). Occultations of M45 (Pleiades) refer to third-magnitude star η Tauri (Alcyone). The last column gives the approximate elongation from the Sun of the event.

**Mercury** returns to the west at sunset in what is the best evening apparition of the year for southern hemisphere observers whereas astronomers in northern temperate latitudes will be lucky to spot the elusive planet at all. Evening apparitions of Mercury are characterised by a fading of brightness; Mercury begins the month at magnitude −1.5 and ends it at −0.1. On 26 September, the planet will outshine first-magnitude Spica (α Virginis) when the two bodies are less than a degree apart some 21° away from the Sun.

**Venus** passes 1.5° south of the first-magnitude star Spica on the second day of the month and the waxing crescent Moon comes close enough to occult the evening star on 14 August. Now past greatest elongation east, the planet presents a crescent shape when observed through a telescope. (**Warning:** To avoid the risk of serious

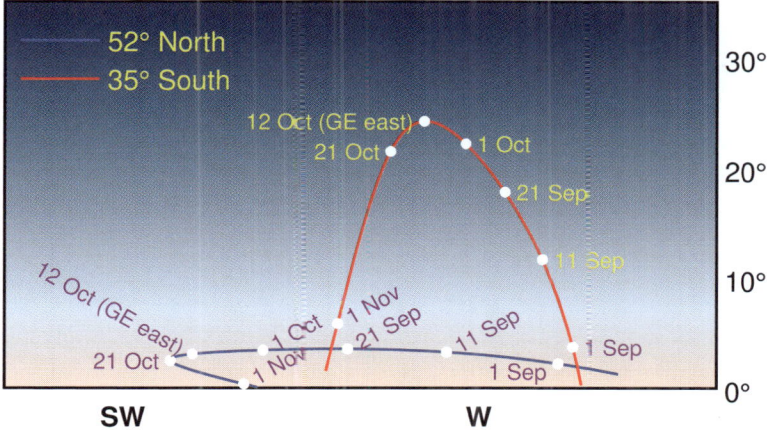

eye damage, *do not attempt this until after the Sun has set.*) Whilst the crescent is getting slimmer (39% illuminated at the beginning of the month, only 14% at the end), Venus is getting closer to Earth and the disk is growing in size (from 30.3″ in apparent diameter to 48.4″). The planet is already quite low in the west at sunset for those looking for it from northern temperate latitudes, with Venus vanishing from view before the end of the month, but it is still quite high above the horizon when seen from southern vantage points.

**Earth** passes through an equinox on 23 September, with astronomical spring bursting forth in the southern hemisphere and autumn beginning in the north.

The **Moon** has it in for the Pleiades this month, occulting the open cluster twice, on 3 September and again on the last day of the month. M44, the Beehive Cluster, is visited by the Moon on 8 September when our satellite passes 0.8° north of the brightest member of that group. Jupiter is occulted on the same day, about 12 hours later, with the occultation of Regulus taking place on 9 September. The crescent Moon makes an attractive sight with Venus on 14 September, with some locations on Earth seeing an occultation. Antares is next, disappearing behind the Moon's disk on 17 September. The nearest apogee this year takes place on 19 September when the Moon is at a distance of 404,221 kilometres.

**Mars** now appears before midnight for observers in northern temperate latitudes but does not rise until early morning for those in the south. It passes 1.0° north of the fifth-magnitude Cepheid variable star ω Geminorum on the second day of the month and 0.8° north of Wasat (δ Geminorum), a fourth-magnitude spectroscopic binary, on 9 September. Mars has a distant encounter with the famous planetary nebula C39 (NGC 2392) on 13 September but only long-exposure photography will bring out the detail in the tenth-magnitude nebula. The red planet moves into Cancer on 25 September and reaches an equinox on 29 September. On this day, spring returns to the northern hemisphere of Mars whilst autumn begins in the south.

**Jupiter** undergoes its first lunar occultation of the year on 8 September when the waning crescent Moon obscures it from around 15:00 UT. The bright planet is visible in the morning sky before sunrise and passes from Cancer to adjacent Leo on 23 September. It will remain in the constellation of the majestic feline for the remainder of the year.

**Saturn**'s retrograde motion brings it out of Pisces and back into Cetus early this month. The rings continue to close, from 8.5° to 7.6°, but the planet gets brighter, from magnitude +0.5 to +0.4, as Saturn approaches opposition and closest approach to Earth early next month. Look for Saturn rising soon after it gets dark.

**Uranus** returns to retrograde motion early this month, reversing course away from the Hyades and back toward the general direction of the Pleiades. Located in Taurus, the faint planet shines at magnitude +5.6 and is best viewed from northern vantage points where it rises in mid-evening. It also rises before midnight for southern observers.

**Neptune** returns to the southern celestial hemisphere mid-month, after straying into positive declinations in April. Opposition takes place on 26 September so Neptune is aloft most of the night. This month the planet is at its closest to Earth – 28.9 au – but is only magnitude +7.8. In a telescope, Neptune appears as a 2.5″ diameter disk. Unfortunately, the light of the Full Moon will flood the sky with light on the date of opposition.

# The Association between Comets and Meteor Showers

## Neil Norman

When our solar system formed some 4.5 billion years ago, some of the leftover debris that did not go on to form the planets and moons clumped together under gravity to form the objects that we see today as asteroids or comets. Comets contain primordial ices and water and differ from asteroids, which are solely dust and rock in composition. Comets do contain rock and dust, although it is the presence of ice and water that makes all the difference.

Images of asteroids reveal little more than a star-like appearance, whereas those of comets, if the comet is primarily gassy in composition, will reveal a 'fuzzy' head or coma and, if we're lucky – as was the case with C/2020 F3 NEOWISE – we can see a tail form. The coma and tail come into being as the frozen nucleus of the comet draws closer to the Sun; the ices on the nucleus have different melting points, the warmth from the Sun causing them to sublimate (convert from a solid to gaseous state without becoming liquid) during this heating process. In addition, if the nucleus contains a lot of dust, this will also be released.

Think of the periodic comet 12P/Pons-Brooks, which last reached perihelion in 2024. This comet has an orbital period of 71 years and, composition-wise, was mostly gassy in nature. This meant that, when observed, it took on the appearance of a "fuzzy star", although on several occasions it experienced an outburst, resulting in its brightness increasing rapidly by a magnitude or more over the course of just a few hours. This increase was due to dust being blown off the surface of the comet, which in turn

Engraving of Denison Olmsted by Scottish-born American artist and engraver Alexander Hay Ritchie (1822–1895). (Wikimedia Commons/ Alexander Hay Ritchie)

made the coma rich in dust, resulting in more sunlight being reflected and the comet appearing brighter. The cause of all this was from jets blowing dust from the nucleus, in much the same way as geysers in Iceland which periodically blow water several meters high into the air. The same thing happens on the nuclei of comets; as the nucleus rotates into sunlight, the pressure of the ices expanding on or just under the surface, as the ices sublimate, causes jets of dust to be expelled from the surface at high speeds.

Comets have two types of tail, one of which is an ion tail, formed from gas and which arcs away from the comet as a result of the pressure of solar radiation. Ion tails do not concern us here, although the other types, the dust tails, do. The dust emanating from the comet actually follows the orbit of the comet – wherever the comet goes the dust tail lags behind – and it is this ejecta that ultimately can, under certain circumstances, create meteor showers that are observable in our skies.

The connection between comets and meteor showers was first mentioned by American astronomer Denison Olmsted (1791–1859). A display of shooting stars in 1799 caught the attention of many, a similar spectacle being observed in 1833. Olmsted witnessed the latter shower, and took it upon himself to research the phenomenon, writing down his findings and asking people to submit their observations to him so that he could gain a bigger picture of the event. He asked for help via a local newspaper, and was surprised to receive a large number of reports from eastern North America.

Olmstead quickly connected a pattern to the occurrences and noted that, in almost every year since 1799, similar events repeated themselves at the same time each

This diagram, taken from the article 'The August and November Meteors' by Dr. H. Schellen in the August 1872 issue of *Popular Science Monthly*, depicts the orbital path of the particles which give rise to the Perseid meteor shower in relation to the Earth's orbit around the Sun. (Wikimedia Commons / Popular Science Monthly / E.L. Youmans / Dr. H. Schellen)

year. Olmsted then asked the New Haven-based amateur astronomer Edward Claudius Herrick (1811–1862) to partake in a study of meteors. After a few years of them working together, Herrick deduced that other meteor displays occurred in April, August and December; today we know these events as the Lyrids, Perseids and Andromedids.

This image, captured on 23 June 2004 by the Multiband Imaging Photometer for Spitzer (MIPS) instrument on the *Spitzer Space Telescope*, depicts Comet 2P/Encke (centre). The picture clearly shows the dusty debris to either side of the parent body. As this material slowly moves away from the comet, it becomes spread out all along the comet's orbital path. (NASA/JPL.Caltech/M.Kelly (Univ.of Minnesota))

With a definitive pattern forming, scientists began working on determining the source of these displays, the man who eventually made a definitive connection being the American astronomer Daniel Kirkwood (1814–1895) in 1861. Kirkwood went on to describe meteors as being cometary in origin. It was the Italian astronomer Giovanni Schiaparelli (1835–1910) who first pointed out the similarity between the orbits of the meteors and that of the recently-discovered (1862) Comet Swift-Tuttle, correctly concluding that the Perseids indeed originated from this comet. These views were shared by both German-American astronomer Christian Heinrich Friedrich Peters (1813–1890) and Austrian astronomer and mathematician Theodor von Oppolzer (1841–1886).

The story now gets a little more complex. Although all comets releasing dust can, in theory, be parent bodies of meteor showers, a certain set of circumstances still have to occur for this situation to be realised. Firstly, the comet has to come close to Earth, either as it approaches perihelion or recedes after perihelion. Even then, the nodes of the orbit, either ascending or descending, must be very close to the Earth's orbital path, thus allowing our planet to scoop up the dusty remnants of the comet and for us to see the consequence as shooting stars at regular intervals each year.

The richness of the stream is also highly relevant. Comet 55P/Tempel-Tuttle, for example, has a stable 33 year period, and good displays of its associated shower (the Leonids) are seen when the comet comes to perihelion as in, for example, 1900, 1933, 1966 and 1999. Because 55P/Tempel-Tuttle is such a regular visitor to the inner solar system, the associated stream is replenished regularly. However, in the case of 109P/Swift-Tuttle, the comet does not frequent the inner solar system as often, taking 133 years to complete an orbit. Comet 109P/Swift-Tuttle is exceptional though, in that it is a large comet which produces more dust, resulting in the Perseid particle stream being adequately filled during each return.

Dozens of meteor showers are currently known, and more could come and go in the future as some become depleted in material. Also, an added mystery to our story is the prolific Geminid meteor shower, visible around mid-December each year. The parent body of this shower is 3200 Phaethon which is not a comet but an asteroid, although that's another article in itself.

# October

| 3 | Moon | Last Quarter Moon | 90° |
|---|---|---|---|
| 4 | Saturn | Opposition | 177° |
| 5 | Moon, Mars | 1.1° apart: occultation | 68° |
| 5 | Moon, M44 (Praesepe) | 0.7° apart: occultation | 65° |
| 6 | Moon, Jupiter | 0.2° apart: occultation | 53° |
| 6 | 2 Pallas | Opposition | 159° |
| 7 | Earth | Arids (ZHR unknown) | |
| 7 | Moon, α Leonis (Regulus) | 0.6° apart: occultation | 44° |
| 7/8 | Earth | Draconids (ZHR 5) | |
| 10 | Earth | Southern Taurids (ZHR 5) | |
| 10 | Moon | New Moon | 4° |
| 11 | Mars, M44 (Praesepe) | 2.6' apart | 71° |
| 12 | Mercury | Greatest elongation east (evening) | 25° |
| 13 | 4 Vesta | Opposition | 168° |
| 14 | Moon, α Scorpii (Antares) | 0.4° apart: occultation | 49° |
| 18 | Moon | First Quarter Moon | 90° |
| 21/22 | Earth | Orionids (ZHR 20) | |
| 24 | Venus | Inferior conjunction: evening → morning | 7° |
| 26 | Moon | Full Moon | 175° |
| 28 | Moon, M45 (Pleiades) | 1.0° apart: occultation | 154° |

Dates are based on Universal Time (UT). Occultations of M44 (Praesepe/Beehive) refer to sixth-magnitude star ε Cancri (Meleph). Occultations of M45 (Pleiades) refer to third-magnitude star η Tauri (Alcyone). Peak activity dates and ZHRs for meteor showers are estimates. The last column gives the approximate elongation from the Sun of the event.

**Mercury** soars above the western horizon for lucky observers in southern latitudes. It is becoming dimmer over the course of the month (magnitude −0.1 down to third magnitude) as its phase decreases (77% down to 5% illumination) even though the planet is getting closer to Earth. Mercury zooms past Venus on 7 October on its way to greatest elongation east when it is 25.2° away from the Sun. It begins to lose altitude about this time. Inferior conjunction is approaching early next month so retrograde motion commences on 24 October.

**Venus** swings into retrograde motion at the beginning of the month and has a distant encounter with Mercury on 7 October when the two inferior planets are 5.1° apart. Venus has already vanished for northern viewers and it plummets toward the western horizon as seen from the favoured southern hemisphere. The planet relinquishes its title of evening star on 24 October when it undergoes inferior conjunction, overtaking Earth in its orbit. It is nearest to our planet the following day when the two worlds are 0.27 au apart. Venus quickly reappears, this time in the morning sky, where it will remain for the rest of the year.

**Earth** is bombarded with comet cast-offs this month, with the Arids and Draconids enjoying very favourable returns early in October. (To be clear, the meteors of the Arid meteor shower emanate from the southern hemisphere constellation of Ara, the altar, not the northern hemisphere constellation of Aries, the ram.) The later Orionids occur much closer to the Full Moon and light pollution may be a factor in observations. The Taurid meteor shower complex is also in force, with activity peaking both this and next month. *Meteor Showers in 2026* provides more information.

Lunar occultations begin with Mars on 5 October although viewing opportunities may be limited. The waning crescent **Moon** also passes through the Beehive Cluster (M44) on the same day. On 6 October, Jupiter vanishes behind the Moon's disk in the hours before dawn for some early risers. Regulus is the next target, with an occultation taking place on 7 October. First-magnitude Antares is occulted by the waxing crescent Moon one week later. Finally, the Pleiades open star cluster is blocked by the Moon on 28 October.

**Mars** rises just before midnight for astronomers in northern temperate latitudes but is visible only during morning hours for observers further south. On 5 October, Mars is occulted by the waning crescent Moon but visibility of this event is largely confined to the Arctic. Then, on 11 October, Mars glides through M44, the famous if faint Beehive Cluster, passing only 2.6′ north of the brightest member of that open cluster, ε Cancri or Meleph. The colour of the red planet should contrast nicely with the predominantly blue-hued stars of the cluster when viewed through binoculars or a telescope. The accompanying diagram shows a binocular view of the progress of Mars through this cluster. Two days later, Mars is found just over a degree north of fourth-magnitude Asellus Australis (δ Cancri). Mars travels into Leo on 30 October, a constellation it will inhabit for the rest of the year.

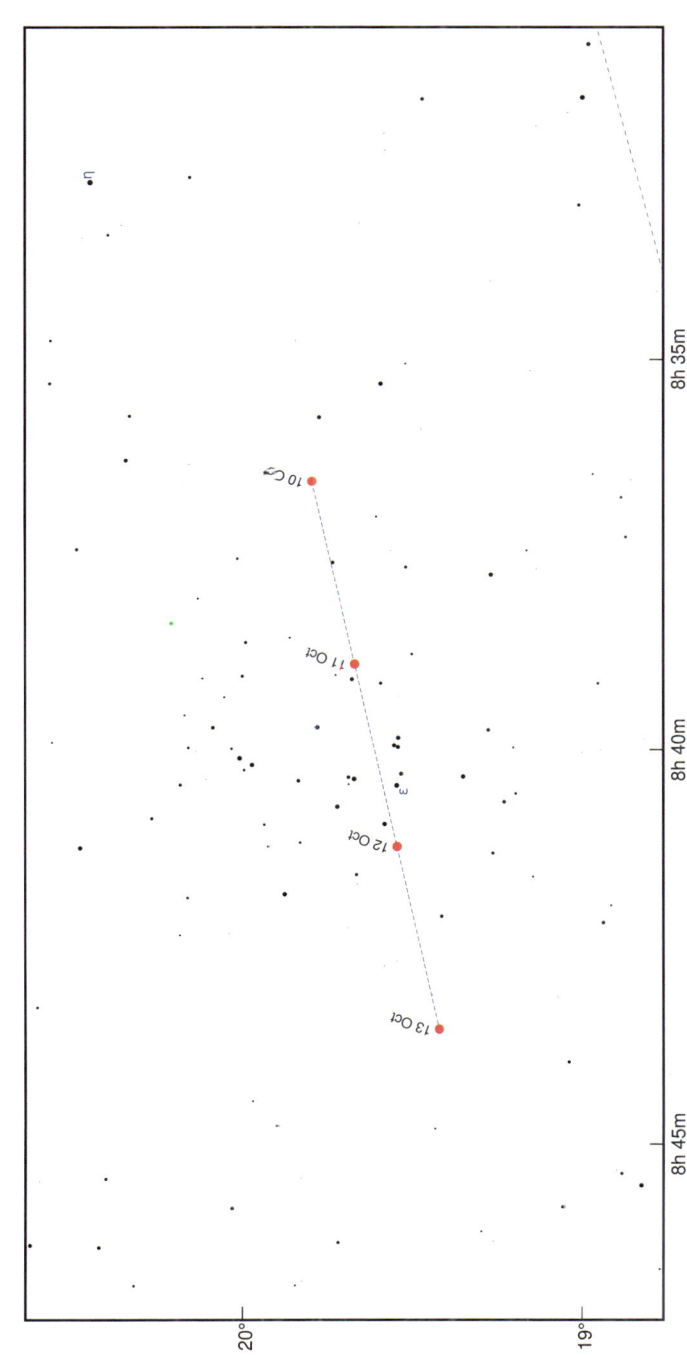

**4 Vesta** is the only asteroid that reaches naked-eye visibility, but this is not a favourable opposition year and the minor planet is barely sixth-magnitude. 4 Vesta attains opposition on 13 October in the constellation of Cetus. It is 1.5 au from Earth and 169° from the Sun. In terms of right ascension, it is opposite to the Sun four days later on 17 October. Optical aids will be necessary to see both this object and 2 Pallas at opposition this month.

**2 Pallas** reaches opposition in Cetus on 6 October when it is 1.9 au from Earth and shining at eighth magnitude. Its elongation from the Sun is 159° and it is opposite to our star in terms of right ascension on 14 October. For more information (including finder charts) for 2 Pallas and 4 Vesta, see *Minor Planets in 2026* later in this volume.

**Jupiter** is again occulted by the Moon this month. On 6 October, beginning at approximately 08:00 UT, the waning crescent Moon will pass in front of the disk of the planet as seen from Earth. Jupiter is magnitude −1.9 in the constellation of Leo and rises during the early morning hours. Northern latitudes are slightly favoured as Leo lies north of the celestial equator.

**Saturn** arrives at opposition on the fourth day of the month. It is at its brightest (+0.3 magnitude) and closest to Earth (8.4 au) on this day. In a telescope, the disk of the planet is 19.8″ across, with the rings open at an angle of 7.5°. The ringed planet is visible for most of the night and is well-placed for viewing from both hemispheres in the non-zodiacal constellation of Cetus.

**Uranus** is approaching opposition next month. Found in Taurus, the planet is best viewed from the northern hemisphere where it rises during early evening hours. It appears mid-evening for those searching for it from southern latitudes.

**Neptune** was at opposition late last month so it already above the eastern horizon at sunset. Choose a dark night and use your telescope to look for the eighth-magnitude planet in Pisces.

# Ussher Chronology: The Age of the Universe

## Jonathan Powell

The necessity to accurately mark the passage of time has entertained the thought of many a scholar throughout history, with the overriding premise that no one person, neither then nor today, has the unchallenged position of saying one interpretation of the passing of time is right, and the other wrong. Whilst time itself is a human-made concept, the need for its existence in the form of, for example, a simple wall clock, acting as a common reference point, far outweighs the debate for the interpretation of its being.

From the most basic clocks to the most ingenious calendars, the need across the globe to produce the ultimate way to reflect the hour by hour passing of time and the changing seasons of the year, has

James Ussher, whose calculations proposed that the creation of the universe occurred on 23 October 4004 BCE. (John McCue)

in turn produced an obsession. The self-invented slave of time has, by our own hand, become our master and ruler, and we are, by its presence, duly humbled.

The need to monitor time has been fulfilled down the centuries by different civilisations in a variety of innovative methods, a number of which have shown how some cultures, originally believed to have not been that scientifically minded, have rightly proved otherwise. However, the premise would suggest the need to record time only in a progressive format, to blueprint an annual picture of lives within a society. This would include, for example, the planting and reaping of crops. So, how would most techniques fare when challenged with working time backward from the present into the past?

The list of those who dedicated a fair proportion of their time in the pursuit of attempting to explain time in reverse, and indeed the very age of the universe, makes for formidable and impressive reading.

Perhaps of greater note from historical records is the realization that what has generally been considered the musings of minds in more recent times (given fresh insight into the cosmos through more advanced technology) was already a consideration from centuries past. Taking a couple of examples we have the Greek philosopher Anaximander of Miletus (611–546 BCE), who proposed the idea of multiple or even infinite universes. Then came the writings of English bishop, scientist and scholar Robert Grosseteste (1175–1253), who described the birth of universe as an explosion followed by the subsequent crystallization of matter.

This poses the question about addressing the age of the Earth, or indeed the creation of the universe itself. One man who has remained very much at the centre of that debate for his theories – even to the present day – is the Irish scholar and Archbishop of Armagh James Ussher (1581–1656). Ussher sought to not only establish the year of the creation, but also the date and time, basing his findings on the text contained within the Old Testament. His work still underpins Young Earth Creationism (YEC), which proposes that the universe was created not billions of years ago, but only thousands of years instead.

Ussher was one of many from in and around the seventeenth century – especially of a religious background – to make such a claim relating to the age of the universe. Jewish scholar and Vice-Chancellor of the University of Cambridge John Lightfoot (1602–1675) produced his own chronology, and whereas it was not the same conclusion as Ussher, it was derived by similar methods and was in the same chronological vicinity. The Franco-Italian religious leader and scholar Joseph Justus Scaliger (1540–1609) was also notable for his work in seeking to revolutionize perceived ideas around the understanding of ancient chronology. Even the great scientist and mathematician Sir Isaac Newton (1642–1727) had a viewpoint on the age of the universe.

The product of his research saw Ussher publish a 1,600-page volume in 1650, an English version of which, entitled *The Annals of the World*, was published in 1658. The text of his work states that the creation occurred on 23 October 4004 BC, the date and year being derived from meticulous and extensive studies of the chronologies used in the Hebrew text of Genesis 5 and 11, coupled with other passages from the bible. According to Lightfoot, the year was 3929; Scaliger 3949; and Newton 4000.

However, Ussher's starting point may well have been the commonly held belief that Earth only had a potential duration of 6,000 years (4,000 years before the birth of Christ, with 2,000 to follow), which would correspond to the six days of creation.

Ussher's work carried great weight because of the high regard in which he was held by many. This was not just for his working life in the Church, but for his

The segment *Tenperantia* (Temperance), from 'The Allegory of Good and Bad Government' by Italian artist Ambrogio Lozenzetti (c.1290–1348), is taken from the scene 'Allegory of Good Government' (1338). Eloquently held and alluded to is the hourglass and the passage of time. Lozenzetti suggests within his series of three fresco panels, painted during 1338 and 1339, that good government has an astral connection with his work depicting along its border the personifications of the planets as well as the seasons. (Ambrogio Lozenzetti / Palazzo Pubblico / The Yorck Project (2002))

academic endeavours, including his mastery of Semitic and classical languages which would have given him great insight into the interpretation of certain texts that were to contribute to his conclusion on the age of the universe.

What is remarkable is that a great number of similar attempts have been made since Ussher to establish such a chronology, and despite many being at odds with Ussher's 4004 BCE, others have been not that dissimilar. More recent years have understandably dictated otherwise. However, at the time it was carried out, Ussher's work commanded a great deal of respect from many quarters, with widespread acceptance of his calculations until well into the eighteenth century.

# November

| | | | |
|---|---|---|---|
| 1 | Moon, M44 (Praesepe) | 0.5° apart: occultation | 92° |
| 1 | Moon | Last Quarter Moon | 90° |
| 2 | Moon, Mars | 1.0° apart: occultation | 81° |
| 2 | Moon, Jupiter | 0.5° apart: occultation | 76° |
| 3 | Moon, α Leonis (Regulus) | 0.8° apart: occultation | 71° |
| 4 | Mercury | Inferior conjunction: evening → morning | 0° |
| 5 | Earth | Southern Taurids (ZHR 5) | |
| 7 | Moon, Venus | 1.0° apart: occultation | 22° |
| 9 | Moon | New Moon | 5° |
| 11 | Moon, α Scorpii (Antares) | 0.3° apart: occultation | 22° |
| 12 | Earth | Northern Taurids (ZHR 5) | |
| 16 | Mars, Jupiter | 1.2° apart | 87° |
| 17 | Moon | First Quarter Moon | 90° |
| 17/18 | Earth | Leonids (ZHR 10) | |
| 18 | Jupiter | West quadrature | 90° |
| 18 | Mars | West quadrature | 90° |
| 20 | Mercury | Greatest elongation west (morning) | 20° |
| 24 | Moon, M45 (Pleiades) | 0.9° apart: occultation | 175° |
| 24 | Moon | Full Moon | 175° |
| 25 | Uranus | Opposition | 180° |
| 29 | Moon, M44 (Praesepe) | 0.2° apart: occultation | 119° |
| 30 | Moon, Jupiter | 1.1° apart: occultation | 101° |
| 30 | Moon, α Leonis (Regulus) | 1.0° apart: occultation | 98° |

Dates are based on Universal Time (UT). Occultations of M44 (Praesepe/Beehive) refer to sixth-magnitude star ε Cancri (Meleph). Occultations of M45 (Pleiades) refer to third-magnitude star η Tauri (Alcyone). Peak activity dates and ZHRs for meteor showers are estimates. The last column gives the approximate elongation from the Sun of the event.

**Mercury** disappears from the evening sky and undergoes inferior conjunction on the fourth. Its last apparition of the year is a morning one, and is a favourable showing for those in northern temperate latitudes. Mercury reaches a stationary point on 13 November and returns to direct motion. Greatest elongation west takes place on 20 November when the tiny planet is 19.6° from the Sun. Mercury brightens from sixth-magnitude around inferior conjunction to magnitude −0.6 by the end of the month.

## Morning Apparition of Mercury
## 4 November to 1 January

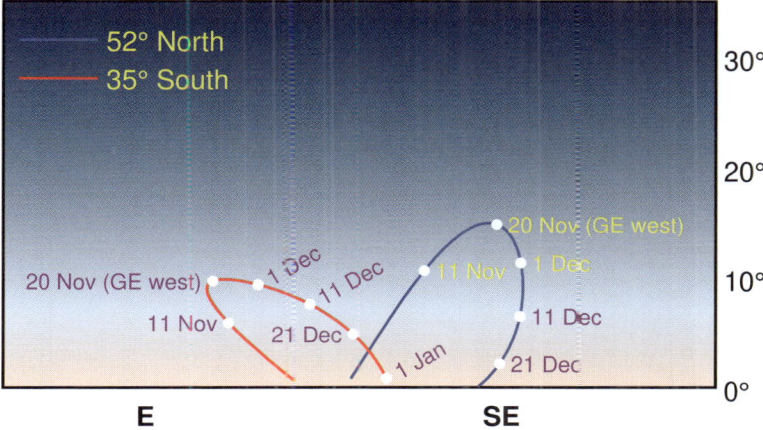

Venus has taken on the mantle of morning star and is well-placed for viewing planet-wide this month. It is occulted by the waning crescent Moon on 7 November when the two celestial bodies are just a degree apart. Venus passes by Spica (α Virginis) for the second time this year on 9 November, when it appears 1.2° south of the first-magnitude star. Retrograde motion ceases mid-month and Venus attains its brightest magnitude of the year, −4.7, a little later.

**Earth** intercepts debris left from comet 2P/Encke, with peaks occurring both early and mid-month. The most famous meteor shower of November, however, is the Leonid shower which has maximum activity around 17–18 November. The Moon is at First Quarter and should not provide an obstacle to post-midnight observations of the meteors. For more information on meteor showers, see *Meteor Showers in 2026*.

The **Moon** moves through M44, an open cluster known both as Praesepe and the Beehive, twice this month, on the first and penultimate days of November. Keep in mind that M44 is only sixth magnitude and the light of the Moon will drown it out when seen by the naked eye. Other lunar occultations follow, with both Mars and Jupiter disappearing behind the Moon's disk on 2 November and first-magnitude Regulus the following day. Venus is also an occultation target, with the bright planet vanishing on 7 November. The Moon claims Antares on 11 November but this will be a difficult event to catch as the Moon is just two days past New and only 22°

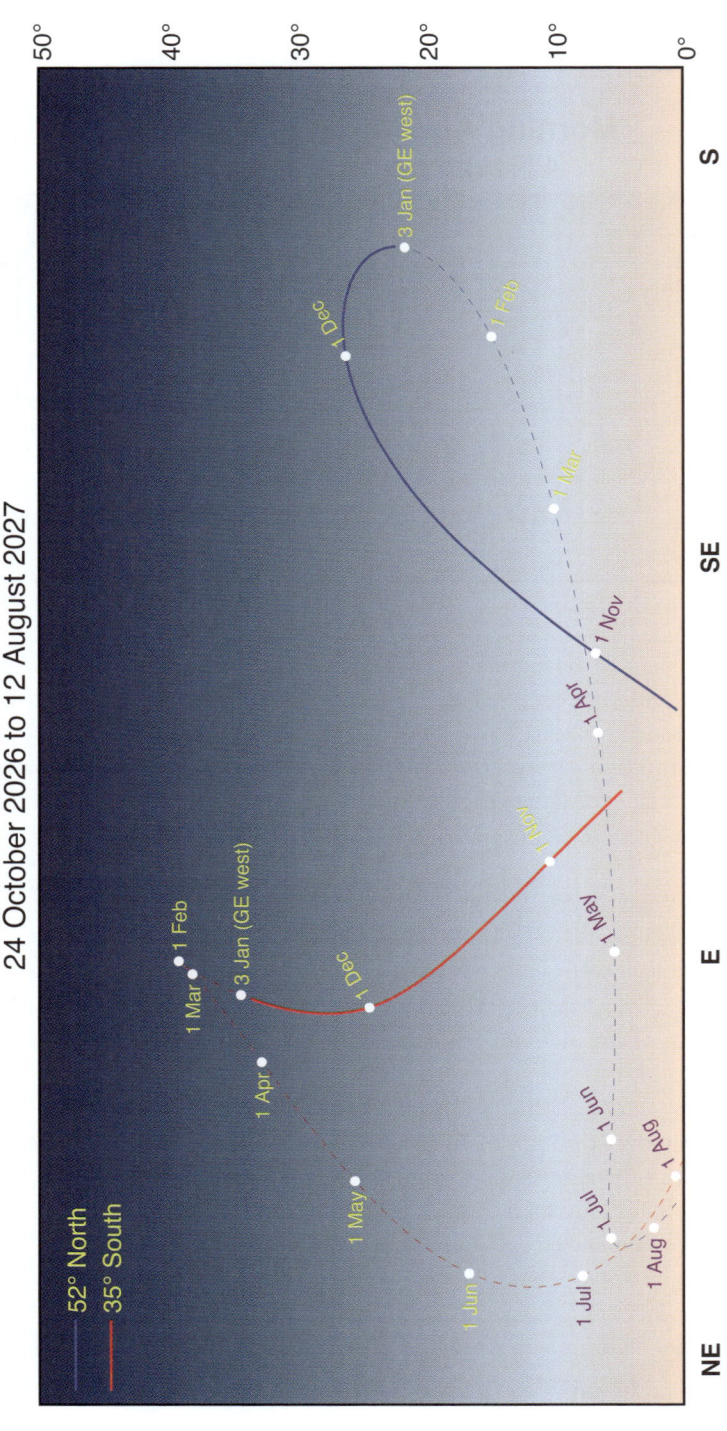

from the Sun. The Full Moon obliterates M45, the Pleiades, on 24 November when our satellite passes less than a degree north of the brightest member of that cluster. Finally, the waning gibbous Moon has another go at both Jupiter and Regulus, occulting them six hour apart on the last day of November.

**Mars** undergoes its final lunar occultation of the year on 2 November when the waning crescent Moon, less than a day past Last Quarter, covers the red planet. Now located in Leo, Mars passes two of the brighter stars of the constellation, including fifth-magnitude $\psi$ Leonis on 11 November and Regulus itself on 26 November. Mars is shining at magnitude +0.7 when it meets Jupiter (a much brighter magnitude −2.1) on 16 November. Mars is at west quadrature on 18 November and in a telescopic view, appears distinctly gibbous as it is only 89% lit. Mars now rises before midnight for all observers.

**Jupiter** is occulted by the Moon twice more, on the second day of the month and again on the last day of November. In the first instance, the Moon is in its waning crescent phase and the occultation event gets underway at around 22:00 UT. The second occultation takes place with a waning gibbous Moon doing the honours, beginning at around 09:00 UT. On 15–16 November, Mars overtakes Jupiter in the constellation of Leo; the two planets are 1.2° apart during this time. Three days later, Jupiter reaches west quadrature. The gas giant rises before midnight for all observers this month.

**Saturn** remains in Cetus this month, shining at magnitude +0.5, and visible in the east as soon as skies darken. The rings will continue to close until next month but are still open to over 6°.

**Uranus** makes another pass by the faint star $\omega^1$ Tauri (8 November) and the brighter star 37 Tauri (22 November). Unlike the July encounters with these stars, however, these take place far from the Sun. Opposition occurs on 25 November. On this date, Uranus shines at magnitude +5.6 and has an apparent diameter 3.7" when viewed through a telescope. It is 18.4 au from Earth. Unfortunately, the Moon is just past Full and will contribute to significant light interference if you attempt to look for Uranus at this time.

**Neptune** is well-placed for viewing in Pisces this month as it is already aloft by the time the sky darkens and doesn't set until after midnight. A telescope is required to view this eighth-magnitude object in dark skies.

# Eta Carinae

## David M. Harland

The early Europeans who ventured into the far southern hemisphere noted the star Eta Carinae at fourth magnitude, but in 1827 the English explorer William John Burchell, in Brazil, saw it at first magnitude. In South Africa, John Herschel made detailed measurements in the 1830s which recorded it consistently at around magnitude 1.4, but in December 1937 it began to brighten, peaking on 2 January 1838 as a rival to Alpha Centauri before fading slightly over the next few months. In March 1843, the Reverend W. S. Mackay in Calcutta wrote that it was first magnitude "... fully as bright as Canopus, and in colour and size very like Arcturus." For the next 15 years its brightness fluctuated but for most of the time

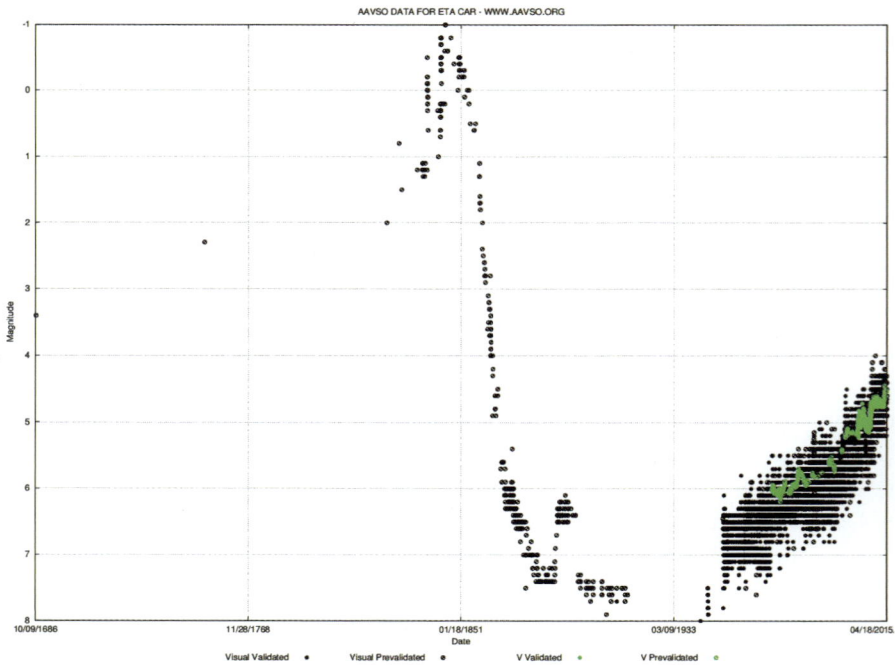

The visual light curve of Eta Carinae from 1686 to 2014. (AAVSO)

was almost as bright as Sirius. By 1868 however, the 'Great Eruption', as it came to be known, was over and the star was lost to naked-eye observers. It spent the first half of the twentieth century at about seventh magnitude and then it started to slowly brighten, fluctuating somewhat.

The spectrum of Eta Carinae showed prominent emission lines whose profiles implied a strong 'stellar wind' blowing material away from the star.

In 1969 astronomers at the California Institute of Technology found Eta Carinae to be extremely luminous at infrared wavelengths. Indeed, at that time it was emitting most of its energy in that part of the spectrum. Infrared spectroscopy showed the presence of a variety of compounds in a dusty circumstellar cloud. Dust grains absorb the stellar radiation and reemit it in the infrared region at wavelengths appropriate to their temperature. It was reasoned that as the Great Eruption faded optically, the infrared increased, with the energy output remaining constant and shifting to longer wavelengths; that is, its return to visual obscurity was merely a veiling effect.

Eta Carinae is in the Carina Nebula (NGC 3372), and appears to be a member of the OB association Trumpler 16, an open cluster roughly 7,600 light years away of very massive blue supergiants that recently formed in the dense gas and dust clouds of the nebula.

At an estimated 160 solar masses, Eta Carinae was initially thought to be the most massive star known, but a study reported in 1996 argued it was a binary system with a period of 5.5 years. It was inferred that the 1827, 1838 and 1843 peaks occurred when the stars were at their closest in a highly elliptical orbit. Radio, optical, X-ray and near-infrared radial velocity and spectral line profile variations at the next such event in late 1997 and early 1998 confirmed Eta Carinae to be a binary.

With the component stars obscured, the orbit can only be estimated, but the eccentricity is 0.9 and with a semi-major axis of about 16 au the separation between the stars varies from around 1.6 au (similar to the distance of Mars from the Sun) to 30 au (the distance of Neptune). But until we have a more accurate orbit and knowledge of the inclination of the orbit relative to our line of sight we cannot accurately determine the masses of the two stars. Nevertheless, the system has been successfully modelled with a primary of between 100 and 120 solar masses and a secondary of between 30 and 60 solar masses. The prodigious mass loss from the primary suggests it may have started out at between 150 and 250 solar masses. The best-fit single-mass-transfer model of the Great Eruption implies that prior to that event the system may have had an overall mass of 250 solar masses.

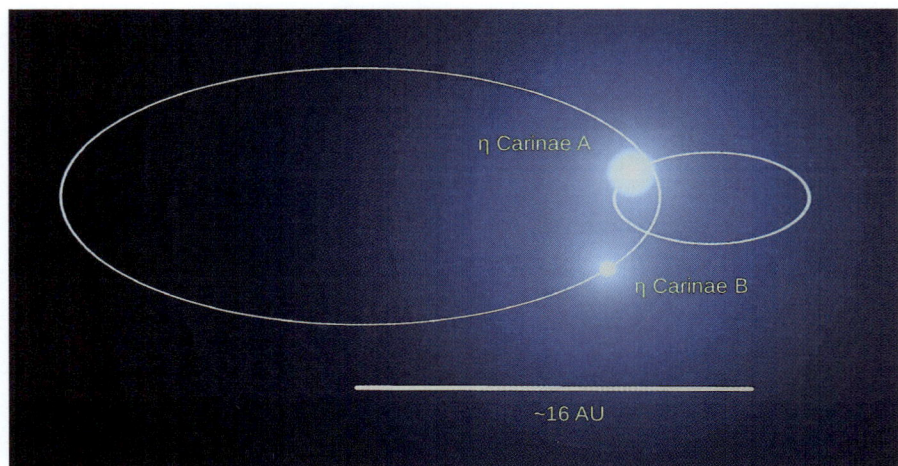

A depiction of the highly eccentric orbit of the stars in the Eta Carinae system. (Wikimedia Commons/Lithopsian derived from the NASA Goddard Space Flight Center movie "NASA Missions Take an Unparalleled Look into Superstar Eta Carinae")

The luminosity of the primary is estimated at five million times that of the Sun, making it one of the most luminous stars. At its peak during the Great Eruption its luminosity may have been as high as fifty million times that of the Sun. Afterward, it returned to its earlier luminosity. In comparison, the secondary may be only several hundred thousand times as luminous as the Sun.

Due to its extreme luminosity, the stellar wind from the primary of Eta Carinae amounts to around one-thousandth of a solar mass per year; one of the greatest known rates of mass loss. During the Great Eruption it was shedding one solar mass per year for a decade or more, much of which is the Homunculus Nebula that hides the stars.

For a portion of its orbit, the secondary may collect material from the primary. In the Great Eruption, it may well have acquired several solar masses of material via an accretion disk whose twin jets would account for the bipolar shape of the Homunculus Nebula.

How can a star undergo a sudden increase in luminosity sufficient to eject as much as 20 solar masses of material in only a couple of decades and then resume its former state? Suggestions for the cause of the Great Eruption include the merger of two components of what was initially a triple star system, an episode of mass transfer from the secondary to the primary, or an internal instability in the primary that ejected a portion of its envelope.

With no very close analogues, Eta Carinae is a rare, and possibly unique, object but we can say with reasonable certainty that it is the progenitor of a future supernova.

A typical core collapse supernova at that distance would peak at an apparent magnitude to rival Venus at its brightest. If the material ejected from the star encounters existing circumstellar material, then the radiation from the shockwave would convert it into a hypernova that was perhaps five magnitudes brighter and much longer-lasting.

The recent visual brightening of Eta Carinae is likely due to a thinning of the dust from the Great Eruption, and astronomers are eager for it to clear sufficiently to finally allow them to inspect the stellar components of the system.

A view of Eta Carinae captured by the *Hubble Space Telescope* showing the bipolar lobes of the Homunculus Nebula. The image is a composite of ultraviolet and visible light data from the High Resolution Channel of the Advanced Camera for Surveys. (ESA/Hubble, NASA)

# December

| | | | |
|---|---|---|---|
| 1 | Moon | Last Quarter Moon | 90° |
| 6 | Saturn | Shallow minimum ring opening (−6.1°) | 114° |
| 7 | Earth | Puppid-Velids (ZHR 10) | |
| 8 | Moon, α Scorpii (Antares) | 0.2° apart: occultation | 8° |
| 9 | Moon | New Moon | 5° |
| 13/14 | Earth | Geminids (ZHR 120) | |
| 17 | Moon | First Quarter Moon | 90° |
| 21 | Earth | Solstice | |
| 21 | Moon, M45 (Pleiades) | 1.0° apart: occultation | 150° |
| 22/23 | Earth | Ursids (ZHR 10) | |
| 23 | Mars, M95 | 0.4° apart | 113° |
| 23 | Neptune | East quadrature | 90° |
| 24 | Moon | Full Moon | 176° |
| 26 | Moon, M44 (Praesepe) | 0.1° apart: occultation | 147° |
| 26 | Mars, M96 | 0.7° apart | 116° |
| 27 | Moon, α Leonis (Regulus) | 1.3° apart: occultation | 126° |
| 29 | Saturn | East quadrature | 90° |
| 30 | Moon | Last Quarter Moon | 90° |

Dates are based on Universal Time (UT). Occultations of M44 (Praesepe/Beehive) refer to sixth-magnitude star ε Cancri (Meleph). Occultations of M45 (Pleiades) refer to third-magnitude star η Tauri (Alcyone). Peak activity dates and ZHRs for meteor showers are estimates. The last column gives the approximate elongation from the Sun of the event.

**Mercury** is visible in east before sunrise, brightening slightly from magnitude −0.6 to −1.3 over the course of the month, but having already attained greatest elongation west last month, is heading back toward the horizon. The best morning apparition of the year for northern temperate latitudes and a fair apparition for those further south ends on the first day of the new year 2027.

**Venus** is the morning star, fairly high in the east before sunrise. It continues to gain altitude for early birds in the southern hemisphere but is already heading back toward the horizon for those spotting it from northern temperate latitudes. It is slowly dimming, from magnitude −4.7 to −4.5, over the course of the month.

Although the illuminated fraction of the planet is increasing, from 28% to 49%, the apparent size of the disk is getting significantly smaller, from 39.4″ to 25.7″. The increasing distance from Earth more than offsets the greater phase, leading to a decrease in brightness.

**Earth** arrives at a solstice on 21 December when the Sun reaches it maximum southerly declination. Astronomical summer begins in the southern hemisphere whilst winter claims the north. December is home to a number of meteor showers (see *Meteor Showers in 2026*) but the major ones are the Geminids mid-month and the Ursids which peak around 22–23 December. This is an excellent year to look for Geminid meteors as the Moon is in its waxing crescent phase and sets by mid-evening. The return of the Ursids is not so favourable, with the Moon nearly in its Full phase during peak activity. The southern hemisphere is treated to a decent return of the Puppid-Velids early in December when the Moon is a waning crescent.

The occultation of first-magnitude Antares by the **Moon** on 8 December takes place too close to the Sun to see, but the obscurations of the Pleiades (21 December), the Beehive Cluster (26 December), and Regulus (27 December) are all well-placed for viewing. Both the most distant apogee of the year (406,419 kilometres on 11 December) and the nearest perigee (356,650 kilometres on 24 December) of the year occur this month. Since the nearest perigee occurs only seven hours after the Moon becomes full, the December Full Moon is the largest of the year in terms of apparent diameter (2010″). Compare this value with that of the 'smallest' Full Moon which occurs at the end of May.

**Comet 161P/Hartley-IRAS** is a Halley-type comet which comes to perihelion this month. Closest approach to Earth took place in October. This faint object, which is unlikely to be brighter than tenth magnitude, is found in the region of the Summer Triangle asterism during December. For more information, see *Comets in 2026*.

**Mars** finishes 2026 in Leo. It rises very late in the evening for all observers and spends the month moving past some interesting deep sky objects. On 23 December, the red planet is 0.4° south of the tenth-magnitude spiral galaxy M95 and three days later, is found 0.7° south of M96, a slightly brighter spiral galaxy. M105, a tenth-magnitude elliptical galaxy, is a little further away on 27 December, when planet and galaxy are 1.5° distant. Mars continues to brighten, beginning the month at magnitude +0.5 and ending the year at −0.1.

**Jupiter** rises before midnight for all planet watchers this month but is best viewed from northern latitudes where it climbs high into the sky. At magnitude −2.2, it is the brightest star-like object in Leo. Jupiter entered the year in retrograde motion and returns to that direction of travel mid-month. The Moon has occulted Jupiter four times in the past three months but misses the gas giant this time, coming closest on 27 December when the two celestial bodies are 1.4° apart.

**Saturn** reaches a shallow minimum ring opening of 6.1° on 6 December after which the rings will begin to open up again. Earth-based astronomers have been observing the southern side of the rings since the ring plane crossing in March 2025 and will continue to do so until the next crossing in October 2038. Retrograde motion ceases mid-month and Saturn reaches east quadrature on 29 December. Currently located in Cetus, Saturn is already aloft by the time skies turn dark, and sets around midnight.

**Uranus** was at opposition late last month and is visible for most of the night throughout December. The sixth-magnitude object is found retrograding through Taurus between the open star clusters of the Hyades and the Pleiades.

**Neptune** returns to direct motion in mid December after spending over five months in retrograde. East quadrature takes place on 23 December and the planet sets around midnight by the end of the year. The blue ice giant is located in Pisces and shines at magnitude +7.8.

# Flammarion and the Comet
## The Sorry 'Tail' of the Downfall of an Honourable Man

### Neil Haggath

"A lie goes around the world while the truth is still pulling up its boots."
Nineteenth-century American proverb.

"Fake news" is nothing new. In the world of journalism, the principle of "Never let the truth get in the way of a good story" was as prevalent a century ago as it is today – and never was it more evident than in the sorry saga of the coverage of the 1910 return of Halley's Comet!

In that year, the comet made a close approach to Earth, at a mere 13 million miles, and was far brighter and more prominent than its next feeble apparition, which many of us (barely) observed in 1986. (However, many elderly people who

Halley's Comet, photographed by E. E. Barnard at Yerkes Observatory, 29 May 1910, and published in the New York Times on 3 July. [Wikimedia Commons/Edward Emerson Barnard/New York Times]

claimed in 1986 to remember how spectacular it was last time were in fact mistaken; they confused it with the much brighter Daylight Comet of January 1910).

In September 1909, after the comet was recovered, astronomers realised that on 19 May 1910, the Earth would pass through the comet's tail. This led to some worrying speculations, as comets were by then known to contain various noxious gases. The subsequent discovery of cyanogen gas in the tail led to scaremongering predictions that the comet was going to poison all of humanity!

Almost all astronomers, of course, assured the public that the matter in a comet's tail is so tenuous that passing through it would have no noticeable effect whatsoever. But saying "Don't worry – nothing is going to happen" is boring, and doesn't sell papers! So the less reputable press, particularly in the United States, milked the scaremongering for all it was worth, and relied on misquotes exaggerations and outright fabrications for their sensational headlines.

Some of their outrageous stories would have put the unlamented *News of the World* to shame. For example, numerous American papers printed the shocking story of how a group of religious fanatics in Oklahoma, believing the comet to be a sign of God's wrath, were prevented, by the last minute arrival of the Sheriff and his posse, from carrying out a human sacrifice (of a virgin, naturally)! But it didn't happen; the story was pure fiction.[1]

But a significant proportion of the population truly believed the claims, and as the dreaded date approached, panic and hysteria became widespread. Many people quit their jobs, withdrew their savings and went on benders, determined to enjoy their last few days. Some even took their own lives, rather than wait to die from cyanide poisoning. One would hope all those editors were rightly proud of themselves!

One man who unwittingly played a central role in this sordid affair was the French astronomer Camille Flammarion (1842–1925). As well as an accomplished astronomer, and founder of the *Société Astronomique de France*, Flammarion was the world's most eminent astronomy writer and populariser; we could rightly call him the Carl Sagan of his time. (See the author's article "Anniversaries in 2025" in the *Yearbook of Astronomy 2025*).

In November 1909, Flammarion published an article in *The New York Herald*, in which he stated that: "The poisoning of humanity by deleterious gases is improbable."[2]

---

1 Richard J. Goodrich, *Comet Madness: How the 1910 Return of Halley's Comet (Almost) Destroyed Civilisation*, Prometheus Books, 2023, p. 240–241.
2 Goodrich, 2023, p.64.

Unfortunately, he went on to indulge in a "thought experiment" about what might happen *if* various substances were to permeate the Earth's atmosphere. He then repeated that it would not happen, and correctly said that in comparison to the density of the tail, the atmosphere would act like a lead shield.

Needless to say, papers across America ignored his assurances, and picked up on the "what if". Flammarion was widely but falsely reported as believing that gases from the tail would "impregnate the atmosphere, and possibly snuff out all life on the planet". While he had said no such thing, the attribution of such a prophecy of doom to such a reputable source played a major part in it becoming widely believed.

Camille Flammarion (John McCue)

On 1 Feb 1910, Flammarion published a follow-up in the *Herald*, warning journalists against "accusing me of announcing the end of the world for May 19 next. The end of the world will not occur on May 19 next."[3] He could not have made his position any clearer – yet the papers ignored his rebuttal, and continued to do exactly that.

After the dreaded date came and went, and nothing happened, the same papers ridiculed Flammarion for the prediction that he had never made, and made him a scapegoat for the panic – suicides and all – created by their own irresponsible fabrications. His reputation lay in ruins; he was the victim of a deliberate and cynical character assassination to sell papers.

Sadly, the myth that Flammarion was responsible for the panic persists to this day. Even Carl Sagan repeated it; in his 1985 book *Comet*, he wrote, "The global pandemonium... was, sadly, fuelled by a few astronomers, who should have known better", and wrongly named Flammarion as one of them.[4] And until 2023, Flammarion's Wikipedia entry repeated the falsehood that he believed that the comet would "possibly snuff out all life"; this has now been rectified, by me.

In the UK, meanwhile, the passage through the tail went almost unnoticed, as it was somewhat overshadowed in the press by another event. It happened just thirteen days after the death of King Edward VII.

---

3 Goodrich, 2023, p.83.
4 Carl Sagan and Ann Druyan, *Comet*, Random House, 1985, p. 143–144.

# Comets in 2026

## Neil Norman

When writing for the comet section of the *Yearbook of Astronomy*, one can only include the objects that are known with certainty to return for any given year; because of this there have been many instances when a bright comet has come along completely by surprise and has not been included within the articles. At the time of writing this section for 2026, the comets destined to break or be very close to magnitude 10 only number four in total. The reason that magnitude 10 is chosen is because this gives the novice with limited equipment a chance to see these wonderful objects. Of course, those of you with larger telescopes and more observing experience can catch comets down to several magnitudes lower, giving you a larger number of comets to seek out and focus upon during the year.

For the very latest comet discoveries and news it is advised that the reader regularly pays a visit to the British Astronomical Association (BAA) Comet Section website at **www.ast.cam.ac.uk/~jds** which is regularly maintained and updated with all the latest discoveries and news.

The usual mix of comets are making a return to perihelion during 2026 of which the vast majority are dim, Jupiter-family comets requiring large aperture equipment to observe; but with that said, there are four comets that are of interest this year.

## 10P/Tempel 2

This comet is a periodic Jupiter family comet with a 5.37 year orbital period that was discovered on 4 July 1873 by the German astronomer Ernst Wilhelm Leberecht Tempel. The 2026 return is very favourable given that the comet reaches perihelion on 2 August at a distance of 1.42 au from the Sun, making its closest approach to Earth at a distance of 0.414 au (6.19 million kilometres) on the following day; the elongation from the Sun at this time will also be large at 163 degrees.

The comet has a similar nuclear size (10.6 kilometres) to 1P/Halley, the albedo of the nucleus being 0.022. This makes the nucleus quite dark, and arises from the fact that 10P/Tempel 2 makes frequent passages around the Sun, the ultraviolet radiation received during these perihelia creating a sticky and tar-like crust made of hydrocarbons.

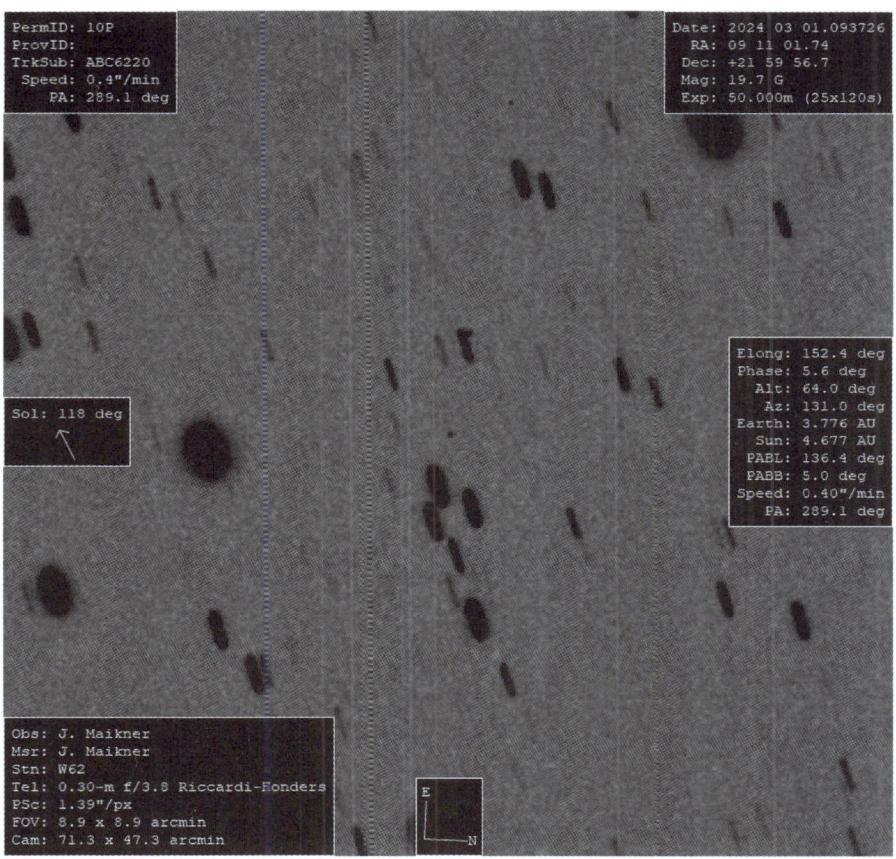

Comet 10P/Tempel 2 (seen here as a dot at centre of image) captured by John Maikner on 3 January 2024 while the comet was 4.67 au from the Sun and almost at its aphelion distance of 4.71 au. (John Maikner)

Given the comets number of returns over the centuries, the comet is quite weak and although the magnitude is at best 7.4, it will be a hard see because the coma will be large and diffuse with little in the way of dust therein. All the same, it is definitely an object to go out and see.

| DATE | RA | DECLINATION | MAGNITUDE | CONSTELLATION |
| --- | --- | --- | --- | --- |
| 15 Jul 2026 | 21 27 45 | −16 40 48 | 8 | Capricornus |
| 2 Aug 2026 | 21 52 24 | −24 01 32 | 7.4 | Capricornus |
| 15 Aug 2026 | 22 07 01 | −29 49 06 | 9 | Piscis Austrinus |

## 24P/Schaumasse

Discovered by French astronomer Alexandre Schaumasse on 1 December 1911, 24P/Schaumasse has an orbital period of 8.24 years, and is also a member of the Jupiter family of comets. The following year the comet was recognised as being periodic with a seven year orbital period, a value later determined to be eight years. The comet was next seen in 1919 when it was recovered by Gaston Fayed from Paris. It was seen again in 1927 but missed in 1935, a subsequent close passage to Jupiter in 1937 increasing its orbital period. During the 1951/1952 approach the comet was far brighter than expected, attaining sixth magnitude. The comet was not seen in 1968, 1976 and 2009, the latter a result of the comet being on the far side of the Sun. while during the return of 2017 it reached magnitude 10, and was last seen on 19 June 2018 when it was 2.7 au from the Sun. During the apparition of 2025/26 the comet is closest to Earth on 4 January 2026 at a distance of 0.597 au (89.3 million kilometres). The comet nucleus is around 2.6 kilometres in diameter.

| DATE | RA | DECLINATION | MAGNITUDE | CONSTELLATION |
|---|---|---|---|---|
| 1 Dec 2025 | 10 20 57 | +19 04 46 | 12 | Leo |
| 15 Dec 2025 | 11 30 39 | +17 16 53 | 11.5 | Leo |
| 1 Jan 2026 | 12 38 04 | +14 42 41 | 10 | Coma Berenices |
| 15 Jan 2026 | 13 47 57 | +11 20 31 | 8 | Boötes |
| 1 Feb 2026 | 14 40 07 | +08 33 55 | 9 | Boötes |
| 15 Feb 2026 | 15 09 19 | +06 57 37 | 10 | Virgo |

## 88P/Howell

Discovered on 29 August 1981 by American astronomer Ellen Howell, 88P/Howell is a Jupiter family comet with a 5.5 year orbital period which will bring it to perihelion on 18 March 2026. The perihelion distance of this comet in 1975 was 1.9 au, although a close approach to Jupiter in 1978 moved the comet closer to the Sun at perihelion to a distance of 1.36 au. The comet will make a close pass of Mars (11.1 million kilometres) on 14 September 2031, the nearest it will come the Earth between the years 2000 and 2050 being 114 million kilometres in June 2042.

Comet 88P/Howell imaged by José J. Chambó Bris on 5 September 2020 in the same field of view as the open star cluster NGC 5897 in Libra. Although seen here alongside NGC 5897, the comet at the time was 1.3 au from Earth as compared to the cluster which lies at a distance of around 41,000 light years. (José J. Chambó Bris, **cometografia.es**)

| DATE | RA | DECLINATION | MAGNITUDE | CONSTELLATION |
|---|---|---|---|---|
| 1 Feb 2026 | 18 43 41 | −24 08 18 | 12 | Sagittarius |
| 15 Feb 2026 | 19 36 10 | −23 09 14 | 11.5 | Sagittarius |
| 1 Mar 2026 | 20 28 30 | −21 06 12 | 11 | Capricornus |
| 15 Mar 2026 | 21 19 13 | −18 17 34 | 10.3 | Capricornus |
| 1 Apr 2026 | 22 17 10 | −13 37 39 | 11 | Aquarius |
| 15 Apr 2026 | 23 01 21 | −09 32 42 | 11.5 | Aquarius |

## 161P/Hartley-IRAS

With an orbital period of 21.5 years, 161P/Hartley-IRAS is defined as a Halley-type comet, this being a comet with a period of between 20 and 200 years. Its orbital path carries it from a closest approach to the Sun of 1.25 au out to a farthest point (aphelion) of 14.19 au, around half-way between the orbits of Saturn and Uranus. The comet was discovered by Malcolm Hartley on 4 November 1983 as a magnitude 15 object on a photographic plate taken by the *Infrared Astronomical Satellite (IRAS)*. Subsequent images obtained by IRAS on 10 November confirmed the discovery, perihelion occurring on 8 January 1984. The comet had brightened quickly prior to

Comet 161P/Hartley-IRAS as imaged by Michael Jager on 8 August 2005. (Michael Jager)

perihelion, attaining magnitude 10 by the end of November and hovering around that magnitude for the rest of the year. 161P/Hartley-IRAS peaked at magnitude 7.4 on 23 February 1984. The comet peaked again at magnitude 10 post-perihelion in July 2005, and during the return of 2026 is expected to peak at magnitude 9.6 during November.

| DATE | RA | DECLINATION | MAGNITUDE | CONSTELLATION |
|---|---|---|---|---|
| 1 Oct 2026 | 23 29 20 | −06 50 49 | 14 | Aquarius |
| 15 Oct 2026 | 21 14 13 | +05 48 57 | 11 | Equuleus |
| 1 Nov 2026 | 19 57 20 | +13 33 06 | 10.5 | Aquila |
| 15 Nov 2026 | 19 31 16 | +16 52 50 | 9.6 | Sagitta |
| 1 Dec 2026 | 19 17 16 | +19 50 16 | 10.5 | Sagitta |
| 15 Dec 2026 | 19 11 14 | +22 22 30 | 11 | Vulpecula |

# Minor Planets in 2026

## Neil Norman

Minor planets – often referred to as asteroids – are a collection of varying sized pieces of rock left over from the formation of the Solar system around 4.6 billion years ago. Millions of them exist, and to date almost 800,000 have been seen and documented, with around 550,000 having received permanent designations after being observed on two or more occasions and their orbits being known with a high degree of certainty. Different family types of asteroids also exist, such as Amor asteroids which are defined by having orbital periods of over one year and orbital paths that do not cross that of the Earth. Apollo asteroids have their perihelion distances within that of the Earth and thus can approach us to within a close distance and Trojan asteroids have their home at Lagrange points both 60 degrees ahead and behind the planet Jupiter respectively. These asteroids pose no problems to the Earth.

Most asteroids travel around the Sun in the main asteroid belt, which is located between the orbits of Mars and Jupiter. However, some asteroids have more elliptical orbits which allow them to interact with major planets, including the Earth. In all, there are around 2,000 which can approach to within a close distance of our planet. These objects are referred to as Potentially Hazardous Asteroids (PHAs). To qualify as a PHA these objects must have the capability to pass within 8 million kilometres of Earth and to be over 100 metres across.

However, some asteroids have orbits which allow them to interact with major planets, including Earth. Over 20,000 asteroids have orbits that can bring them into close proximity to our planet. Such an object is referred to as a Near Earth Asteroid (NEA). To be classed as an NEA, an object must have an orbit which allows it to pass within a distance of 1.3 au of the Earth. A Potentially Hazardous Asteroid (PHA) on the other hand is defined as being one which can approach to within 0.05 au (19.5 lunar distances) of our planet, and have a diameter of at least 140 metres.

Objects of this size could pose a serious threat if on a collision course with Earth. It is estimated that several thousand exist with diameters of over 100 metres, and with around 150 of these being over a kilometre across. A large number of smaller asteroids, measuring anything between just a few meters in diameter to several tens of meters wide, pass close to our planet on a regular basis, with considerable numbers of smaller ones entering the Earth's atmosphere every day, burning up harmlessly as meteors.

Those observers with a particular interest in following these objects should go to the home page of the Minor Planet Center. It is their job to keep track of these objects and determine orbits for them. This page can be accessed by going to **www.minorplanetcenter.net** where you will find a table of newly discovered minor planets and Near Earth Objects (NEOs). At the top of the page is a search box that you can use to locate information on any object that you are interested in, and from this you can obtain ephemerides of the chosen subject. The Minor Planet Center site is the one that all dedicated asteroid observers should consult on a regular basis.

Observers who are beginning the hobby of asteroid hunting should remember that these objects will appear as nothing more than points of light in your field of view. You therefore need to compare your observations with detailed star charts (especially for the dimmer objects) to ascertain that you have indeed observed your intended target. Details of a selection of minor planets observable during 2026 are given below. Some of the objects described are visible in binoculars or small telescopes, and – because they arrive at opposition during this year – are particularly good targets for the backyard astronomer. Others are much fainter, and will therefore present greater challenges to the would-be observer.

Right ascensions and declinations tabulated in this section are referred to the mean equator and equinox of J2000. This allows them to be plotted directly on modern star charts such as *Sky Atlas 2000.0* by Wil Tirion and Roger Sinnott. All positional and magnitude data is taken from the Minor Planet & Comet Ephemeris Service at **minorplanetcenter.net/iau/MPEph/MPEph.html**

## 2 Pallas

Pallas has a diameter of 512 kilometres and was the second asteroid to be discovered when first spotted by the German astronomer Heinrich Wilhelm Matthias Olbers on 28 March 1802. He named the object after the Greek goddess of wisdom and warfare Pallas Athena, an alternative name for the goddess Athena. Pallas travels around the Sun once every 1,686 days, its orbit being highly eccentric, and its path around the Sun steeply inclined to the main plane of the asteroid belt, rendering it fairly inaccessible to spacecraft. The accompanying chart shows the positions of Pallas (and Vesta) for the three months around their October 2026 oppositions.

Pallas comes to opposition on 6 October in the constellation of Cetus.

Minor Planets in 2026   195

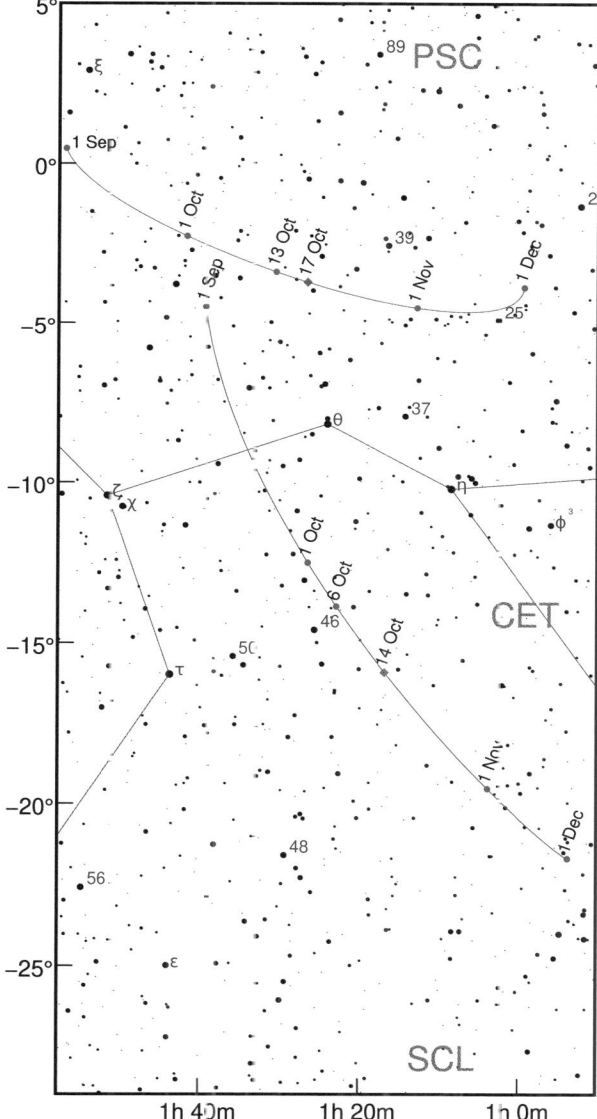

The paths of 2 Pallas (blue) and 4 Vesta (magenta) between 1 September and 1 December 2026. Pallas is at opposition on 6 October, when it is magnitude +8.2. Vesta is at opposition on 13 October, when it is magnitude −6.5. Dates of opposition in Right Ascension are 14 and 17 October respectively, and are indicated by a diamond. Dot sizes do not denote magnitude. Stars are shown to magnitude +9.0. (David Harper)

| DATE | RA | DECLINATION | MAGNITUDE | CONSTELLATION |
|---|---|---|---|---|
| 1 Sep 2026 | 01 39 04 | −04 25 44 | 8.8 | Cetus |
| 15 Sep 2026 | 01 35 28 | −08 02 45 | 8.5 | Cetus |
| 1 Oct 2026 | 01 26 28 | −12 31 04 | 8.2 | Cetus |
| 15 Oct 2026 | 01 15 53 | −16 09 43 | 8.2 | Cetus |
| 1 Nov 2026 | 01 03 09 | −19 30 24 | 8.5 | Cetus |
| 15 Nov 2026 | 00 55 33 | −21 04 53 | 8.7 | Cetus |

## 3 Juno

With a mean diameter of around 270 kilometres, Juno ranks as the twelfth largest asteroid and is one of the two largest stony (S-type) asteroids, comprising around 1% of the mass of the entire asteroid belt and being the second largest stony asteroid after 15 Eunomia. Discovered by the German astronomer Karl Ludwig Harding on 1 September 1804, its mean distance from the Sun is 2.6 au, its journey around our star taking 4.36 years to complete.

Juno comes to opposition on 26 July in the constellation of Aquila.

| DATE | RA | DECLINATION | MAGNITUDE | CONSTELLATION |
|---|---|---|---|---|
| 15 Jun 2026 | 20 35 39 | −03 34 39 | 10.0 | Aquila |
| 1 Jul 2026 | 20 28 52 | −03 31 54 | 9.6 | Aquila |
| 15 Jul 2026 | 20 18 57 | −04 05 06 | 9.3 | Aquila |
| 1 Aug 2026 | 20 04 21 | −05 29 20 | 9.1 | Aquila |
| 15 Aug 2026 | 19 53 02 | −07 04 46 | 9.3 | Aquila |
| 1 Sep 2026 | 19 43 44 | −09 11 41 | 9.5 | Aquila |

## 4 Vesta

Discovered by German astronomer Heinrich Wilhelm Matthias Olbers on 29 March 1807, Vesta is one of the largest of the asteroids with a diameter of 525 kilometres and an orbital period of 3.63 years. Vesta holds the distinction of being the brightest minor planet visible from Earth. With a maximum magnitude of 6, Vesta is the only one of the minor planets which is regularly bright enough to be glimpsed with the naked eye. The preceding chart shows the positions of Vesta (and Pallas) for the three months around their October 2026 oppositions.

Vesta comes to opposition on 13 October in the constellation of Cetus.

| DATE | RA | DECLINATION | MAGNITUDE | CONSTELLATION |
|---|---|---|---|---|
| 1 Sep 2026 | 01 56 18 | +00 42 01 | 7.2 | Cetus |
| 15 Sep 2026 | 01 52 14 | −00 31 09 | 6.9 | Cetus |
| 1 Oct 2026 | 01 41 21 | −02 10 33 | 6.6 | Cetus |
| 15 Oct 2026 | 01 28 25 | −03 30 20 | 6.5 | Cetus |
| 1 Nov 2026 | 01 13 06 | −04 28 17 | 6.8 | Cetus |
| 15 Nov 2026 | 01 04 04 | −04 31 15 | 7.2 | Cetus |

## 134340 Pluto

Discovered on 18 February 1930 by American astronomer Clyde Tombaugh, and orbiting the Sun once every 248 years at a mean distance of around 5.9 billion kilometres, Pluto was considered as being the ninth fully-fledged planetary member of the solar system until 2006, at which point it was downgraded to the status of dwarf planet by the International Astronomical Union.

This year Pluto comes to opposition on 27 July in the constellation of Capricornus. For those equipped with large aperture telescopes, and perhaps a great deal of patience, Pluto may be located by using the information given here.

| DATE | RA | DECLINATION | MAGNITUDE | CONSTELLATION |
|---|---|---|---|---|
| 15 Jun 2026 | 20 32 56 | −23 07 16 | 15.2 | Capricornus |
| 1 Jul 2026 | 20 31 43 | −23 13 59 | 15.1 | Capricornus |
| 15 Jul 2026 | 20 30 28 | −23 20 08 | 15.1 | Capricornus |
| 1 Aug 2026 | 20 28 49 | −23 27 23 | 15.0 | Capricornus |
| 15 Aug 2026 | 20 27 29 | −23 32 46 | 15.1 | Capricornus |
| 1 Sep 2026 | 20 26 02 | −23 38 06 | 15.2 | Capricornus |

## 136108 Haumea

This year's challenging object is the minor planet 136108 Haumea. Discovered in December 2004 by American astronomers Mike Brown, Chad Trujillo and David Rabinowitz, it was classified as a dwarf planet on 17 September 2008. This is a trans-Neptunian object with aphelion and perihelion distances of around 52 au and 35 au respectively. Named after the Hawaiian goddess of childbirth and fertility, Haumea is ellipsoidal in shape with a mean radius of a little less than 800 kilometres, and holds the distinction of having two moons and of being the first object of its type to have a ring system around it.

This year 136108 Haumea comes to opposition on 23 April in the constellation of Boötes.

| DATE | RA | DECLINATION | MAGNITUDE | CONSTELLATION |
| --- | --- | --- | --- | --- |
| 1 Apr 2026 | 14 43 10 | +14 26 31 | 17.2 | Boötes |
| 15 Apr 2026 | 14 42 19 | +14 33 27 | 17.2 | Boötes |
| 1 May 2026 | 14 41 14 | +14 39 22 | 17.2 | Boötes |
| 15 May 2026 | 14 40 17 | +14 42 27 | 17.2 | Boötes |

# Meteor Showers in 2026

## Neil Norman

A shooting star dashing across the sky is a wonderful sight that often captures the imagination of many young minds and perhaps sparks an interest for astronomy in them. On any given night of the year you can expect to see several of these, and they belong to two groups – sporadic and shower. Quite often the ones you see will be 'sporadic' meteors, that is to say they can appear from any direction and at any time during the observing session. These meteors arise when a meteoroid – perhaps a particle from an asteroid or a piece of cometary debris orbiting the Sun – enters the Earth's atmosphere and burns up harmlessly high above our heads, leaving behind the streak of light we often refer to as a 'shooting star'. The meteoroids in question are usually nothing more than pieces of space debris that the Earth encounters as it travels along its orbit, and range in size from a few millimetres to a couple of centimetres in size. Meteoroids that are large enough to at least partially survive the passage through the atmosphere, and reach the Earth's surface without disintegrating, are known as meteorites.

At certain times of the year the Earth encounters more organised streams of debris that produce meteors over a regular time span and which seem to emerge from the same point in the sky. These are known as meteor showers. These streams of debris follow the orbital paths of comets, and are the scattered remnants of comets that have made repeated passes through the inner solar system. The ascending and descending nodes of their orbits lie at or near the plane of the Earth's orbit around the Sun, the result of which is that at certain times of the year the Earth encounters and passes through a number of these swarms of particles.

The term 'shower' must not be taken too literally. Generally speaking, even the strongest annual showers will only produce one or two meteors a minute at best, this depending on what time of the evening or morning that you are observing. One must also take into account the lunar phase at the time, which may significantly influence the number of meteors that you see. For example, a full moon will probably wash out all but the brightest meteors.

The following is a table of the principle meteor showers of 2026 and includes the name of the shower; the period over which the shower is active; the Zenith Hourly Rate (ZHR); the parent object from which the meteors originate; the date of peak shower activity; and the constellation in which the radiant of the shower is located.

Most of the information given is self-explanatory, but the Zenith Hourly Rate may need some elaborating.

The Zenith Hourly Rate is the number of meteors you may expect to see if the radiant (the point in the sky from where the meteors appear to emerge) is at the zenith (or overhead point) and if observing conditions were perfect and included dark, clear and moonless skies with no form of light pollution whatsoever. However, the ZHR should not be taken as gospel, and you should not expect to actually observe the quantities stated, although 'outbursts' can occur with significant activity being seen.

The observer can make notes on the various colours of the meteors seen. This will give you an indication of their composition; for example, red is nitrogen/oxygen, yellow is iron, orange is sodium, purple is calcium and turquoise is magnesium. Also, to avoid confusion with sporadic meteors which are not related to the shower, trace the path back of the meteor and if it aligns with the radiant you can be sure you have seen a genuine member of the particular shower.

## Meteor Showers in 2026

| SHOWER | DATE | ZHR | PARENT | PEAK | CONSTELLATION |
|---|---|---|---|---|---|
| Quadrantids | 1 Jan to 5 Jan | 120 | 2003 EH$_1$ (asteroid) | 3/4 Jan | Boötes |
| Lyrids | 16 Apr to 25 Apr | 18 | C/1861 G1 Thatcher | 22/23 Apr | Lyra |
| Pi Puppids | 16 Apr to 30 Apr | Varies | 26P/Grigg-Skjellerup | 23/24 Apr | Puppis |
| Eta Aquariids | 19 Apr to 28 May | 50 | 1P/Halley | 6/7 May | Aquarius |
| Tau Herculids | 19 May to 19 Jun | Varies | 73P/Schwassmann–Wachmann | 9 Jun | Hercules |
| Delta Aquariids | 12 Jul to 23 Aug | 25 | 96P/Machholz | 28/29 Jul | Aquarius |
| Perseids | 17 Jul to 24 Aug | 100 | 109P/Swift–Tuttle | 12/13 Aug | Perseus |
| Kappa Cygnids | Jun to Sep | 3 | Unknown | 17 Aug | Cygnus |
| Arids | 28 Sep to 14 Oct | ? | 15P/Finlay | 7 Oct | Ara |
| Draconids | 6 Oct to 10 Oct | 5 | 21P/Giacobini–Zinner | 7/8 Oct | Draco |
| Southern Taurids | 10 Sep to 20 Nov | 5 | 2P/Encke | 10 Oct | Taurus |
| Orionids | 2 Oct to 7 Nov | 20 | 1P/Halley | 21/22 Oct | Orion |
| Southern Taurids | 10 Sep to 20 Nov | 5 | 2P/Encke | 5 Nov | Taurus |
| Northern Taurids | 20 Oct to 10 Dec | 5 | 2004 TG$_{10}$ (asteroid) | 12 Nov | Taurus |
| Leonids | 6 Nov to 30 Nov | 10 | 55P/Tempel–Tuttle | 17/18 Nov | Leo |
| Puppid-Velids | 1 Dec to 15 Dec | 10 | Not Known | 7 Dec | Vela |
| Geminids | 4 Dec to 17 Dec | 120 | 3200 Phaethon (asteroid) | 13/14 Dec | Gemini |
| Ursids | 17 Dec to 26 Dec | 10 | 8P/Tuttle | 22/23 Dec | Ursa Minor |

## Quadrantids

The parent object of the Quadrantids has been identified as the near-Earth object of the Amor group of asteroids 2003 EH$_1$ which is likely to be an extinct comet. With peak rates known to exceed 100 meteors per hour, the Quadrantids rivals the August Perseids, although there is a drawback in that the period of maximum activity takes place over a very short period of between two and three hours. The radiant lies a little to the east of the star Alkaid (η Ursae Majoris) and the meteors are fast moving, reaching speeds of 40 km/s. Maximum activity occurs on the night of 3/4 January although this year a Full Moon will hinder observations.

## Lyrids

Produced by particles emanating from the long-period comet C/1861 G1 Thatcher – which last came to perihelion on 3 June 1861 – these are very fast moving meteors that approach speeds of up to 50 km/s. The peak falls on the night of 22/23 April with the radiant lying near the prominent star Vega in the constellation of Lyra. This year a First Quarter Moon will be out of the way to leave perfect conditions.

## Pi Puppids

Discovered in 1972, the Pi Puppids are associated with the periodic comet 26P/Grigg-Skjellerup. The shower favours southern observers and is only really detectable when the parent comet is approaching perihelion. Hourly rates of these relatively slow moving meteors vary, and are often quite low which, when taken into account with the fact that the parent comet was last at perihelion on Christmas Day 2023, reduces hopes for a decent display in 2026. A First Quarter Moon on the night of peak activity (24 April) may assist with observations of this elusive shower.

## Eta Aquariids

One of the two showers associated with 1P/Halley, the Eta Aquariids are active for a full month between 19 April and 28 May. The radiant lies just to the east of the star Sadalmelik (α Aquarii), from where up to 50 meteors per hour are normally expected during the period of peak activity, although displays of up to 60 meteors per hour can occasionally be seen. Maximum activity occurs in the pre-dawn skies of 7 May, although in 2026 a waning Moon six days after full phase may interfere with observations.

## Tau Herculids

Appearing to originate from the star Tau (τ) Herculis, this shower runs from 19 May to 19 June with peak activity taking place on 9 June. The Tau Herculids were first recorded in May 1930 by observers at the Kwasan Observatory in Kyoto, Japan. The parent body has been identified as the periodic comet 73P/Schwassmann–Wachmann – also known as Schwassmann–Wachmann 3 – discovered on 2 May 1930 by the German astronomers Arnold Schwassmann and Arno Arthur Wachmann during a photographic search for minor planets being carried out from Hamburg Observatory in Germany.

73P/Schwassmann–Wachmann has an orbital period of 5.36 years. During 1995 the comet began to fragment, and at the time of its 2006 return, the *Hubble Space Telescope* observed at least eight fragments of varying size. Within the vicinity of

On 18 April 2006, the *Hubble Space Telescope* observed at least eight fragments of varying size emanating from the head of 73P/Schwassmann–Wachmann. Within the area of these fragments were hundreds of pieces of smaller debris that will litter the comet's orbit and eventually intersect with the orbit of Earth and create a true spectacle. (NASA/ESA/H. Weaver (APL/JHU)/ M. Mutchler and Z. Levay (STScI))

these fragments were hundreds of pieces of smaller debris which will litter the orbit of 73P/Schwassmann–Wachmann and eventually intersect with the Earth's orbit, potentially creating a spectacular display. From these observations it appears that 73P/Schwassmann–Wachmann is close to total disintegration.

The observed rate of meteors from the Tau Herculid shower is generally low. However, the above-mentioned observations suggest that Earth will be getting closer to the streams of debris left behind by the fragmenting comet. Consequently, heavier meteor showers can be expected in the near future, with a hefty shower possible for 2049. Peak activity in 2026 occurs during a Last Quarter Moon resulting in smaller meteors being dimmed out, although it may be worth looking for enhanced activity up to two weeks before the peak date in case an outburst occurs early.

## Delta Aquariids

Probably linked to the short-period sungrazing comet 96P/Machholz, the Delta Aquariids is a fairly average shower which coincides with the more prominent Perseids. However, Delta Aquariid meteors are generally much dimmer than those associated with the Perseids, making their identification somewhat easier. The radiant lies to the south of the Square of Pegasus, close to the star Skat ($\delta$ Aquarii). Located to the north of the bright star Fomalhaut in Pisces Austrinus, the radiant is particularly well placed for those observers situated in the southern hemisphere. The shower peaks during the early hours of 29 July which this year ties in with a Full Moon resulting in observing conditions that will be far from ideal.

## Perseids

Associated with the parent comet 109P/Swift-Tuttle – and radiating from a point in northern Perseus, close to the border with the adjoining constellation Cassiopeia – the Perseids are a beautiful sight, with fast moving meteors appearing as soon as night falls and up to 100 meteors per hour often recorded. This is usually one of the best meteor showers to observe, with large numbers of very bright meteors often seen. There will be a New Moon at the time of peak activity on the night/morning of 12/13 August, and viewing conditions are set to be perfect.

## Kappa Cygnids

The Kappa Cygnids is a weak meteor shower, the radiant of which starts at the antihelion point in late June and then locates to Cygnus in July before taking up a position just west of the bright star Vega in early August. The ZHR is rather poor at just 3 meteors per hour. The shower is named after the position of the radiant

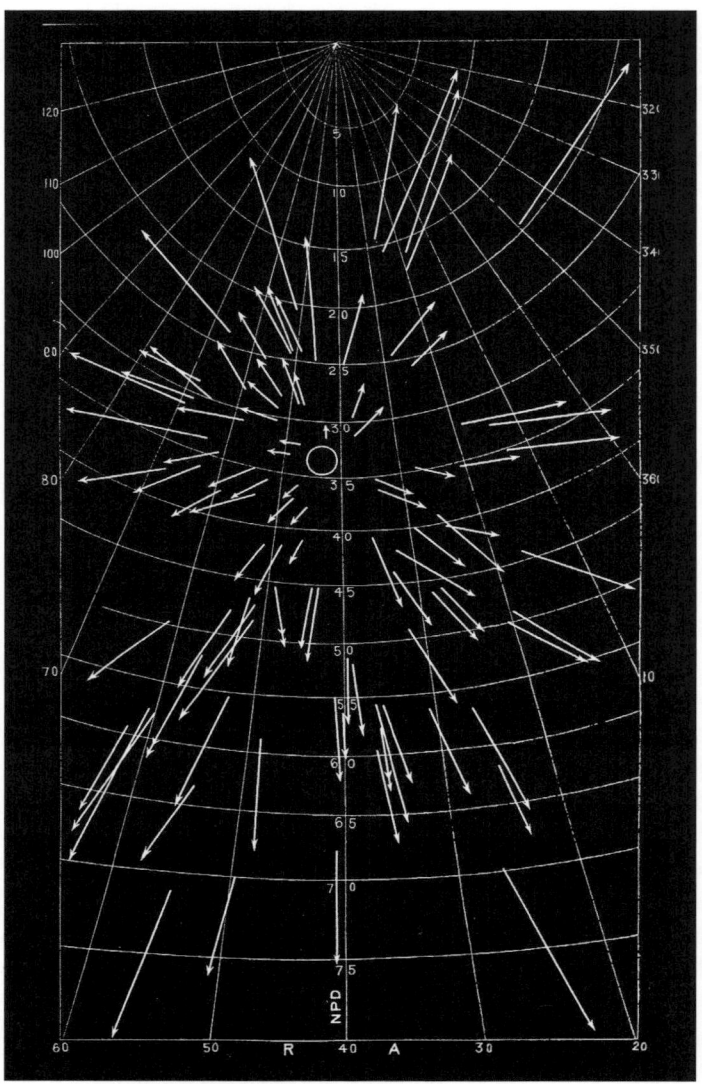

This drawing illustrates the meteor shower radiant for the Perseids. It was made during the period 8 August to 18 August 1880 and featured in the article 'The August Meteors' by W. F. Denning published in Volume 18 of *Popular Science Monthly*. The drawing depicts the paths of numerous meteors that can be traced back to the single point in the sky known as the radiant, and enables an observer to determine whether an observed meteor is truly a meteor of the shower or simply a sporadic meteor happening to burn up during the same period. (Wikimedia Commons/ W. F. Denning/*Popular Science Monthly*)

at the peak of shower activity, which is near the star Kappa (κ) Cygni. One point of interest is the shower is usually devoid of activity, bursting into what little activity it has once every seven years, with the last recorded activity being that of 2021. In 2026, on the night of supposed maximum activity, a waxing crescent Moon will give conditions that should be reasonably good for observing this shower.

## Arids

In 1999 astronomers began speculating that a new meteor shower could be forming, with a radiant in the southern constellation Ara (the Altar) and resulting due to the Earth encountering the debris left behind by the periodic comet 15P/Finlay. In September 2021, the Central Bureau for Astronomical Telegrams (CBAT) released a circular announcing the discovery of the new meteor shower as predicted.

Peter Jenniskens (SETI Institute and NASA Ames Research Center); Timothy Cooper (Astronomical Society of Southern Africa); and Dante Lauretta (University of Arizona), reported that Cameras for Allsky Meteor Surveillance (CAMS) video-based meteoroid orbit survey networks in New Zealand and Chile detected the new meteor shower. Fireballs resulting from the Earth running into the stream left behind from the 1995 return of the comet were first observed on the evening of 28/29 September 2021. Activity lasted into October with our planet encountering the debris left behind by the comet during its 2014/15 return. A total of thirteen slow moving meteors were detected during the period of visibility.

The Arids is a new shower and activity will be unpredictable. Some years will see a limited number of meteors and other years perhaps showing greater activity. This is certainly a meteor shower for observers in the southern hemisphere to monitor closely. As far as the 2026 shower is concerned, at the time of predicted peak activity on 7 October the Moon will be approaching new phase, resulting in potentially excellent observing conditions.

## Draconids

Also known as the Giacobinids, the Draconid meteor shower emanates from debris left behind by the periodic comet 21P/Giacobini-Zinner. The duration of the shower is just four days, from 6 October to 10 October, with the shower peaking on 7/8 October. The ZHR of this shower varies with poor displays in 1915 and 1926 but stronger displays in 1933, 1946, 1998, 2005 and 2012. Radiating from the 'head' of Draco, the meteors from this shower travel at a relatively modest 20 km/s. The Draconids are generally quite faint, although the fact that the Moon will be approaching new phase at the time of peak activity, observing conditions should be quite favourable.

## Southern Taurids/Northern Taurids

Running collectively from 10 September to 10 December, the Taurids are two separate showers, with southern and northern components, the Southern Taurids linked to the periodic comet 2P/Encke and the Northern Taurids to the asteroid 2004 $TG_{10}$. Meteors from both components emanate from the western regions of Taurus, the radiant for the Southern Taurids being located a little to the north of the star Xi (ξ) Tauri – close to the border with neighbouring Cetus – and that for the Northern Taurids immediately to the south east of the Pleiades open star cluster. The southern hemisphere encounters the first part of the stream, followed later by the northern hemisphere encountering the second part. The ZHR of these showers is low (between 5 and 10 per hour), although they can be beautiful to watch as they glide across the sky. The Southern Taurids first peak on 10 October when the Moon will be at new phase allowing for ideal conditions. The main peak for the Southern Taurids is on 5 November, about a week before the Northern Taurids, when a waning crescent Moon should provide good observing conditions. The Northern Taurids will also benefit from ideal conditions, with a waxing but slender crescent Moon setting well in advance at the time of peak activity on 12 November.

## Orionids

The second of the meteor showers associated with 1P/Halley, the Orionids radiate from a point a little to the north of the star Betelgeuse in Orion. Best viewed in the early hours when the constellation is well placed, the shower takes place between 2 October and 7 November with a peak on the night of 21/22 October. The velocity of the meteors entering the atmosphere is a speedy 67 km/s. This year, the presence of a waxing gibbous Moon will hinder observations.

## Leonids

Running from 6 November to 30 November, this is a fast moving shower with particles varying greatly in size and which can create lovely bright meteors that occasionally attain magnitude −1.5 (or about as bright as Sirius) or better. The radiant is located a few degrees to the north of the bright star Regulus in Leo. The parent of the Leonid shower is the periodic comet 55P/Tempel-Tuttle which orbits the Sun every 33 years. It was last at perihelion in 1998 and is due to return in May 2031. The Zenith Hourly Rates vary due to the Earth encountering material from different perihelion passages of the parent comet. For example, the storm of 1833 was due to the 1800 passage, the 1733 passage was responsible for the 1866 storm and the 1966 storm resulted from the 1899 passage (for additional information see Courtney Seligman's article *Cometary Comedy and Chaos* in the *Yearbook of Astronomy*

2020). The Leonid shower peaks on the night of 17/18 November when the light from a First Quarter Moon will interfere with observations, although as the Moon will be setting around midnight, observing conditions after this time will improve.

## Puppid-Velids

This is another shower for the southern hemisphere, albeit a weak one, the parent body of which has also yet to be identified conclusively. A low ZHR of 10 means that meteors will be few and far between even on the night of maximum activity 7 December. However, the presence of a very thin waning crescent Moon at this time may greatly aid observation of this elusive shower.

## Geminids

The Geminid meteor shower was originally recorded in 1862 and originates from the debris of the asteroid 3200 Phaethon. Discovered in October 1983, this rocky five kilometre wide Apollo asteroid has an unusual orbit that carries it closer to the Sun than any other named object of its type. Classified as a potentially hazardous asteroid (PHA), 3200 Phaethon made a relative close approach to Earth

In this long exposure image of the region around the constellation Gemini, captured in 2017 by Alan Dyer, the radiant of the Geminid meteor shower can be clearly identified. (Alan Dyer/ **Amazingsky.com**)

on 10 December 2017, when it came to within 0.069 au (10.3 million kilometres) of our planet. The Geminid radiant lies near the bright star Castor in Gemini and the shower peaks on the night of 13/14 December. This is considered by many to be the best shower of the year, and it is interesting to note that the number of observed meteors appears to be increasing annually. This year a waxing crescent Moon setting early in the evening will leave dark skies which, when coupled with a high ZHR of around 120 for the Geminids, should ensure an excellent show.

## Ursids

Discovered by William Frederick Denning during the early twentieth century, this shower is associated with comet 8P/Tuttle and has a radiant located near Beta ($\beta$) Ursae Minoris (Kochab). With relatively low speeds of around 33 km/s, the Ursids are seen to move gracefully across the sky. Research jointly carried out by Dutch/American astronomer Peter Jenniskens and his colleague, the Finnish astronomer Esko Lyytinen (1942–2020), revealed that outbursts may occur when 8P/Tuttle is at aphelion due to some meteoroids being trapped in a 7/6 orbital resonance with Jupiter. The Ursids run from 17 December to 26 December with peak activity taking place on the night of 22/23 December, although this year an almost fully illuminated Moon will severely hinder observations.

# Article Section

# Recent Advances in Astronomy

## Rod Hine

### Gravitation Lensing of Distant Galaxies and Individual Stars

The principle of 'gravitational lensing' arises from the way that time and space are fused into a quantity known as 'spacetime'. Light travelling from a far distant object is curved by passing close to massive objects because the mass has curved spacetime. In this picture, light from a distant quasar has been deflected by the mass of an intervening galaxy forming multiple images of just one object. In some cases, if the alignment is almost perfect, the distant object is seen as a partial or even complete ring – the so-called 'Einstein Ring'.

The *James Webb Space Telescope (JWST)*, launched on 25 December 2021, has captured numerous examples of gravitational lensing and has given us unprecedented views of objects that existed during the very earliest time in the Universe. The light from such objects is redshifted well from the visible spectrum into the infrared, so the key *JWST* instruments are the Near Infrared Camera (NIRCam) and Near Infrared Spectroscope (NIRSpec). Stars and galaxies all have

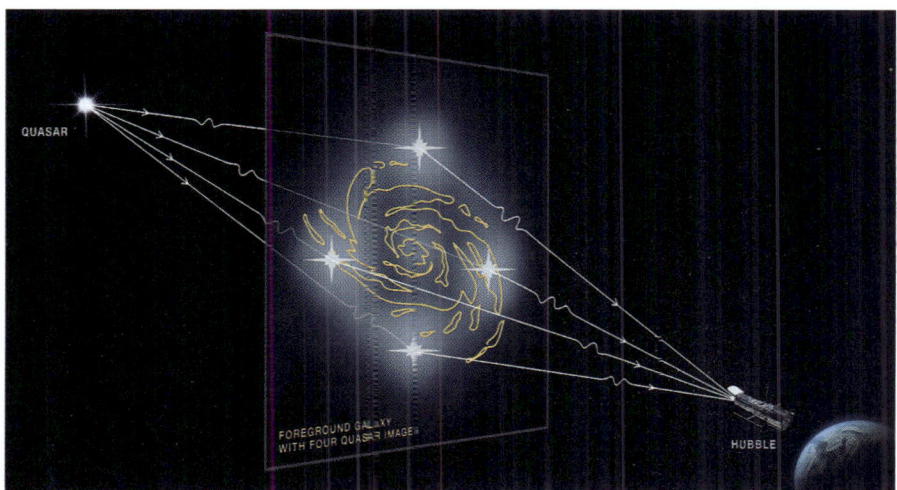

Light from a distant quasar is deflected by the mass of the intervening galaxy and arrives at Earth giving four images of the same object. (NASA, ESA and D. Player (STScI))

Earendel is seen as a tiny spot on the arc of the ring-shaped image of the distant galaxy. (NASA, ESA, CSA, D. Coe (AURA/STScI for ESA), Z. Levay)

their own very specific spectral fingerprint so that data from NIRSpec can be used to classify and identify any interesting images recognised by NIRCam.

Many distant galaxies had indeed already been identified from *Hubble Space Telescope (HST)* images, but *JWST* now extends this to yet further distant galaxies and even individual stars. At the time of writing, the oldest individual star appears to be WHL0137-LS, or Earendel, which was first discovered in *HST* images by a team led by Brian Welch of The Centre for Astrophysical Sciences, John Hopkins University, Baltimore. It was described as 'A highly magnified star at redshift 6.2' in a paper published in 2022.[1] The work was based on a series of observations over three and a half years and verified by four independent lensing modelling processes. The fact that the observations have been consistent for so long has increased confidence that it really is a star and not a transient event, and initial results indicate that it is likely to be a massive star, probably of 50 or more solar masses and observed just 900 million years after the Big Bang. Further observation may reveal more details and enable it to be given a place in the Hertzsprung-Russell diagram.

The *JWST* has also revealed gravitational lensed images of galaxies from a period just a few hundred million years after the Big Bang. The massive gravity of galaxy cluster MACS0647 acts as a cosmic lens to bend and magnify light from the more distant MACS0647-JD system. It also triply lensed the JD system, causing its

---

1. Welch, B., Coe, D., Diego, J.M. et al., 2022, 'A highly magnified star at redshift 6.2', *Nature*, **603**, 815–818. **arxiv.org/pdf/2209.14866.pdf**

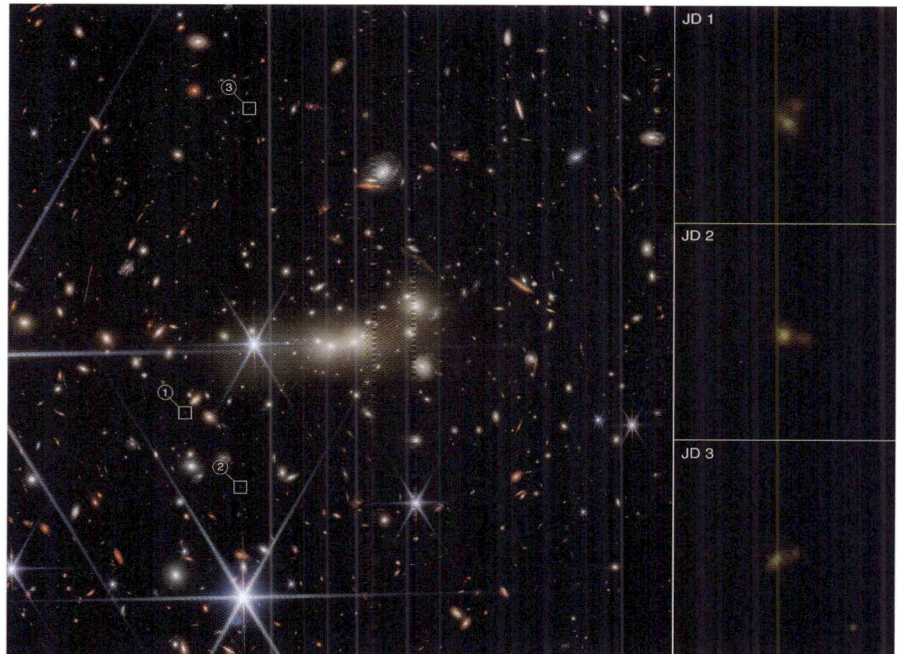

The massive gravity of galaxy cluster MACS0647 has formed three gravitationally lensed images of the distant JD galaxy. (NASA, ESA, CSA, Dan Coe (STScI), Rebecca Larson (UT Austin), Yu-Yang Hsiao (JHU), Alyssa Pagan (STScI))

image to appear in three separate locations. These images, which are highlighted with white boxes on the accompanying picture, are marked JD1, JD2, and JD3. With a redshift of $z = 10.7$, this galaxy was formed about 426 million years after the Big Bang. It had been observed by *HST* in 2012 but these new images from the *JWST* NIRCam have improved the resolution and it's likely that it may be a pair of galaxies rather than the single galaxy seen in just a few pixels by *HST*.

## More Results from Event Horizon Telescope Collaboration

Building on the success of the Event Horizon Telescope (EHT) in producing the first images of black holes in April 2019, evidence has now emerged to show jets of matter ejected at almost the speed of light coming from the supermassive black hole at the heart of radio galaxy 3C84, located in the region known as Perseus A. The Perseus super cluster lies about 230 million light years from Earth so that places the object as one of our closest supermassive black holes.

The team led by Georgios F. Paraschos of the Max Planck Institute for Radio Astronomy, Bonn, Germany, have used the observations of the jets to deduce the likely structure of the intense magnetic fields surrounding the black hole. Understanding how the magnetic fields shape and direct the jets will help to understand the gravitational effects of a spinning black hole and improve the modelling of jet formation.

## Closer to Home – Periodic Radiation from Centre of Milky Way

In 2021, repetitive bursts of gamma-ray pulses were detected from the region known as Sagittarius A* at the heart of our Milky Way galaxy, about 26,700 light years from Earth. Occurring every 76 minutes, the gamma-ray pulses appear to be in harmony with X-ray pulses of twice that period from the same location. This mystery may have been solved by Gustavo Magallanes-Guijón and Sergio Mendoza of the National Autonomous University of Mexico. They have analysed several sets of publicly available data and deduced that a highly magnetised blob of matter may be orbiting around the supermassive black hole.

## Brightest Quasar Seen So Far

The object known as J0529-4351 showed up first in images from the ESO Schmidt Southern Sky Survey around 1980. Finding quasars, or quasi-stellar objects, requires detailed examination of the object and its immediate surroundings to establish that the emissions arise from a supermassive black hole at the centre of a galaxy surrounded by an accretion disk where gas and dust are being consumed – in other words, an 'active galactic nucleus' or AGN. The emissions are seen across nearly all of the electromagnetic spectrum, making them distinct from other objects such as stars, nebulae and ordinary galaxies.

Searching through huge databases to identify quasars has been done very successfully by machine learning methods, but such methods do have limitations. Machine learning works by 'training' a computer program with numerous examples of data from known quasars. Then, the computer searches through new data, trying to find a similar pattern. The danger is that new data with some extreme values, not seen in the training data, will be rejected or re-classified as a different object. This appears to be the case with J0529-4351.

Automated analysis of the data from the European Space Agency *Gaia* satellite considered the image of J0529-4351 to be so bright that it was most likely a bright star close to Earth. However, a team led by Christian Wolf, from the Australian National University (ANU) used new data from the 2.3 metre telescope at Siding Spring Observatory in Australia to identify it as a quasar of hitherto unknown

Region around quasar J0529-4351 showing the location of the quasar previously thought to be a bright nearby star. (ESO/Digitized Sky Survey 2/Dark Energy Survey)

brightness. It took further observations from the X-Shooter Spectrograph on the ESO's Very Large Telescope (VLT) on Cerro Paranal in the Atacama Desert in northern Chile to confirm the properties, making it the brightest quasar and fastest-growing black hole ever seen to date.

Its mass is estimated at 17 billion solar masses and it appears to be consuming the equivalent of about one solar mass every day. The accretion disk is seven light years in diameter and the energy released makes it 500 trillion times more luminous than our Sun, the light having travelled more than 12 billion years to reach us. But this may not be the end of the story. An upgrade called GRAVITY+ is being incrementally rolled out at the VLT site to improve the existing interferometry capabilities, where the light from all four 8-metre VLT instruments is combined to improve resolution. New adaptive optics will better compensate for blurring effects in the atmosphere and will give resolution down to milliarcseconds, equivalent to a 130-metre telescope. Future observations of J0529-4351 with GRAVITY+ may help us to understand exactly how such incredibly massive black holes came into existence so early in the history of the Universe and just how they may have evolved since then.

# Recent Advances in Solar System Exploration

## Peter Rea

This article was written during the spring of 2024. As the missions mentioned are either active or due for launch imminently, the status of some missions may change after the print deadline. The mission websites are shown in bold in each section and these can be visited for the very latest information.

### Seeking Answers at the Sun

Regular readers of the *Yearbook of Astronomy* who choose to read my articles on solar system exploration will recall that the NASA mission *Parker Solar Probe* is slowly getting closer to the Sun's corona. The corona is the outermost part of the Sun's atmosphere and is usually hidden by the bright light of the Sun's surface. That makes it difficult to see without using special instruments. However, this feature can be viewed during a total solar eclipse, manifesting itself as streamers

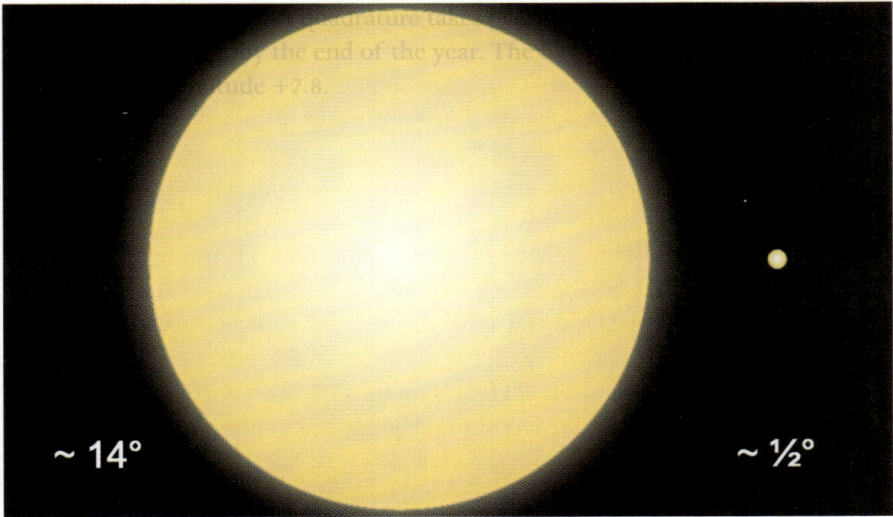

Depiction of the Sun's apparent size as seen from Earth (half a degree) on the right, against the apparent size it would appear to someone looking at it from the Parker Solar Probe's closest distance to it (14 degrees at perihelion of the probe). (Wikimedia Commons/Maringaense)

emanating from the Sun. The corona is composed of plasma with a temperature in excess of one million degrees Celsius, far hotter than the visible surface of the Sun – the photosphere – which is around 6,000 degrees Celsius. It is into this region that the Parker Solar Probe will venture to perform its measurements.

Launched on 12 August 2018 into a heliocentric (sun-centred) orbit, it has over the past few years used the gravity of Venus to slowly lower the perihelion (the point in its orbit at which it is closest to the Sun). The seventh and final gravity assist occurred on 24 November 2024 which set up the first of several very close passes to our star on 24 December 2024. It will race past the Sun at 702,000 kilometres per hour, approaching to within 0.04 au (around six million kilometres) of the solar surface. A special heat shield pointing toward the Sun protects the spacecraft from the intense 1,400 degrees Celsius temperatures.

There will be two more scheduled passes before nominal end of mission. These will be in March and June 2025. Interested readers should check the mission website listed below for news of any possible mission extension unknown to the author at the time of writing.

Parker Solar Probe (NASA): **parkersolarprobe.jhuapl.edu**

The European Space Agency's *Solar Orbiter* mission was launched on 10 February 2020 into a highly elliptical heliocentric orbit. At its closest to the Sun it will be at 0.28 au, which is well within the orbit of Mercury. At its furthest from the Sun it will be 0.91 au, almost back at the orbital distance of the Earth. Its mission complements that of the *Parker Solar Probe* by focusing on the heliosphere and solar wind. The mission goals will be attempting to answer the following questions:

- What drives the solar wind and where does the coronal magnetic field originate from?
- How do solar transients drive heliospheric variability?
- How do solar eruptions produce energetic particle radiation that fills the heliosphere?
- How does the solar dynamo work and drive connections between the Sun and the heliosphere?

These are vital questions, and the *Solar Orbiter* is currently addressing them. Solar Orbiter is in a highly elliptical orbit around the Sun and which initially had zero inclination, placing it within the plane of the ecliptic along with the Earth and the rest of the planets. Over the past few years that inclination has been increasing, up

This image captures 'totality' during the 2 July 2019 total solar eclipse, the moment that the Moon passes directly in front of the Sun from Earth's perspective, blocking out its light and allowing the Sun's extended atmosphere – the corona – to be seen. (ESA/CESAR/Wouter van Reeven/ CC BY-SA 3.0 IGO license)

to 24° and hopefully up to 33° in an extended mission, allowing observations of the polar regions of the Sun.

Solar Orbiter (ESA): **esa.int/Science_Exploration/Space_Science/Solar_Orbiter**

Our Sun, a typical G-type star, gives the Earth the light and heat that keeps all species on Earth alive, and has done so for around 4.6 billion years. It gives us life and it can take it away. Massive Coronal Mass Ejections (CMEs) can play havoc with life on Earth, and in the modern world can disrupt power grids and orbiting satellites. Scientists have long studied the Sun to try and understand what powers our parent star. In the last few articles I have focused on the *Parker Solar Probe* and *Solar Orbiter*. This is not meant to imply there are no other missions studying our Sun. In fact there are many, but space does not allow me to mention them all. Readers interested in past and current missions to study the Sun can check out 'List of Heliophysics Missions' available at: **en.wikipedia.org/wiki/List_of_heliophysics_missions**

Readers interested in an overview of the different aspects of the Earth-Sun environment that affect our planet may want to view NASA's 'Living with a Star' program at: **science.nasa.gov/heliophysics/programs/living-with-a-star**

## Mercury Within Grasp

The joint ESA/JAXA mission *BepiColombo* launched on 20 October 2018 is well into its seven year cruise to Mercury. An explanation on why it takes seven years to reach Mercury rather than six to nine months for Mars was discussed in the *Yearbook of Astronomy 2025*. The spacecraft should be inserted into orbit around Mercury during November 2026, before which it will split into its two components, these being the ESA-built Mercury Planetary Orbiter (MPO) and the JAXA (Japanese Aerospace Exploration Agency) Mercury Magnetospheric Orbiter (MMO). Once in orbit the two separate spacecraft will conduct observations of both the tenuous exosphere and magnetic field as well as study Mercury's interior structure and surface geology.

BepiColombo: **esa.int/Science_Exploration/Space_Science/BepiColombo_overview2**

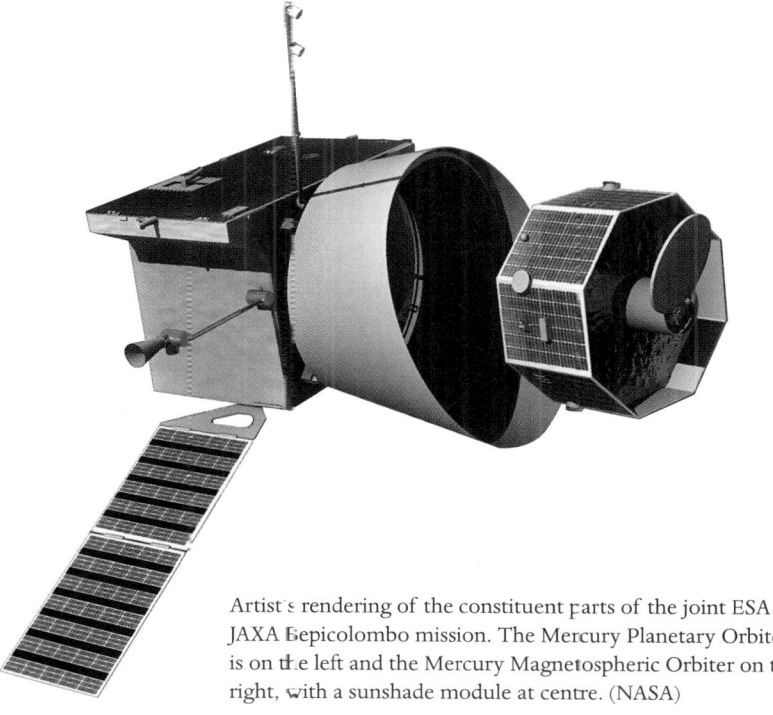

Artist's rendering of the constituent parts of the joint ESA/JAXA Bepicolombo mission. The Mercury Planetary Orbiter is on the left and the Mercury Magnetospheric Orbiter on the right, with a sunshade module at centre. (NASA)

## Decade of the Moon

As at time of writing, it has been 55 years since the first crewed mission – Apollo 11 – famously landed on the lunar surface, and 52 years since the last Apollo mission returned to Earth. I remember it all vividly. Since then, missions to the Moon have been few and far between. The current decade has seen a renewed interest in exploring our nearest interplanetary neighbour. The American *Artemis* programme plans to return human footprints to the Moon within the next few years. There is also much interest from other nations in sending spacecraft to the lunar surface. In recent years the Indian Space Research Organisation (ISRO) were successful in landing *Chandrayaan 3* near the south polar region in August 2023, making India the first nation to land in that region. The Russians were less successful with *Luna 25*, launched in August 2023, their first mission to the Moon since *Luna 24* in 1976. An anomaly caused *Luna 25* to crash onto the lunar surface. The Japanese had a close call with their *Smart Lander for Investigating Moon (SLIM)* lunar lander. Although it successfully landed on the Moon, it was not at the correct orientation, resulting in the solar panels not facing the Sun to recharge the batteries. The lander was put into a low activities state to conserve battery power. A few days later, with the Sun higher in the sky, the solar panels were able to recharge the batteries and the lander was re-activated. Astrobotic – an American private company working with NASA and their Commercial Lunar Payload Services (CLPS) programme – had built with their own funds the *Peregrine* lander on which were mounted various experiments including some CLPS experiments supplied by NASA. A propulsion anomaly shortly after launch precluded any landing attempt. Another commercial company – Intuitive Machines – landed their *Nova-C* lander, named *Odysseus*, on the Moon on 22 February 2024. *Odysseus* was carrying six payloads under the CLPS contract. As if to prove how difficult it is landing a spacecraft on the Moon, the lander tipped over on touchdown and came to rest on its side, this as the result of it having some residual forward motion instead of coming straight down as intended.

Readers interested in current and future missions to the Moon are invited to check out the 'List of Missions to the Moon' website at: **en.wikipedia.org/wiki/List_of_missions_to_the_Moon**

## Mars – The Red Planet

Look up into the night sky when Mars is at opposition, and you will see a very bright and distinctly red object. This is Mars – also known as the Red Planet – its unmistakable colour emanating from the presence of iron oxides on its surface. The planet is named for Mars, the Roman God of War. The Greek equivalent is Ares,

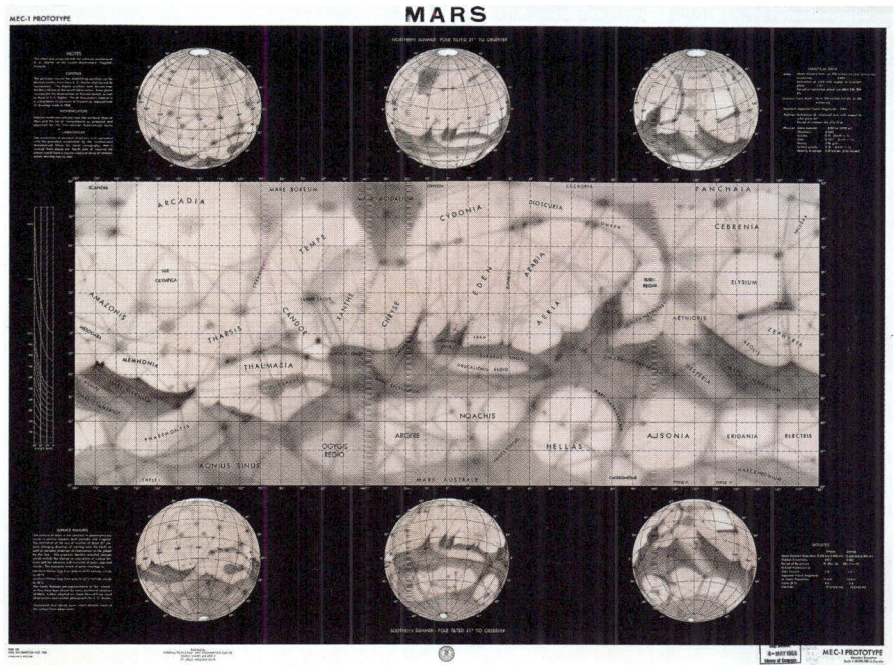

Map of Mars showing dark albedo features with lines connecting some of these features which Italian astronomer Giovanni Schiaparelli described as *canali* (channels), but which Percival Lowell in the USA described as canals. These dark lines were only seen in poor or low resolution telescope and are illusionary; they disappeared when imaged by spacecraft with high resolution cameras. (Aeronautical Chart and Information Center, USA)

from which we derive the word *areology*, the study of Mars. Mars has fascinated astronomers both professional and amateur for a long time. Maps drawn in 1877 by the Italian astronomer Giovanni Schiaparelli (1835–1910) during a favourable opposition showed dark lines connecting features on the surface of Mars. Percival Lowell (1855–1916), observing at his observatory at Flagstaff in Arizona, also saw these dark lines, which he referred to as 'canals' a mistranslation of the Italian word *canali*, or channels, which Schiaparelli had used on his maps. Our view of Mars has changed significantly since the dawn of the Space Age. The dark albedo features have been revealed to us, and show no sign of any 'canals'. Percival Lowell would doubtless be disappointed. Instead of the inhabited planet, pictured by Lowell as having 'canals' stretching across the globe, Mars is seen to be a desiccated world. Amongst its most notable features are several huge volcanoes which are far larger

Global mosaic of 102 *Viking 1 Orbiter* images of Mars taken on orbit 1,334, 22 February 1980. The images are projected into point perspective, representing what a viewer would see from a spacecraft at an altitude of 2,500 kilometres. At centre is Valles Marineris, over 3,000 kilometres long and up to eight kilometres deep. Note the channels running up (north) from the central and eastern portions of Valles Marineris to the dark area, Acidalia Planitia, at upper right. At left are the three Tharsis volcanoes and to the south is ancient, heavily impacted terrain. A far cry from the drawing of Percival Lowell. (NASA/Viking 1 Orbiter/MG07S078-334SP)

than any volcano on Earth. Running just south of the equator within the Tharsis Montes region situated on the western hemisphere of Mars, is a huge canyon system, which was formed by rifting when the Tharsis Bulge rose up. It is named Valles Marineris, and was named after the *Mariner 9* orbiter, an American spacecraft that first saw it in 1972.

There have been active missions at Mars for the last twenty years. I would like to mention two in particular, because of their age. Launched on 7 April 2001, the *Mars Odyssey* spacecraft has been exploring Mars since its arrival on 24 October 2001. The name *Odyssey* is significant, tying in as it does with the year of launch, and is a tribute to the Arthur C. Clark book, and the subsequent Stanley Kubrick film *2001: A Space Odyssey*. The *Mars Odyssey* probe is using a camera that provides visible and infrared imaging to characterize how minerals are distributed on the surface of Mars. It also provided a relay of signals back to Earth for the *Curiosity* rover, as well as the two *Mars Exploration Rover (MER)* mission rovers *Spirit* and *Opportunity*, and the *Phoenix* Mars lander mission. Attitude control propellant used for keeping the spacecraft correctly orientated is running low, and by the end of 2025 could become depleted, leading to the end of the mission. Also worthy of note is the ESA mission *Mars Express*. Launched on 2 June 2003 it carried the *Schiaparelli* lander. Arriving at Mars on Christmas Day 2003 the lander was released and correctly entered the atmosphere. Unfortunately the landing was not successful, although the *Mars Express* mission has been working continuously since then returning outstanding stereo images as well as other data. ESA has not given a potential end date for this mission.

## Table of Current Mars Missions

| Mission | Origin | Type | Orbital Insertion | Landed |
|---|---|---|---|---|
| Mars Odyssey | USA | Orbiter | 24 Oct 2001 | |
| Mars Express | Europe | Orbiter | 25 Dec 2003 | |
| Mars Reconnaissance Orbiter | USA | Orbiter | 10 Mar 2006 | |
| Mars Science Laboratory – Curiosity | USA | Lander | | 6 Aug 2012 |
| Maven | USA | Orbiter | 22 Sep 2014 | |
| ExoMars – Trace Gas Orbiter | Europe | Orbiter | 19 Oct 2016 | |
| Hope Mars Mission | UAE | Orbiter | 9 Feb 2021 | |
| Tianwen-1 | China | Orbiter/Lander | 10 Feb 2021 | 14 May 2021 |
| Perseverance Rover | USA | Lander | | 18 Feb 2021 |

The table of missions reproduced here shows which spacecraft are currently operational as of 2024. Interested readers can enter the spacecraft name into their preferred search engine for further information. It may help to add the word 'mission' after the spacecraft name. (Peter Rea)

Spacecraft that should be operational by the time this edition of the *Yearbook of Astronomy* is published include the dual-spacecraft *Escape and Plasma Acceleration and Dynamics Explorers (EscaPADE)* mission consisting of a pair of spacecraft designed to complement the work done by the American *Mars Atmospheric and Volatile EvolutioN (MAVEN)* mission already returning data on Mars' atmospheric loss. Both of the *EscaPADE* spacecraft will carry similar instruments and, by flying two spacecraft, *EscaPADE* can do what the *MAVEN* mission cannot; fly in two places around Mars at once and gather data with greater geographic resolution. They will launch on a Blue Origin *New Glenn* reusable rocket in the spring of 2025.

The Indian Space Research Organization (ISRO) is returning to Mars with their *Mars Orbiter Mission 2 (Mangalyaan 2)* spacecraft, which continues the work of *Mangalyaan 1* which arrived at Mars in September 2014.

Launching in 2026 the Japanese Aerospace Exploration Agency (JAXA) hopes to land on the largest of Mars' two moons Phobos to collect samples to bring back to Earth. (NASA/JPL-Caltech/University of Arizona)

The Japanese Aerospace Exploration Agency (JAXA) are planning a Mars orbiter that will make close flybys of the two moons of Mars called Phobos and Deimos. It is their intention to deliver a small rover to the surface of Phobos. Details of these three missions can be found using the web links at the end of this section.

There are other planned missions to Mars, and these will be discussed in future articles.

If the 2020s is a 'Decade of the Moon' then the last couple of decades have been a 'Golden Age' of Martian exploration. One day in the future human footprints will be placed on the sands of Mars. I hope I live to see it.

EscaPADE (NASA): **science.nasa.gov/mission/escapade**

Mangalyaan 2 (ISRO): **www.isro.gov.in/MarsOrbiterMissionSpacecraft.html**

Martian Moons Exploration (JAXA): **mmx.jaxa.jp/en**

New Glenn: **blueorigin.com/new-glenn**

## Rock On!

As of this writing there are two missions en route to asteroids; *Lucy*, launched on 16 October 2021 and *Psyche*, launched on 13 October 2023. Both of these are part of NASA's Discovery program, being the thirteenth and fourteenth respectively to be launched. Named for the 3.2 million year old hominid fossil of an Australopithecus Afarensis, *Lucy* has been launched toward the Trojan asteroids co-orbiting with Jupiter. They are situated at the Lagrange points $L_4$, located 60 degrees ahead of Jupiter and $L_5$, situated 60 degrees behind Jupiter. In the early history of the Solar System the dust and rocky material clumped together to form these ancient rocky bodies. These asteroids, rather than the main belt asteroids, could be "fossils of planet formation", hence the link to that early fossil. *Lucy* will encounter the first trojan asteroid, 3548 Eurybates, in August 2027. Before then the probe will have visited two main belt asteroids, the first of which will be 152830 Dinkinesh, named for the local language name for *Lucy* in the Afar region of Ethiopia where the fossil was discovered. When this object was visited on 1 November 2023 there was a major surprise in store; Dinkinesh was discovered to have a companion, seen quite clearly behind the main asteroid. As *Lucy* then flew past the pair its field of view was more side-on, and yet another surprise was revealed. Selam, as the companion was later called, was found to be a contact binary, making it the first known example of a contact binary satellite of an asteroid. Contact binaries are two asteroids that are actually touching but not fused together, being held together by very weak gravitational forces. The second asteroid to be visited by *Lucy* will be

This image shows the asteroid 152830 Dinkinesh and its satellite as seen by the *Lucy Long-Range Reconnaissance Imager (L'LORRI)* as NASA's *Lucy* Spacecraft departed the system. This image was taken at 1 pm EDT (17:00 UTC) on 1 November 2023, about six minutes after closest approach, from a range of approximately 1,630 kilometres. From this perspective, the satellite is revealed to be a contact binary; the first time a contact binary has been seen orbiting another asteroid. (NASA/Goddard/SwRI/Johns Hopkins APL)

52246 Donaldjohanson (named for the discoverer of the Lucy fossil) which will be encountered in April 2025. As this has yet to happen from my perspective, I will have more to say in the forthcoming *Yearbook of Astronomy 2027*.

The second asteroid mission is *Psyche*, launched to the metallic asteroid of the same name. As *Psyche* will not arrive until 2029, I will limit my remarks to one particular experiment carried onboard the spacecraft. Currently spacecraft return data using radio frequency (RF) wavelength. This does tend to limit the amount of data that can be returned in a given time. *Psyche* is carrying an experiment that will make use of optical frequencies, using a laser to increase data rates by at least ten times, with the possibility of 100 times in the future. This is the Deep Space Optical Communications (DSOC) experiment, and is NASA's first demonstration of optical communications beyond the Earth-Moon system. An article about DSOC will appear in a future edition of the *Yearbook of Astronomy* and will go into more detail.

As part of NASA's Planetary Defence Program, the *Double Asteroid Redirection Test (DART)* mission was launched toward the double asteroid Didymos and its companion Dimorphos. The purpose of the mission was to collide part of the *DART* spacecraft with Dimorphos to see if the kinetic energy of the impact could alter the orbital velocity of the asteroid. Results from this study would determine whether an asteroid on a collision course with Earth could be deflected. The velocity of Dimorphos, and hence shape of its orbit around Didymos, was significantly altered. The impact left a significant crater, and the ESA mission *HERA*, scheduled for launch in October 2024, will be sent to observe the asteroid and the effect of the impact. Further details can be found at the mission website listed below.

ESA's *Hera* mission concept, currently under study, would be humanity's first mission to a binary asteroid; the 800-metre diameter Didymos is accompanied by a 170-metre diameter secondary body. *Hera* will study the aftermath of the impact caused by the NASA spacecraft *DART* on the smaller body. (ESA–Science Office / CC BY-SA 3.0 IGO license)

Lucy (NASA): **lucy.swri.edu**

Psyche (NASA): **jpl.nasa.gov/missions/psyche**

Hera (ESA): **esa.int/Space_Safety/Hera**

Dart (NASA): **dart.jhuapl.edu/Mission/index.php**

## The Medician Stars are Expecting Visitors

In early-January 1610, using the recently invented telescope, the Italian astronomer Galileo Galilei (1564–1642) made observations of Jupiter that would change for ever our view of the solar system, discovering previously unknown natural satellites, or moons, of Jupiter. There were four in total, observations carried out on subsequent nights revealing that they were continually changing position. For example, on some nights he would observe three on one side, sometimes two on either side. Galileo correctly concluded that these "stars" were revolving around the planet. He initially referred to them as the *Cosmica Sidera* (Cosimo's Stars) in honour of the high status banker and politician (and patron of Galileo) Cosimo II de' Medici. After all, if you need a patron, it is always a good idea to ingratiate oneself with the great and good (and wealthy). The four moons were subsequently renamed the Medicean Stars, thereby honouring all four of the Medici brothers (Cosimo, Francesco, Carlo, and Lorenzo). Nowadays we refer to them collectively as the

This composite image includes the four largest moons of Jupiter which are known as the Galilean satellites, having been first seen by the Italian astronomer Galileo Galilei in 1610. Shown from left to right in order of increasing distance from Jupiter is the closest Io, followed by Europa, Ganymede, and Callisto. (NASA/JPL-Caltech/PIA01299/DLR)

Galilean satellites of Jupiter, as well as by their individual names Io, Europa, Ganymede, and Callisto.

Initially imaged by *Pioneer 10* and *Pioneer 11* at low resolution, it was the *Voyager 1* and *Voyager 2* missions that revealed these four moons in much greater detail. I recall seeing these pictures for the first time during the late-1970s and early-1980s, not via the World Wide Web (this was in the future) but by news sheets posted to me from the Jet Propulsion Laboratory. How things have moved on in recent years! The Galilean moons were fascinating worlds in their own right, and each was quite different from the rest in surface detail. They were visited again by the *Galileo* mission, which arrived at Jupiter in December 1995. For some time now it has been speculated in the science community that Europa, Ganymede, and Callisto have sub surface oceans. Recent observations by the *Juno* spacecraft, in polar orbit around Jupiter since July 2016, have shown the presence of salts and organic compounds on the surface of Ganymede; further evidence for water below the icy crust.

Further exploration of Europa, Ganymede, and Callisto is a high priority for NASA and the European Space Agency. To that end, ESA launched the *Jupiter Icy Moons Explorer (JUICE)* mission on 14 April 2023. When it arrives at Jupiter in 2034 it will fly by Europa and Callisto a few times before going into orbit around Ganymede for long term observations. Rather than going into orbit around any of the moons, the NASA *Europa Clipper* mission will make many close flybys of Europa. As mentioned in the introduction, this article was written in the spring of 2024, with launching scheduled for October 2024. If the launch is successful, *Europa Clipper* is scheduled to arrive at the Jovian system in April 2030. Interested readers can follow the missions at the websites listed below.

Ganymede, larger than even Mercury and Pluto, has an icy surface speckled with bright young craters overlying a mixture of older, darker, more cratered terrain laced with grooves and ridges. The cause of the grooved terrain remains a topic of research, with a leading hypothesis relating it to shifting iced plates. Ganymede is thought to have an ocean layer that contains more water than Earth, and which might contain life. Just like our own Moon, Ganymede keeps the same face turned towards its parent planet, in this case Jupiter. The image seen here was captured by NASA's robotic *Juno* spacecraft as it passed only about 1,000 kilometres above Ganymede's surface, this close pass reducing *Juno's* orbital period around Jupiter from 53 days to 43 days. *Juno* continues to study the giant planet's high gravity, unusual magnetic field, and complex cloud structures. (NASA/JPL-Caltech/SwRI/MSSS; processing by Kevin M. Gill)

JUICE (ESA): **esa.int/Science_Exploration/Space_Science/Juice**
Europa Clipper (NASA): **jpl.nasa.gov/missions/europa-clipper**
JUNO (NASA): **missionjuno.swri.edu**

## Welcome Back Old Friend

*Voyager 1* was launched on 5 September 1977, which, as at time of writing (May 2024) was nearly 47 years ago. Like its sister spacecraft *Voyager 2*, launched earlier on 20 August 1977, it performed impeccably. *Voyager 1* passed the planet Jupiter on 5 March 1979 and Saturn on 12 November 1980, the latter encounter including a flyby of Saturn's largest moon Titan. Its primary mission now concluded, the spacecraft continued on a trajectory that would take it out of the solar system and into interstellar space. In 2012 *Voyager 1* crossed the heliopause (located at a distance of around 123 au from the Sun) and entered interstellar space, becoming the first human-made object to exit the solar system.

In November 2023 *Voyager 1* experienced issues with one of its onboard computers, resulting in the spacecraft being unable to send usable data back to Earth. The problem was found to be in the Flight Data System (FDS) computer. Engineers surmised that a memory module was corrupted, possibly as a result of having being struck by a high-energy particle, or simply due to age. Fixing the problem was not going to be an easy task. *Voyager 1* was so far away at the time that it was taking 22½ hours for commands sent from Earth to reach it, and for the team having to wait a total of 45 hours to receive a response. However, by March 2024 communication with *Voyager 1* was restored. This was achieved by deleting old, no longer needed code, and moving code away from the defective FDS memory chip into these freed up memory areas. *Voyager 1* was back in business … for the time being …

Voyager Mission Website: **voyager.jpl.nasa.gov**

Voyager Mission Status: **voyager.jpl.nasa.gov/mission/status**

Voyager 1 Real Time Tracker: **theskylive.com/voyager1-tracker**

Voyager 2 Real Time Tracker: **theskylive.com/voyager2-tracker**

As always, Solar System exploration continues to excite and inspire, and next year promises to be no different.

# Anniversaries in 2026

## Neil Haggath

### Halley on St. Helena

Edmond Halley (1656–1742) was destined for great things from an early age. In 1673, aged only 16, he began studying at The Queen's College, Oxford, and while an undergraduate, published papers on the solar system and sunspots.

After John Flamsteed (1646–1719) became the first Astronomer Royal in 1675, he began producing his famous star catalogue of the northern sky. Halley proposed to do the same for the region of the southern sky which is never visible from Europe. In late 1676, still aged only 20, he dropped out of Oxford to do so, with the support of King Charles II.

He sailed to the South Atlantic island of St. Helena, from where he could still see some of the northern stars, against which to cross-reference his observations

Edmond Halley. Portrait by Thomas Murray, 1690. (Wikimedia Commons / Thomas Murray)

of the southern. He set up an observatory with a large sextant with telescopic sights, and observed over the course of a year – the first telescopic survey of the southern sky.

Halley returned to England in May 1678, and produced a map of the southern sky. Oxford would not let him return, as he had violated the residence requirements; the King intervened, and he was awarded his master's degree in December. In the same month, he was elected a Fellow of the Royal Society, when still only 22.

The following year, he published *Catalogus Stellarum Australium* (Catalogue of the Stars of the South), which was presented to the Royal Society by Robert Hooke.

While on St. Helena, Halley made perhaps his greatest contribution to astronomy. In 1677, he observed a transit of Mercury, and it occurred to him that observations of transits from widely separated locations could be used to refine the measurement of the Earth-Sun distance. Having read Jeremiah Horrocks' (1618–1642) paper on the first observed transit of Venus in 1639, Halley realized that transits of Venus would be far more suitable for the purpose. Despite knowing that no such event would happen in his lifetime – the next one would not occur until 1761 – he wrote a paper laying down the challenge to astronomers of a future generation.

When the time came, both British and French astronomers took up the challenge in earnest – but that's another story.

## Joseph Ritter von Fraunhofer (1787–1826)

Two hundred years ago, a man died, who made remarkable contributions to astronomy in a tragically short life.

Joseph Fraunhofer was born on 6 March 1787 in Straubing, Bavaria. His childhood was like something from a Dickens novel; he was orphaned at the age of 11, had little education, and at 14, was apprenticed to a glassmaker, and was cruelly treated.

After a few months, the workshop suddenly collapsed, and Joseph was buried in the rubble. His rescue was witnessed by the Prince-Elector of Bavaria, who took an interest in the boy, giving him money to buy his freedom and pay for an education. Joseph Utzschneider, a partner in a reputable glassmaking company, also became a benefactor, allowing him to combine his education with practical training.

Joseph Ritter von Fraunhofer. Portrait by an unknown artist taken from *Die grossen Deutschen im Bilde* by Michael Schönitzer, 1936. (Wikimedia Commons)

He later joined the firm of the renowned Swiss glass and lens maker Pierre-Louis Guinand, where he proved exceptionally skilled. In 1806, he joined the Munich Optical Institute, later becoming its director when only 31. Under his leadership, Bavaria overtook England as the centre of optical instrument making.

In 1822, Fraunhofer was awarded an honorary doctorate by the University of Erlangen. Two years later, he was given the highest honour of the new Kingdom of Bavaria; he was made Knight of the Order of Merit of the Bavarian Crown by King Maximilian I, becoming a member of the nobility, with the title "Ritter von".

Fraunhofer had a passion for making lenses – with a grinding machine of his own invention – and refracting telescopes. His finest creation was a 9½-inch achromatic refractor, which was bought by the Russian government for the Dorpat Observatory in Estonia. This was then the world's biggest refractor but more importantly, it was the first telescope with an equatorial mount and clock drive.

However, Joseph von Fraunhofer is best known for his work in spectroscopy. He invented the modern spectroscope and, in 1814, discovered over 500 dark lines in the spectrum of the Sun. He later found similar lines in the spectra of several bright stars. Their cause was not explained until 1859, by Kirchoff and Bunsen, but they are still known as Fraunhofer lines.

Like many glassmakers of the era, Fraunhofer's health was damaged by heavy metal vapours. He contracted tuberculosis, and died in Munich on 7 June 1826, aged only 39.

## Giuseppe Piazzi (1746–1826)

Another eminent astronomer also died in 1826 who, unlike Fraunhofer, had a long life. This was Giuseppe Piazzi, the Italian priest, best known for his discovery of the first asteroid on the first day of the nineteenth century.

Piazzi was born on 16 July 1746 in Ponte in Valtellina, Lombardy. Little is known about his early life and education, but he reached academic heights at an early age.

In 1770, aged only 24, he took the Chair of Mathematics at the University of Malta. In 1781, after several moves, he became Lecturer in Mathematics at the University of Palermo in Sicily, becoming Professor of Astronomy six years later. In 1790, he established an observatory there, the Osservatorio Astronomico di Palermo.

Portrait of Giuseppe Piazzi by Italian artist Costanzo Angelini. c. 1825. (Wikimedia Commons / Costanzo Angelini / Osservatorio Astronomico di Capodimonte / INAF)

In 1803, Piazzi published the *Palermo Catalogue*, which listed the positions of 7,646 stars with unprecedented precision. He discovered that 61 Cygni had the largest proper motion of any star then known – it became known as Piazzi's Flying Star – and correctly reasoned that it must be relatively close. It later became the first star whose parallax, and thus distance, was measured, this by German astronomer Friedrich Bessel (1784–1846) in 1838.

On 1 January 1801, Piazzi found that an eighth-magnitude object had moved with respect to background stars. He initially announced it as a comet, but soon realized that it was something else – the first known and largest asteroid, though it was initially called a new planet. He named it Ceres, after the Roman goddess of grain, who was the patron goddess of Sicily.

Some decades earlier, the so-called Titius-Bode Law – a supposed mathematical relationship between the distances of the planets from the Sun, now known to be completely unfounded – had "predicted" the existence of an unseen planet between Mars and Jupiter. A society of astronomers known as the "Celestial Police" had been founded in 1800 to hunt for it – but Piazzi had discovered Ceres by chance before receiving his invitation to join! Ceres was initially thought to be the "missing" planet, but it was soon found to be tiny compared to the other planets.

Three more small bodies – Pallas, Juno and Vesta – were found over the next six years, and were also initially called planets. Then in 1845, a fifth, much smaller one was found, and it was realized that they were simply the biggest objects of many. Then the term "asteroid" was coined, and Ceres was designated as 1 Ceres.

In 2006, the International Astronomical Union reclassified it under the new category of "dwarf planet".

In 1817, Piazzi oversaw the completion of the Capodimonte Observatory in Naples. He died in Naples on 22 July 1826, aged 80.

## Richard Carrington (1826–1875)

In the same year in which Fraunhofer died, another astronomer was born, one who also achieved great things in a life cut short.

Richard Christopher Carrington was born in Chelsea, Middlesex on 26 May 1826. He graduated from Cambridge in 1848, and spent an unsatisfactory four years working as an observer at Durham University.

With a comfortable income from his father's brewery business, which he would later inherit, he was able to devote himself to astronomy as an amateur, becoming a highly accomplished one. In 1853, he built a house and observatory on Furze Hill, Redhill, Surrey, equipped with a 4½-inch refractor by William Simms. He compiled

Richard Carrington's house and observatory on Furze Hill, Redhill. (Wikimedia Commons/ Science Museum)

a catalogue of circumpolar stars, published in 1857, for which he was awarded the Gold Medal of the Royal Astronomical Society. He later served as Secretary of the RAS, and was elected Fellow of the Royal Society in 1860.

However, Carrington is best known for his studies of the Sun. Between 1853 and 1861, he made over 5,000 observations of sunspots, and discovered the Sun's differential rotation, with different rotation periods at different latitudes. He also independently discovered Spörer's Law, concerning the distribution in the latitudes of sunspots.

He was a pioneer of eclipse travel; he observed a total eclipse in Sweden in 1851, and later wrote a guide for people intending to travel to South America for another in 1858.

On 1 September 1859, Carrington observed a huge solar flare, which in the following days disrupted telegraph communications across Europe and America. This has been known ever since as the Carrington Event.

A severe illness in 1865 curtailed Carrington's career. He retired to Churt, Surrey, where he died suddenly on 27 November 1875, aged only 49.

## Allan Sandage (1926–2010)

This year sees the centenary of perhaps the most eminent cosmologist of the late-twentieth century, Allan Sandage. His life is described in David Harland's article, 'Allan Rex Sandage', elsewhere in this volume.

## Vikings on Mars

Fifty years ago, NASA achieved a major "first" – the landings of the two *Viking* probes on the surface of Mars.

Prior to 1976, three NASA *Mariner* probes had made successful flybys of Mars, with *Mariner 9* being the first to orbit the planet. Meanwhile, three Soviet missions had orbited Mars, but three landers had all failed.

The *Viking* program, which began in 1968, was the most expensive US planetary mission to date. *Viking 1* and *Viking 2* were launched on 20 August and 9 September 1975 respectively, by Titan IIIE-Centaur rockets, and each reached Mars and went into orbit about ten months later.

Each spacecraft consisted of an orbiter, with a launch mass of 2,328 kg, and a lander of 663 kg. The orbiters were based on a bigger version of the *Mariner* design, which had achieved several flybys of both Venus and Mars, and one of Mercury.

Carl Sagan poses with a full scale model of a *Viking* lander. (NASA/JPL)

Each was powered by solar panels with a combined area of 15 square metres, and carried cameras to map the planet with unprecedented resolution.

The landers, powered by radioisotope thermoelectric generators, would be deployed from orbit, enclosed in aeroshells during entry to the atmosphere, and would land using a combination of parachutes and retro-rockets. They had been sterilized before launch, to avoid contaminating Mars with terrestrial microbes.

Each orbiter spent about a month photographing the planned landing areas, to enable mission controllers to select the best landing site, before deploying the lander. It had been planned for *Viking 1* to land on the momentous date of 4 July 1976, but this was delayed due to difficulties in selecting the landing site. *Viking 1* eventually landed on 20 July in Chryse Planitia, and *Viking 2* on 3 September in Utopia Planitia.

The landers could transmit data both directly to Earth and via the orbiters. Each was equipped with a robotic arm and scoop, to deposit samples of surface material into a number of instruments to analyse it. These included a miniature biology lab, with three experiments designed to detect any signs of biological activity. No such signs were found; there were a couple of false alarms, which were explained by purely chemical processes.

The *Viking 2* lander operated until April 1980 and *Viking 1* until November 1982. In total, the orbiters returned over 51,000 images, mapping 97% of the planet at a resolution of 300 metres, and the landers returned over 4,500 images from the Martian surface.

# The Astronomers' Stars
## Taking It to Extremes

### Lynne Marie Stockman

Extremes in colour are the hallmarks of some stars. Extremes in motion set others apart. But extremes in age or size or magnetism or just unconventionality are what make some stars stand out from the rest.

American astronomer **Paul Willard Merrill** (1887–1961) was born in Minnesota, USA, but spent most of his life in California. His father was a minister who, it was said, was somewhat antagonistic toward the sciences but that did not stop Merrill from earning a bachelor's degree in mathematics from Stanford University and later a PhD in astronomy at Berkeley. It was at Berkeley and Lick Observatory that Merrill became interested in the astronomical applications of spectroscopy. After several years as an instructor at the University of Michigan, he was employed as a physicist at the National Bureau of Standards in Washington, DC, during the First World War. Here he worked on problems associated with infrared and aerial photography. Merrill finally arrived at Mount Wilson Observatory in 1919 where he would remain for the rest of his professional life. His passion was for the spectroscopy of unusual stars, particularly long-period variables, and also for infrared spectroscopy, a field still in its infancy. In 1922, he devised a new stellar type, the S class, for carbon stars defined by the bands of zirconium monoxide appearing in their spectra. Shortly before his retirement in 1952, he also discovered the element technetium in the spectra of some S-type stars. Technetium is the lightest element with no stable isotopes; its maximum half-life is just over four million years. The discovery of this short-lived element in the atmospheres of stars proved beyond doubt that heavier elements were being created through nucleosynthesis inside stars. He was also a pioneer in observational infrared astronomy, discovering diffuse interstellar absorption lines. The origin of many of these absorption bands remains unknown. Merrill was a member of numerous learned and professional societies, and was awarded the Henry Draper Medal of the National Academy of Sciences in 1945 for his contributions to spectroscopy. He died unexpectedly following surgery just nine years after retiring (Wilson 1964).

Four short paragraphs announced the discovery of an unusual star that was later to bear his name (Merrill 1938):

An emission line (D3) in the yellow portion of the spectrum of a tenth-magnitude star was detected by Miss Cora G. Burwell on objective-prism spectrograms taken by William C. Miller at Mount Wilson on September 26 and October 30, 1937. It is the northern of two small stars which precede the bright pair BD +16°3774 and BD +16°3775 by 1.0 minutes of time…It is star No. 209, *Bordeaux Astrographic Chart* 516….

Merrill's Star (WR 124) blazes forth from the centre of this James Webb Space Telescope false-colour image which combines near- and mid-infrared wavelengths. The surrounding nebula is composed of gas blown off from the central Wolf-Rayet star, demonstrating that past ejections have been both random and asymmetric (NASA/ESA/CSA/STScI/Webb ERO Production Team)

Follow-up spectrograms revealed that **Merrill's Star** was a Wolf-Rayet star with an unusually large radial velocity.

Wolf-Rayet stars were first identified in 1867 at the Observatoire de Paris by French astronomers Charles Joseph Étienne Wolf and Georges-Antoine-Pons Rayet. They found that the spectra of three unusual stars in the constellation of Cygnus had broad, bright emission lines rather than narrow absorption lines. Today astronomers subdivide these kinds of stars into three subtypes: WN (strong nitrogen lines), WC (strong carbon lines), and WO (strong oxygen lines). The broad emission lines are due to incredibly powerful stellar winds which push the outer layers of the star away at high speed. The emission features seen in spectrograms are actually formed in the gas which is already far from the central star. Wolf-Rayet stars are an evolved form of hot, massive O-type stars. They have short lifetimes, well under a million years, and have been theorised to be the progenitors of gamma ray bursts, supernovae, magnetars, and even collapsars. They remain an active topic of research today (Crowther 2007).

Even amongst Wolf-Rayet stars, those of type WN8 are peculiar. They are variable, they tend to avoid clusters and stellar associations, they are rarely found in binaries, and they exhibit unusually large space velocities. Some researchers conjecture that a nitrogen-rich Wolf-Rayet star is an evolutionary stage of a Thorne-Żytow Object or TŻO, a hypothetical hybrid star which forms when a giant star collides with a neutron star, creating a red giant or supergiant with the neutron star at its core! Merrill's Star is of this spectral type and has an astonishingly large radial velocity of nearly 200 kilometres per second, making it the fastest runaway Wolf-Rayet star in the galaxy. This object weighs in at around 30 solar masses, has a radius about 16 times that of the Sun, and is six times more luminous than our star. It is also surrounded by a nebula. Over the years, this nebula has been described as an H II (ionised hydrogen) region, a planetary nebula, a ring nebula, and even a bubble blown in the interstellar medium by strong stellar winds. It is moving in tandem with Merrill's Star and is most likely material ejected from the surface of the star. The nebula is also very young, only tens of thousands of years old (Toalá et al 2018).

Like most stellar objects, Merrill's Star has a host of other names, including the variable star designation QR Sagittae and the Wolf-Rayet number WR 124. Its surrounding nebula is known variously as (Minkowski) M 1-67 and (Sharpless) Sh 2-80, amongst other catalogue identifiers.

Although the star is named for Merrill, two other people, both frequent collaborators with Merrill in the early and mid-twentieth century, were instrumental in identifying the unusual characteristics of this object. **Cora Gertrude Burwell**

(1883–1982) was one of the Mount Wilson 'computers', women who measured and counted and calculated, relieving their male colleagues of the tedious work. She joined the staff at Mount Wilson in 1907, working with eminent astronomers George Ellery Hale and Walter Sydney Adams. In 1922 she began her long and fruitful collaboration with Merrill, finally retiring in 1949 as a recognised expert in the measurement and interpretation of stellar spectra. **William Curt Miller** (1910–1981) was an expert photographer who was responsible for some of the most outstanding astronomical images of his time, including the first colour portrait of M31, the Andromeda Galaxy. He also developed an interest in archaeology and in 1955, proposed that two petroglyphs recently discovered in northern Arizona represented the supernova of 1054 CE which we now know as the Crab Nebula.

**Georges-Achille Van Biesbroeck** (1880–1974) was born and raised in Ghent, Belgium. He developed an early interest in mathematics and the sciences, and upon completion of his studies, joined the Brussels Department of Roads and Bridges as an engineer. However, he carried out observational work at a variety of European observatories, chiefly of variable stars, visual binaries, and comets, publishing his findings in the international astronomical journal *Astronomische Nachrichten*. He was appointed to a position at Yerkes Observatory in Chicago in 1916, his family travelling through war-torn Europe to join him in the United States. Despite a heavy observing schedule, he found time to edit and publish much of the work left unfinished by the death of his colleague Edward Emerson Barnard in 1923. Mandatory retirement came in 1945 but he continued his observational work in a number of observatories around the world. In 1961, he published the *Van Biesbroeck Star Catalogue* of low luminosity stars discovered during observations made at McDonald Observatory in Texas. Two years later, he took up an appointment at the University of Arizona's Lunar and Planetary Laboratory. He died in 1974 (Hardie 1974).

BD+04°4048 is a ninth-magnitude red dwarf star with a large annual proper motion of well over an arcsecond. In 1944, Van Biesbroeck uncovered a faint companion star:

> In 1940 the writer started at the prime focus of the 82-inch reflector of the McDonald Observatory a systematic search for faint companions to known proper motion stars in order to extend our knowledge toward the lowest luminosity stars. Comparison of two plates of the field of BD +04°4048 revealed the presence of a very faint companion to that star, sharing its motion…This will make the photovisual magnitude of the companion 19.2, or three magnitudes fainter than the lower limit known up to now, which was

held by Wolf 359 at 16.5 according to Kuiper…The companion seen from a distance of one astronomical unit would have magnitude $-14$, which is very nearly the brightness of the full moon.

**Van Biesbroeck's Star** was, at the time of its discovery, the least luminous star known. Modern values from *Gaia* give a distance of 5.92 pc which, when combined with a revised apparent visual magnitude of $+17.3$, gives an absolute visual magnitude of $+18.7$. This would result in an apparent magnitude of $-12.8$ at a distance of 1 au which is indeed very close to the brightness of the full moon. The tiny red dwarf star is right at the lower end of the mass limit needed to initiate nuclear fusion and become a star rather than a brown dwarf. Only just larger than the planet Jupiter, Van Biesbroeck's Star is still one of the smallest main sequence stars ever discovered.

**Horace Welcome Babcock** (1912–2003) was always destined to become an astronomer. His father was an electrical engineer and physicist by training who was working at the newly opened Mount Wilson Solar Observatory when Babcock was born. Babcock grew up surrounded by astronomers and worked as a volunteer at the observatory during his youth. Particularly interested in astronomical instrumentation, he graduated from the California Institute of Technology with a degree in structural engineering and earned his doctorate whilst working at Lick Observatory. Following his time at Lick he worked at the McDonald and Yerkes observatories under the direction of Otto Struve. The Second World War intervened and he worked on radar at MIT and aircraft rocket launchers at Caltech. Babcock returned to astronomy in 1946 when he joined the staff at the Mount Palomar and Mount Wilson observatories. He headed the Mount Wilson grating ruling laboratory (earlier directed by his father) and developed a number of innovative instruments such as automatic telescope guiders and electronic exposure meters for spectroscopy. He also explored the idea of using adaptive optics to sharpen astronomical images, an idea which was four decades ahead of its time. His studies in stellar and solar magnetism were groundbreaking, measuring for the first time magnetic field strengths in the Sun and other stars. In collaboration with his father, he successfully demonstrated that the Sun's magnetic field reverses polarity every other sunspot cycle. He was made director of the Mount Palomar and Mount Wilson observatories in 1964 and was instrumental in choosing the site for what became the Las Campanas Observatory in Chile. He retired in 1978. Babcock was a member of the American National Academy of Sciences and the recipient of many awards, including the Eddington Medal (1958) and Gold Medal (1970) of the Royal Astronomical Society (Vaughan 2003).

The star HD 215441 is an Ap star, an A-type star with a peculiar spectrum. Babcock studied the star from 1958, taking numerous spectrograms with the Hale 200-inch telescope at Mount Palomar in 1959 and 1960. In the presence of a strong magnetic fields, the spectral lines will be split into components and this is what caught Babcock's eye (Babcock 1960):

> The A0p star HD 215441 has a magnetic field sufficiently strong and uniform to show distinct resolution of many spectrum lines into the π and σ components of the Zeeman patterns. The mean field intensity derived from four plates is +34400±266 gauss…and the magnetic pressure is sufficient to dominate over the gas pressure in the outer layers of the star.

At 34,400 gauss (3.44 Tesla in more modern SI units), this was by far the strongest magnetic field yet discovered on a non-degenerate star. By comparison, the

In 1908, American astronomer George Ellery Hale published the first evidence of extraterrestrial magnetism, specifically, in sunspots. In this image, green computer-generated magnetic field lines associated with solar features such as sunspots are created using data from the Solar and Heliospheric Observatory (SOHO) spacecraft. However, the magnetic field strength of Babcock's Magnetic Star far outstrips that of the Sun. (NASA/Goddard Space Flight Center)

magnetic field strength on the surface of the Earth varies between 0.25 gauss and 0.65 gauss; on the surface of the Sun it ranges from around one gauss at the poles to several thousand in a sunspot. Even magnetic resonance imaging (MRI) scanners on Earth tend to operate at lower field strengths than **Babcock's Magnetic Star**.

Also known by the variable star designation GL Lacertae, the magnetic star is about 500 pc distant and is four times the mass of the Sun. Although classified as type A0p, some argue that it is closer to a B spectral type. Babcock thought that the magnetic field strength varied, although not the polarity, with later researchers finding a period of approximately 9.5 days. The light output and details of the spectrum of the star also vary on this time scale. This is the rotational period of the star and not the result of any sort of pulsation. Thus, Babcock's Magnetic Star is a natural laboratory for researchers experimenting with magnetic stellar models with rotation.

Babcock discovered the first magnetic star, 78 Virginis, in 1947, and went on to compile a catalogue of magnetic stars in the 1950s. Magnetic stars continue to be discovered; most of them are type B or Ap, but Babcock's Magnetic Star remains the main sequence star with the strongest magnetic field. The origin of these massive magnetic fields, however, is a question yet to be answered.

The closest region of massive star formation to Earth, the Orion Nebula (M42 or NGC 1976) is actually visible to the naked eye as a glowing fuzzy patch south of the three bright stars marking Orion's Belt. Within the nebula is a very young open cluster known as the Trapezium and it was near this cluster in 1965 that American astrophysicist **Eric E. Becklin** (bn 1940) and German astronomer **Gerhart Neugebauer** (1932–2014) found an infrared source that had no visual counterpart (Becklin and Neugebauer 1967):

> In January, 1965, an intensity map of the Orion Nebula in the wavelength region from 2.0 to 2.4 µ[m] was made using a dual-beam photometer…On the survey seven point sources were detected which could be identified positively with photographically visible stars, and one source…was found which could not be identified.

Becklin and Neugebauer decided that, on the balance of probabilities, the infrared source was located within the nebula, rather than in front of it (making it extremely cool and dim) or behind it (making it very bright and reddened), and that it was likely a protostar of around six solar masses and with a luminosity a thousand times that of the Sun. The **Becklin-Neugebauer Object** was one of the brightest infrared objects in the sky and attracted the immediate attention of the

The Trapezium Cluster was discovered in 1617 by Galileo. It is an open cluster of young stars embedded in nebulosity, with the brightest member being fifth-magnitude $\theta^1$ Orionis C. The Becklin-Neugebauer Object was ejected from this cluster several thousand years ago. This image is a mosaic of 81 images taken at the European Southern Observatory (ESO) Very Large Telescope at the Paranal Observatory in Chile. (ESO/Mark J. McCaughrean et al/AIP))

astronomical community. Within two years the Kleinmann-Low infrared nebula was discovered in the vicinity of the BN Object, followed by other infrared sources. Investigations at other wavelengths followed, with the BN Object detected at radio wavelengths and found to be weakly variable with a period of just over 8 days in the near-infrared. The reason for the variability is not understood but could be a result of pulsation, bright star spots rotating in and out of view, or an eclipsing binary.

Another possibility is occultation by an asymmetric circumstellar disk (Hillenbrand et al 2001). Later studies have suggested that the BN Object is, in fact, a zero-age main sequence star of spectral type B, with a mass of 8–12 solar masses, a radius about four times that of the Sun, and a brightness of 10,000 solar luminosities. Furthermore, it appears to be a runaway star, ejected some 4000 years ago from the Trapezium and triggering a violent explosive outflow event about 500 years ago when it passed close by another radio source, the massive protostar Source I (Tan 2004).

Discovery of the BN Object was one of the first major finds in the field of infrared astronomy. Investigations into this star continue to make important contributions to the theories of early stellar evolution and to stellar ejection models.

American astronomer **Nicholas Sanduleak** (1933–1990) was born in New York, USA, to Romanian immigrant parents. The family soon moved to Cleveland, Ohio, where Sanduleak eventually attended Case Institute of Technology (now Case Western Reserve University), earning his doctorate in 1965. After short tenures at Kitt Peak and Cerro Tololo observatories, he joined the staff of Warner and Swasey Observatory (the observatory of Case Western Reserve University) as a research

Sanduleak's Star resides in the Large Magellanic Cloud, a satellite galaxy to our own Milky Way. This ground-based image of the LMC was taken with the Blanco 4-metre telescope at the Cerro Tololo Inter-American Observatory in Chile. (NOIRLab/AURA/NSF)

associate, a post he held for the rest of his life. Sanduleak specialised in objective prism spectroscopy and had a special affinity for the Magellanic Clouds. One of the stars that he studied, Sanduleak −69° 202, fulfilled its destiny in 1987 when it exploded as SN 1987A, thus becoming the first supernova to have spectroscopic data about it prior to its destruction. He was also a co-author of a list of Hα-emission stars, the most famous being the enigmatic SS 433. Sanduleak died unexpectedly of cardiac arrest in 1990 (Stephenson 1991).

Sanduleak spotted an interesting object whilst surveying the region in and around the Large Magellanic Cloud (Sanduleak 1977):

A new emission-line object ($\alpha = 5^h 45^m.7$, $\delta = -71°17'$, 1975), possibly associated with the Large Magellanic Cloud, appears to have shown strongly variable Hα emission…The suspected spectral variability would suggest that this is some type of eruptive variable star rather than a planetary nebula.

**Sanduleak's Star** (it also appears in the literature as **Sanduleak Anonymous** and 'LMC Anonymous') has proven to be a bit of a puzzle to astronomers. Some classify it as a symbiotic star, a binary system containing a white dwarf and a cool red giant embedded in an emission nebula, where the giant star loses mass to the white dwarf, either directly or via an accretion disk. Other astronomers point out that its spectrum is similar to those of the nebular features found near the massive hot star η Carinae or even supernova SN 1987A. Then, in 2011, a giant collimated jet was found to be emanating from Sanduleak's Star. Extending 14 parsecs, it was the largest stellar jet discovered and the first one seen beyond our galaxy (Angeloni et al 2011). What is Sanduleak's Star? It might be symbiotic, despite the lack of evidence of a cool red giant companion. It might be a massive eruptive variable. It has been observed to be slowly fading in brightness over several decades so it might be a nova, now recovering after a recent thermonuclear event, or it might be the progenitor of a future supernova. It might be something we have yet to envisage. Even today, Sanduleak's Star remains a mystery.

## Acknowledgements

This research has made use of NASA's *Astrophysics Data System Bibliographic Services*, operated at the Harvard-Smithsonian Center for Astrophysics, Cambridge, Massachusetts, USA, and the SIMBAD astronomical database, operated at CDS, University of Strasbourg, France. The author would like to thank Dr David Harper for his enthusiastic encouragement and helpful comments.

## References

Angeloni, R., Di Mille, F., Bland-Hawthorn, J., & Osip, D. J. (2011). 'Discovery of a Giant, Highly Collimated Jet from Sanduleak's Star in the Large Magellanic Cloud'. *The Astrophysical Journal Letters*, **743** (1), L8. doi.org/10.1088/2041-8205/743/1/L8

Babcock, H. W. (1960). 'The 34-Kilogauss Magnetic Field of HD 215441'. *The Astrophysical Journal*, **132** (3), 521–531. doi.org/10.1086/146960

Becklin, E. E., & Neugebauer, G. (1967). 'Observations of an Infrared Star in the Orion Nebula'. *The Astrophysical Journal*, **147**, 799–802. doi.org/10.1086/149055

Crowther, P. A. (2007). 'Physical Properties of Wolf-Rayet Stars'. *Annual Review of Astronomy and Astrophysics*, **45**, 177–219. doi.org/10.1146/annurev.astro.45.051806.110615

Hardie, R. H. (1974). 'Georges Van Biesbroeck (1880–1974)'. *Journal of the Royal Astronomical Society of Canada*, **68**, 202–204.

Hillenbrand, L. A., Carpenter, J. M., & Skrutskie, M. F. (2001). 'Periodic Photometric Variability in the Becklin-Neugebauer Object'. *The Astrophysical Journal*, **547**, L53–L56. doi.org/10.1086/318884

Merrill, P. W. (1938). 'A Wolf-Rayet Star with High Velocity'. *Publications of the Astronomical Society of the Pacific*, **50** (298), 350–351. doi.org/10.1086/124982

Sanduleak, N. (1977). 'A Suspected Variable Emission-line Object in the Direction of the Large Magellanic Cloud'. *Commission 27 of the I.A.U. Information Bulletin on Variable Stars*, No. 1304.

Stephenson, C. B. (1991). 'Nicholas Sanduleak, 1933–1990'. *Bulletin of the American Astronomical Society*, **23** (4), 1491–1492.

Tan, J. C. (2004). 'The Becklin-Neugebaur Object as a Runaway B Star, Ejected 4000 Years Ago from the $\theta^1$ Orionis C System'. *The Astrophysical Journal*, **607**, L47–L50. doi.org/10.1086/421721

Toalá, J. A., Oskinova, L. M., Hamann, W.-R., Ignace, R., Sander, A. A. C., Shenar, T., Todt, H., Chu, Y.-H., Guerrero, M. A., Hainich, R., & Torrejón, J. M. (2018). 'On the Apparent Absence of Wolf-Rayet+Neutron Star Systems: The Curious Case of WR124'. *The Astrophysical Journal Letters*, **869** (1), L11–L15. doi.org/10.3847/2041-8213/aaf39d

van Biesbroeck, G. (1944). 'The star of lowest known luminosity'. *The Astronomical Journal*, **51**, 61–62. doi.org/10.1086/105801

Vaughan, A. H. (2003). 'Horace Welcome Babcock, 1912–2003'. *Bulletin of the American Astronomical Society*, **35** (5), 1454–1455.

Wilson, O. C. (1964). 'Paul Willard Merrill 1887–1961'. *Biographical Memoirs*, National Academy of Sciences (Washington, DC).
nasonline.org/wp-content/uploads/2024/06/merrill-paul.pdf

# Hawking Stars

## Andrew D. Santarelli and Matthew E. Caplan

Have you heard the song *Black Hole Sun* by Soundgarden come on at the pub, and wondered with your mates what would happen if there was literally a black hole in the Sun? If you haven't had this experience, you could at least imagine how the conversation would go. But I guarantee you that the conversation is completely different when your mates have PhDs in astrophysics, and that one song could turn an otherwise ordinary conversation into a search for the solution to some of the biggest problems in astrophysics.

In 1971 it occurred to Stephen Hawking that the early universe, filled with plasma denser than the cores of today's neutron stars, may have been an ideal location for black holes to form. These 'primordial black holes' would have formed in the first fractions of a second, even before the first protons, and could have masses anywhere from trillions of times larger, or smaller, than our Sun. So who's to say a small one, if it exists, couldn't find its way into the centre of a star? Hawking did indeed do the maths on it; if such black holes are abundant, one might even expect one as massive as an asteroid in the centre of the Sun. Such a black hole would be tiny, smaller even than an atom.

This seemingly outlandish idea captured the attention of many astrophysicists as it offered a solution to a major problem in solar physics. As we understood it, the core of the Sun fuses hydrogen into helium at a rate high enough to power itself under the pressures and temperatures that we knew it had. As a by-product, these nuclear reactions shower an immense number of neutrinos into space, whizzing straight out through the sun and washing over the solar system. As a literal rule of thumb, about a hundred billion of these solar neutrinos pass through your thumbnail every second like light through glass. This is nothing to worry about, neutrinos react with other particles so rarely that you're unlikely to have even one neutrino collision in your body in a decade.

Measuring and counting these neutrinos proved hard in the late-1900s but was the key to confirming that our model of the sun's core was correct. However, in the mid-1960s the first neutrino detectors were born and the measured number of neutrinos was not quite the flood that was predicted. In fact, it wasn't even close, at almost exactly one-third of the expected value. One proposed solution to this 'solar neutrino problem' was that there is another source of energy within the sun in addition to fusion that doesn't emit neutrinos. When Hawking's primordial black hole proposal came along a few years later, it provided a neutrino-less means of making the other two-thirds of the Sun's energy.

You're probably thinking, "Wouldn't a black hole suck energy out of a star, not give it more?", but the opposite is actually true. Black holes are only black when they are alone. If you feed them, they become some of the brightest things in the universe. While light and energy are lost when crossing the event horizon, matter approaching but just outside of it is accelerated to tremendous speeds, close to the speed of light. Friction between infalling streams drives the matter around the black hole to extreme temperatures that shine brilliantly with high energy radiation. A small black hole at the centre of the Sun would certainly feed on some matter but in the process it would produce an enormous amount of radiation. Enough to,

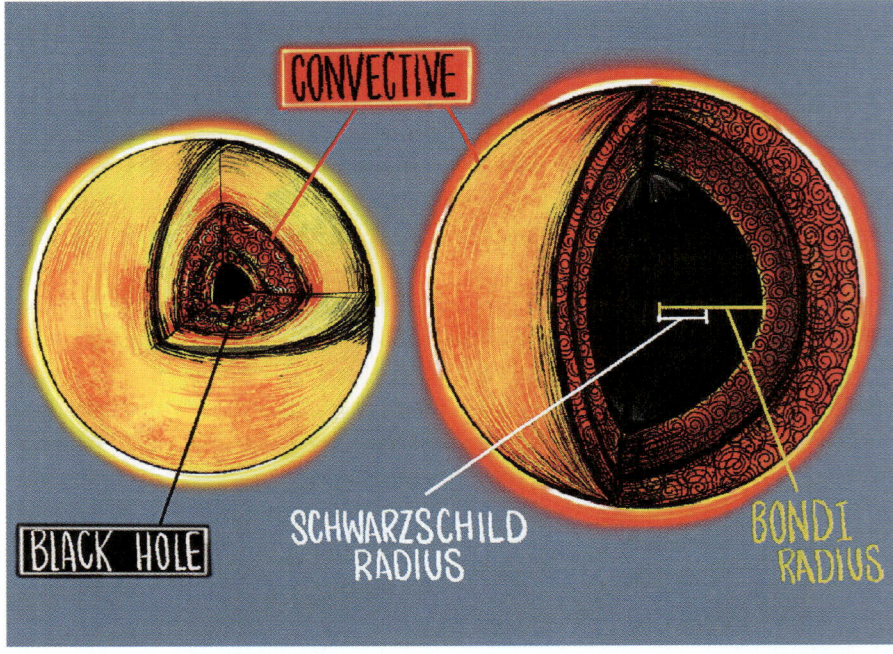

Here we show the Hawking star cross section and compare between early and late lifetimes. In the left figure, earlier in the star's life, we see that the normally radiative core is now convective due to the presence of the black hole. Then, as the black hole grows later in the star's life, the star becomes completely convective throughout and puffs up into a red straggler. (Illustration by Naomi Satoh)

say, compensate for fusing less hydrogen and thus emitting fewer neutrinos. Great, we've solved the solar neutrino problem!

Well, not quite.

The Sun is actually producing exactly as many neutrinos as we originally expected, although our neutrino detectors just were not advanced enough yet to pick up the other two types in the 1960s. So the Sun is fusion powered, and there is no need for a black hole engine in its core. But that doesn't rule out the fact that a primordial black hole could wind up in the centre of a star and go completely unnoticed, and in the past decade the interest in primordial black holes has once again soared.

Primordial black holes could be the solution to one of the biggest outstanding problems in astrophysics: dark matter. It turns out we can't see the vast majority of the matter in the universe, and the most common explanation is that it is made up of some kind of new, undiscovered electrically neutral particle. The electric neutrality would explain the darkness without electric charge there is no interaction with electromagnetic radiation, meaning light, and such a particle would be invisible to us (just like neutrinos!). However, primordial black holes present an alternate explanation, as they easily pack colossal amounts of mass into easy to miss specks. This idea has been around for a while, giving astronomers plenty of time to look for, and rule out, the existence of primordial black holes of many different masses. Today, there is still one mass window that hasn't been constrained and it is that of medium to large asteroids, coincidentally the exact masses that Hawking thought could get captured in stars. If dark matter does consist of primordial black holes in this mass range, there would be so many of them that suggesting they could end up in the Sun wouldn't get you too many strange looks.[1] Thought about this way, stars could be used as dark matter detectors, so long as that dark matter is primordial black holes. This brings us back to our question: what happens if there's a tiny black hole in the centre of a star? We have understood how stars evolve for decades now: a massive gas cloud collapses on itself due to its own gravity until a core with a suitably high pressure and temperature is formed and begins to undergo hydrogen fusion. After billions of years, all of this fuel gets used up and the star goes through a series of death-throes. For our sun, this means puffing up into a red giant, fusing helium, and shedding the gas around its core leaving behind a 'dead' white dwarf.

---

1. There are other motivators as well, such as gravitational waves. The merging black holes seen by the Laser Interferometer Gravitational-Wave Observatory (LIGO) are likely produced in supernova; however, some observations could be explained by some mergers being primordial black holes.

When adding in a black hole to the mix it can only end up in a star if it is initially in the gas cloud that forms the star, sinking to the centre of the star shortly after it was born. Primordial black holes with the mass of a medium sized asteroid are about the size of an atom, and so would have little effect on the formation or early life of the star. As the black hole starts feasting on the matter around it, the energy created begins to radiate outward and creates a pocket about the size of the Earth where things begin to behave differently from normal. While this region is normally mostly static, the newly added radiation causes it to whirl and seethe as it becomes convective. This region gets pushed out as the black hole grows and, eventually, devours the entire star. This doesn't happen as rapidly as one might think though.

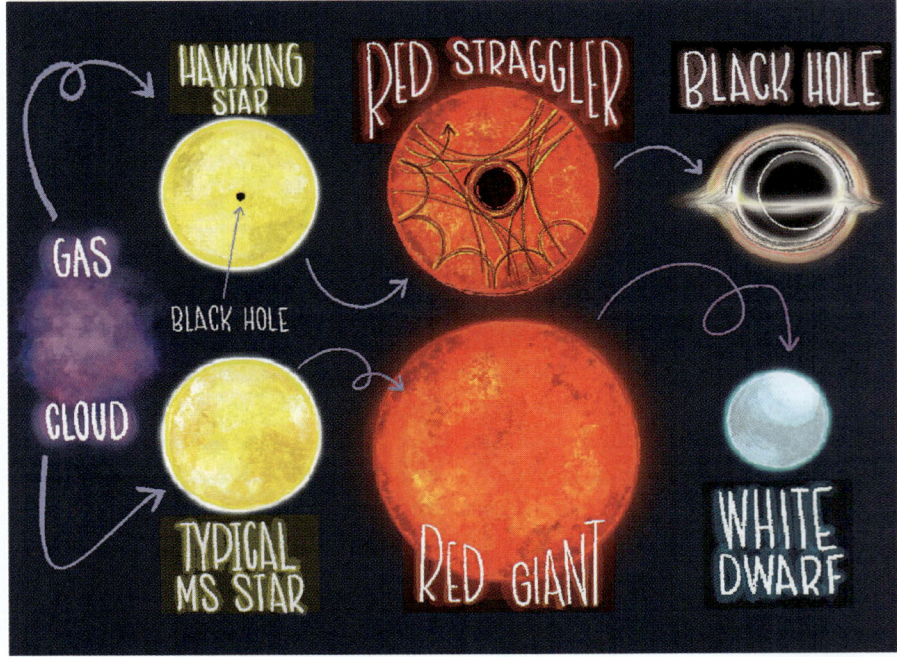

Here we see the evolution of a Hawking star (top) versus that of a typical star (bottom). Beginning with a cloud of gas, the stars appear to be the exact same. It is not until they begin their post-main sequence lifetimes where things change. The Hawking star becomes a red straggler, and is less bright and cooler than the red giant that the typical main sequence star becomes. At the end of their lives, the Hawking star becomes a lower mass black hole than should be allowed by typical stellar evolution. The typical red giant instead sheds its outer layers and becomes the dense white dwarf one would expect. (Illustration by Naomi Satoh)

The effects of this outward radiation are actually two-fold. Not only does it create a new convective zone in that earth-sized pocket, it also fights back against the black hole feeding frenzy. Additionally, the tiny size of the black hole also severely limits its growth rate. An atom sized black hole can only consume a couple hundred tonnes of matter per second. This may seem like a lot but is it is peanuts in comparison to the full mass of the Sun. For reference, the Sun loses about a million tonnes per second due to solar wind, and even that is mostly negligible over the lifetime of the star. Eventually though, the black hole will grow to a more substantial size, eating faster as it grows larger. This also means that it begins to emit more and more energy, which does put a damper on the black hole's accretion rate but leads to it becoming a bigger player in the energy production of the star.

In a few billion years, depending on the exact mass of the star and initial mass of the primordial black hole, the black hole will have grown to be about the mass of Uranus while only being a handful of centimetres across. While small in size, it is at this point that the black hole takes centre stage. The energy produced from the infalling matter is now greater than that put out by the star's typical fusion processes. This excess energy disrupts the delicate balance supporting the star, and ultimately is responsible for the star evolving into something completely different – a 'Hawking star'.

The star's outer shell now bloats outward resembling a 'normal' star puffing up to enter its giant phase, except several billion years too soon. But the expansion of a Hawking star is very different than in a typical giant. While a typical giant has a very dense helium burning core in the very heart of an otherwise sparse ball of gas, the Hawking star has no such inner structure. Deep inside, the heat and energy from the black hole has pushed the star's core out, mixing it away into the gas of the outer layers. Ultimately, this stops fusion all together making the Hawking star entirely black hole powered.

The expansion also further stalls the black hole's feeding, causing the Hawking star to not quite get to the peak size of a typical dying star. It will only reach about five times the size of the Sun as opposed to 100 times for a proper red giant, and only about 10 times brighter as opposed to 1,000 times. Instead, it would look like a very rare star called a sub-subgiant or red straggler. Over billions of years the planet-massed black hole feasts, growing from the size of a coin to the size of a small city when it has finished consuming its star.

This is of course the result of simulations and calculations. Hawking stars are, at present, hypothetical. We don't know if Hawking stars or primordial black holes exist. For this to be science, we must test the hypothesis. We must go looking for

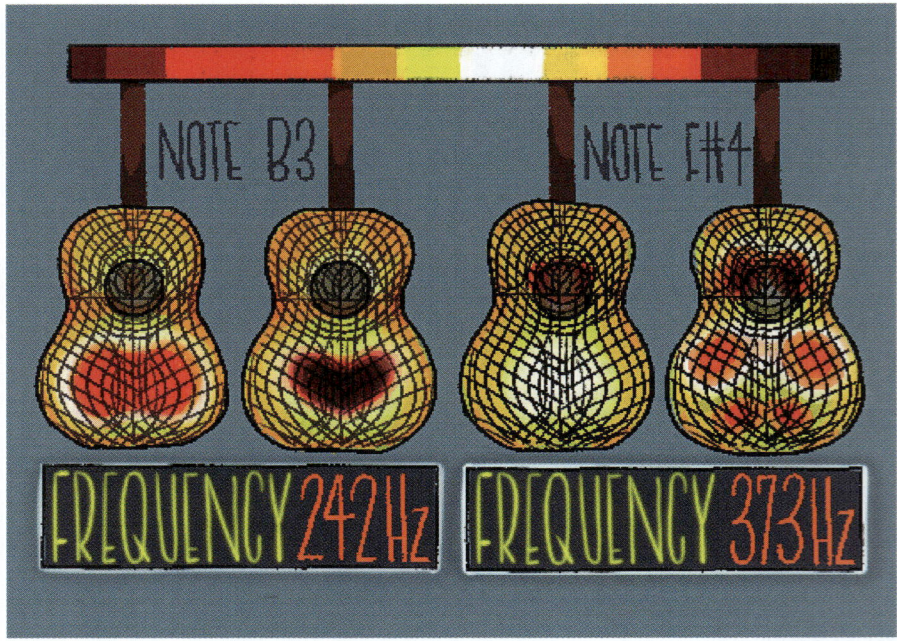

Just like guitars, stars vibrate and oscillate in different ways, resulting in different frequencies and modes. In a guitar, each note makes the guitar vibrate in a different way, and these vibrations are what create the music we enjoy. For a star, we don't hear the music it plays but we are still able to observe their oscillations. (Illustration by Naomi Satoh)

Hawking stars in the sky, and find ways to distinguish stars with black holes at their cores from sub-subgiants with more mundane origins. To find a Hawking star is to solve the mystery of dark matter, but to find none is also valuable as it allows us to close the book on primordial black holes as dark matter.

It is hard to peer inside a star to know if it is fusion powered or black hole powered, but not impossible. Returning to our Sun, we know that the black hole would have to be less massive than Uranus as that is the mass threshold for the Hawking star to transition into its giant phase. It would also need to be small enough to not account for too much of the total energy output, as a lower fraction coming from fusion would mean that not enough neutrinos are being produced to match our measurements. The uncertainty in current measurements would allow for up to 1/1000 of the Sun's energy to come from a non-fusion source. Taking this into account along with the fact that the temperature and pressure changes from the black hole would affect fusion rate, crunching the numbers shows that the black hole would have to be less than the mass of Mercury for us to not have noticed its effects on neutrino output.

The odds of our Sun being a Hawking star are admittedly slim to none, but that's not to say that one of the hundreds of billions of stars in our galaxy isn't. Assuming primordial black holes exist in the correct mass window, at least some of these stars would be entering their black hole powered red straggler stages. From the outside, measurements of their luminosity, temperature, and radius are good for finding candidates, but it's not enough to verify our exotic origin story. For that we need more detailed observations.

An obvious place to start is a measurement of the star's chemical composition with its spectrum. Since the Hawking star mixes the formerly fusing core into the outer layers of the star, an obvious signature is an excess of helium in the spectrum and is a relatively straightforward observation to make. This only gives us the surface though, and to peer inside a distant star we need to measure one final thing: the sound of the star.

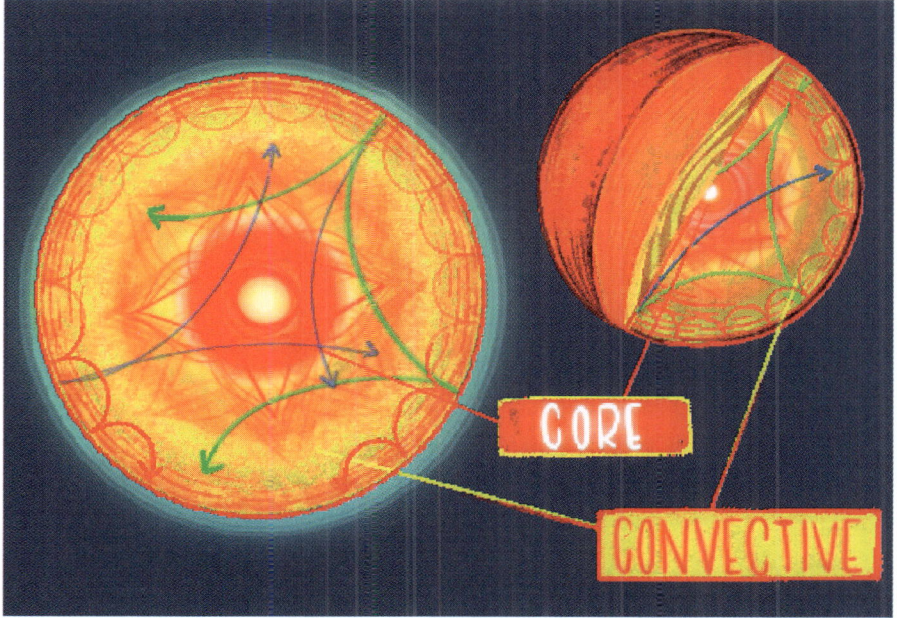

Here we take a look at the cross section of a typical star and show the vibrational modes within. The most important of these that are the most promising in distinguishing Hawking stars are the G-modes, represented by the flower petal-like lines around the core of the star. (Illustration by Naomi Satoh)

All stars vibrate. As the plasma inside churns and lurches, different waves bounce around the stellar interior and cause global vibrations much like a guitar. These vibrations dictate the sound of the guitar, and are defined by the structure: the shape, the thickness and hardness of the material, the size, etc. The same is true for stars, each subtly vibrating on its surface at hundreds of different frequencies determined by the exact structure deep inside. Once measured, these vibrations can be reverse engineered to reconstruct a map of their inner structure. This concept is at the heart of an entire field called 'asteroseismology'. A Hawking star would have a distinct asteroseismological signature, both from the additional vibrations caused by the sloshing powered by the black hole and also the absence of the vibrations characteristic of a star with a dense fusing core.

Edison famously found hundreds of ways not to make a light bulb, and astrophysics is similar. Science, and astrophysics, often works by process of elimination. Most hypotheses are wrong and will die under testing, but we test them anyway. Eventually there will be a hypothesis that our experiments and observations fail to kill because that hypothesis may be the right one. In either case, a search for Hawking stars will yield an interesting result. With non-detection, we have yet another limit on the number and mass of primordial black holes, potentially ruling them out as a dark matter candidate entirely. But more optimistically, with detection of a Hawking star we know that primordial black holes are real and that they could be the solution to the dark matter mystery.

Who would have thought that Soundgarden was on to something the entire time?

# Subrahmanyan Chandrasekhar and Professor A. S. Eddington

## David M. Harland

Subrahmanyan Chandrasekhar was born on 19 October 1910 into a Tamil family in Lahore, the capital of the province of Punjab in North India (now Pakistan). His name, which meant 'Moon' in Sanskrit, was informally shortened to 'Chandra'.

His father, Chandrasekhara Subrahmanya Ayyar (known as Ayyar), was the eldest member of his generation. After excelling at *his* father's college and then at Presidency College in Madras, Ayyar took the All India Examination to gain entry into the Indian Audits and Accounts Office of the government. His posting to the railways demanded a lot of travelling across the subcontinent.

When Chandra was eight years old the family returned to its roots in the city of Madras (now Chennai), capital of Madras Province (now known as Tamil Nadu) in South India, where he enjoyed the tropical climate of the Bay of Bengal.

Presidency College, Madras, seen here shortly after its construction. (Wikimedia Commons/ Illustrated London News)

It was a Tamil Brahmin household which in the Hindu caste system was upper class, well to do, and valued education. Chandra was eldest of ten children born to Sitalakshmi Balakrishna. As a youngster he was closer to his mother than to his authoritarian father. It was the norm in those days for a family of such status to provide early education in the home. Chandra's parents soon became aware that he was exceptionally bright, particularly in mathematics. His grandfather was Ramanathan Chandrasekhar, a professor of mathematics in Visakhapatnam, several hundred kilometres north of Madras. He died the year that young Chandra was born, leaving a library of books on mathematics (some written by himself) which Chandra devoured.

The family resolved to make a permanent home in Madras in 1923 by beginning construction of a mansion in the neo-colonial style in an affluent suburb of the city. Home tutoring continued until the age of 12, following which he attended a nearby Hindu High School until 15. After that, two years earlier than usual due to his mathematical skills, he was accepted by Presidency College. Soon after being established in 1840, Presidency College was made the nucleus of the University of Madras. It was the best college in South India, with most of the teachers being British. In tests, Chandra's answers tended to be more detailed than his lessons, with the result that he usually scored in excess of 100%. He could read mathematical expressions as easily as prose, and absorb text books in a single reading. After two years, he opted for a course that would result in an honours degree in physics.

In the summer of 1928 Chandra visited his uncle Raman, who was a professor of mathematics at the University of Calcutta (now Kolkata) in West Bengal. Raman lent him a copy of *The Internal Constitution of the Stars* by Arthur Stanley Eddington of Cambridge University. Published in 1926, the book was a review of what was known, or suspected, about the physics of stars. Chandra was fascinated by the process of 'radiative transfer' as atoms absorb and emit light, and the fact that this generates a pressure that helps a star to stabilise itself against the inward pressure of gravity.

English astrophysicist Arthur Stanley Eddington. (Wikimedia Commons/ Cambridge University)

Shortly thereafter, back home, Chandra learned that the renowned German physicist Arnold Sommerfeld was in Madras during a lecture tour. In 1919 Sommerfeld had published *Atomic Structure and Spectral Lines*, and, by chance, Chandrasekhar had read the English translation. Emboldened despite being only 17 years old, he paid the man a visit. Sommerfeld gave him a crash course on Quantum Theory, emphasising the fact that in certain circumstances electrons can create a pressure to resist compression. This was a consequence of the Exclusion Principle discovered in 1925 by Wolfgang Pauli, who was one of Sommerfeld's former students.

Combining his recent reading influences, Chandra set himself the task of figuring out how light (radiation), electrons, and atomic nuclei interact in a star. He made such rapid progress that later in the year he published his first research paper.[1] In January 1929 he presented this work at a meeting of the venerable Indian Science Congress in Madras, where his uncle Raman was presiding. Chandra was 18 years old, still only in his second year at college, and had neither specialised tuition nor supervision; a fact the audience of professional scientists appreciated.

In the spring of 1929 Chandra happened across a research paper in the *Monthly Notices of the Royal Astronomical Society* by Ralph Howard Fowler at Cambridge, published in 1926, which considered whether the pressure of electrons could assist a very dense type of star known as a white dwarf in resisting gravitational contraction. Chandra boldly sent Fowler a copy of his *Indian Journal* paper. Fowler replied, making some suggestions. Chandra revised the paper and sent it back to Fowler, who arranged for its publication by the Royal Society of London.[2] Chandra followed up the next year with a paper in *Philosophical Magazine*.[3]

When another of Sommerfeld's former students, Werner Heisenberg, now at Leipzig University, visited Madras in October 1929 on a lecture tour, Presidency College asked Chandra to show their esteemed visitor around. It gave the youngster an opportunity to discuss the Uncertainty Principle that Heisenberg had discovered two years earlier.

In January 1930 Chandra attended a meeting of the Indian Science Congress where he met Meghnad Saha, the Indian scientist who had, a decade earlier,

---

1. Chandrasekhar, S., (1928), 'The Thermodynamics of the Compton Effect with Reference to the Interior of Stars', *Indian Journal of Physics*, **3**, 241–250.
2. Chandrasekhar, S., (1929), 'The Compton Scattering and the New Statistics', *Proceedings of the Royal Society* (A), **125**, 231–237.
3. Chandrasekhar, S., (1930), 'The Ionisation Formula and the New Statistics', *Philosophical Magazine*, **9**, 292–299.

Subrahmanyan Chandrasekhar at Cambridge in 1934. (University of Chicago/CXC/NASA)

undertaken work on how ionisation in the atmospheres of stars influences their spectra.[4]

Quite remarkably, the college student who was regarded as a genius was meeting leading physicists during visits to Madras, and soaking up their ideas. By the spring though, he had to put his personal activities aside in order to concentrate on his final examinations. He passed with distinction, of course.

Being educated under the British Raj led naturally to his desire to study in England. Even prior to sitting his finals, he was unofficially informed that a scholarship to Trinity College at Cambridge University was reserved for him.

Spending several days in Bombay (now Mumbai) awaiting his ship, Chandra, now 19 years old, gave a lecture at the Royal Institution of Science.

The *Lloyd Triestino* sailed on the afternoon of 31 July 1930. To keep himself occupied, Chandra had a stock of books and research papers. Conditions crossing the Arabian Sea were hot, so he spent most of his time in a recliner on deck.

On re-reading Fowler's 1926 paper on the role of an electron gas in white dwarfs, he was surprised that even though Eddington's book had explained the procedures for calculating the internal temperatures and pressures of a star in terms of its

---

4. Saha, M. N., (1921), 'On a Physical Theory of Stellar Spectra', *Proceedings of the Royal Society*, **99(A)**, 135–153.

surface temperature, mass and radius, neither Eddington nor Fowler had done this for Fowler's solution for a white dwarf. Since it was a trivial task, Chandra made the necessary calculations and found the density at the core of a white dwarf to be a million times that of water. This high density prompted him to wonder how fast the electrons might be travelling. Fowler had not said, so, just as he often did, Chandra improved someone else's work. Heisenberg's Uncertainty Principle meant that the more the positions of the electrons were constrained by squeezing them together, the greater would be the uncertainty of their velocities. Chandra found that in a white dwarf the electrons would be travelling at speeds that were a substantial fraction of light itself. It would be necessary to incorporate the Theory of Special Relativity discovered by Albert Einstein in 1905, whereby the masses of 'relativistic' electrons would be greatly increased.

By *not* allowing for Special Relativity, Fowler's calculations were an approximation (in the sense that Newtonian physics addresses objects which are not travelling at speeds approaching light). In exploring how this might alter the outcome, Chandra discovered that there would be an upper limit to the mass of a white dwarf. To calculate the *value* of the limiting mass would require a detailed analysis involving

Great Court, Trinity College, Cambridge University, pictured here in around 1870. (Wikimedia Commons/William Winfield/Kimberly Blaker)

a great many factors. That would have to wait. What was significant was there *was* a limiting mass for a white dwarf. And, of course, it begged the question of what would happen to a star that exceeded this mass.

After the ship docked at Genoa, Italy, Chandra travelled by train to France, then by ferry to England, arriving in London on 19 August.

Eddington, who was appointed the Plumian Professor of Astronomy and Experimental Philosophy at Cambridge in 1915, now led the field in astrophysics and considered himself to be the expert on white dwarf stars. By claiming there was a maximum mass for such stars, Chandra had upset the applecart.

## The White Dwarf Dilemma

In the decades spanning the end of the nineteenth century and the start of the twentieth century, the Harvard College Observatory directed by Edward Charles Pickering assembled a vast collection of photographic plates of stellar spectra. This enabled it to create a classification system with the letter sequence OBAFGKM. Other observatories had been determining distances to stars using a variety of methods. By knowing the apparent brightness of a star on the sky and its distance, it was a simple matter to calculate its true luminosity.

Henry Norris Russell at Princeton plotted all the stars for which he had absolute magnitudes (i.e. luminosity) and spectral types (seen as a proxy for temperature).[5] The stars were strongly concentrated along the line that sloped from hot-and-luminous to cool-and-

Henry Norris Russell in 1913. (Wikimedia Commons/The World's Work/Princeton University)

faint which became known as the 'main sequence'. There was also a populated zone above this line. Some years earlier, the Danish astronomer Ejnar Hertzsprung had discovered that red stars come in two forms: highly luminous 'red giants' and very

---

5. Russell, H. N., (1914), 'Relations Between the Spectra and Other Characteristics of the Stars', *Popular Astronomy*, **22**, 275–294.

dim 'red dwarfs'.[6,7] The Hertzsprung-Russell diagram, as this plot became known, was a major advance in the study of stellar evolution.

As more stars were added to the plot, it was discovered that there were several stars of type 'A' situated far below the line, one being the very faint companion to Sirius in the constellation of Canis Major. Its existence was inferred from periodic 'wiggles' in the proper motion of Sirius. It was a binary system. Despite the primary being the brightest star in the night sky, on 31 January 1862 Alvan Graham Clark observed the companion visually while testing an 18.5-inch refractor, the largest such telescope at the time. He reported it to George Phillips Bond at the Harvard College Observatory who, after additional observations, made the official announcement.[8] Walter Sydney Adams at the Mount Wilson Observatory in California found in 1914 that the two stars in the system

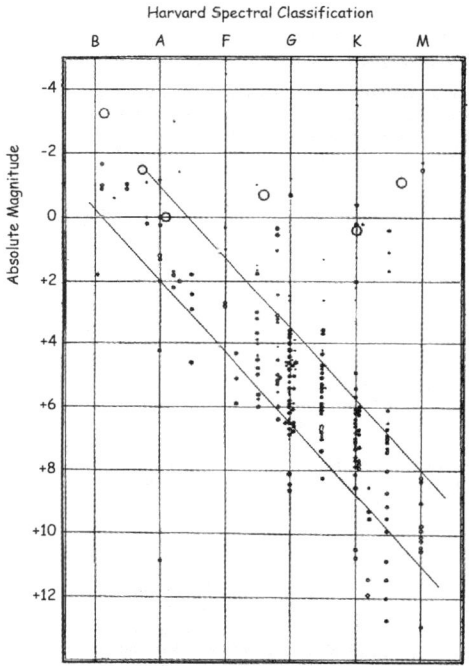

The stars plotted by Russell in terms of absolute magnitudes and spectral types. (Based on Figure 1 of his paper 'Relations Between the Spectra and Other Characteristics of Stars' (1914), *Popular Astronomy*, **22**, 285)

had similar spectra, implying their temperatures were comparable.[9] Yet the low luminosity of the companion meant it must be very small. The parameters of the orbit enabled its mass to be calculated, revealing an incredibly high density. By some mysterious process, a mass similar to the Sun was occupying a volume comparable to Earth!

---

6. Hertzsprung, E., (1908). 'About the Stars of the Subdivisions "c" and "ac" According to the Spectral Classification of Antonia C. Maury', *Astronomische Nachrichten*, **179**, 373–380.

7. H. N. Russell, H. N., (1913), '"Giant" and "Dwarf" Stars', *The Observatory*, **36**, 324–329.

8. Bond, G. P., (1862), 'Letter to Editor', *Astronomische Nachrichten*, **58**, 85–90.

9. Adams, W. S., (1915), 'The Spectrum of the Companion of Sirius', *Publications of the Astronomical Society of the Pacific*, **27**, 236–237.

An artist's impression of the Sirius system. The primary star (blazing on the left) is on the main sequence with a mass twice that of the Sun, a temperature of 9,940 K, and a luminosity 25 times that of the Sun. Although the mass of the white dwarf companion (the smaller blue star to the right) is comparable to the Sun and its temperature is 25,200 K, the fact that its diameter is similar to that of Earth means it is only 0.06 as luminous as the Sun. If it was on its own, it would appear in the sky as an 11th magnitude star. (NASA/ESA/STScI/G. Bacon)

The term 'white dwarf' for such a star was coined in 1922 by Willem Jacob Luyten, who was at that time working at the Lick Observatory in California.[10]

In 1917 Eddington realised that the environment within a star must be so intense that 'radiation pressure' must be acknowledged.[11] That is, a stable star represents a delicate balance between the inward pressure of gravitation and the *combined* outward pressures of the gas consisting of atoms and ions and the radiation (photons of light) that are absorbed and emitted by the electrons.

One of Eddington's most significant discoveries about stars was the mass-luminosity relationship, which states that the more massive a star, the more

---

10. Holberg, J. B., (2005), 'How Degenerate Stars Came to be Known as White Dwarfs', *Bulletin of the American Astronomical Society*, **37**, 1503.
11. Eddington, A. S., (1917), 'The Radiation of the Stars', *Nature*, **99**, p. 445.

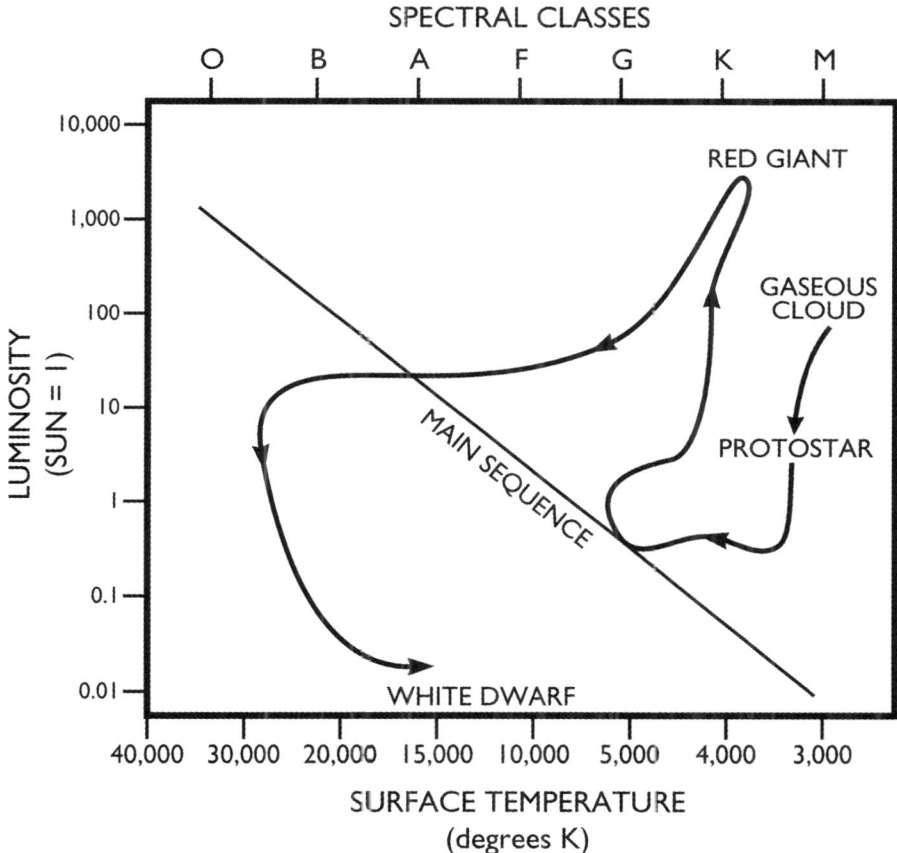

When a star like the Sun is formed by the collapse of an interstellar cloud of gas and dust it sits on the main sequence. After billions of years it evolves into a red giant with a cool surface and high luminosity before ending its life as a hot but faint white dwarf. (Based on Figure 3 of the paper 'Stellar Evolution and Social Evolution: A Study in Parallel Processes' (1995) by Robert L. Carneiro, *Vistas in Astronomy*, **39**, #4, 711)

luminous it is.[12] To his surprise, it applied not only to stars on the main sequence but also to the giants, implying their interiors could all be described by the 'perfect gas law' that links pressure, volume and temperature.

---

12. Eddington, A. S., (1924), 'The Relation Between the Masses and Luminosities of the Stars', *Nature*, **113**, 786–788.

At the time of Eddington's 1926 book only a few white dwarfs were known, but he believed they marked the end point of stellar evolution and hence must be abundant. It did not seem likely that the perfect gas law would apply to their interiors. Whatever law was applicable was a mystery.

In wondering how a white dwarf could have such a high density as to shrink it to a radius comparable to Earth, Eddington proposed that in such an environment virtually every atom would have been stripped of its electrons. Since the nucleus of an atom is so tiny, most of its volume is occupied by the orbiting electrons. When atoms are fully ionised, the nuclei can be packed together more closely before their positive charges generate a repulsive force. In that case, a much greater density could be achieved. He speculated that as a star grows old, its radiation pressure diminishes and the task of resisting gravity is increasingly borne by the gas pressure, causing the star to progressively contract. This posited a dilemma. Could the density attain a value sufficient to cause the star to succumb to gravity? Eddington dismissed as ridiculous the possibility of a star collapsing in on itself. He thought the inevitable fate would be an inert solid body, a 'dead star'.

Shortly after Eddington published his 1926 book, Fowler, who was aware that an incredibly dense gas of electrons would *not* behave as a perfect gas, suggested that applying Quantum Theory might rule out the possibility of a white dwarf completely collapsing. In such a gas, the negatively charged electrons would remain as far apart as possible. In 1925 Wolfgang Pauli found that electrons cannot be packed together arbitrarily closely. To be precise, Pauli's Exclusion Principle states that no two electrons can share the same 'quantum state'. Taking this into account Fowler realised that, as the density increases, the electrons will accommodate compression until they occupy all of the lowest energy states. With the electrons so constrained that they are no longer able to interact by absorbing and emitting light, there is no radiation pressure. Eddington had presumed that at this point the task of resisting gravity would have to be borne by the gas pressure alone. But at precisely this point in stellar evolution the electrons exert a different kind of pressure that resists further compression. An electron gas in which the electrons occupy all of the lowest energy states is 'degenerate', and compression is resisted by 'degeneracy pressure'. Because this pressure is the result of the higher kinetic energy of the electrons, it depends only on density, not, as in the case of a perfect gas, on temperature.[13]

---

13. Fowler, R. H., (1926), 'On Dense Matter', *Monthly Notices of the Royal Astronomical Society*, **87**, 114–122.

By eliminating gravitational collapse, Fowler's analysis meant white dwarfs would continue to shine to a grand old age.[14]

Then came Chandra, pointing out in 1930 that Fowler's method was flawed because there was an upper limit to the mass of a white dwarf, and a larger star would indeed collapse. At Cambridge, he lost no time in writing up his discovery. When he showed his paper to Fowler (his supervisor) in November it garnered a lukewarm response. Regardless, Chandra sent it off for publication.[15] He followed up with papers on the detailed physics right through to 1934.[16] Although Eddington provided him with a powerful calculator to help in determining the precise value of the limiting mass, Eddington seemed to be indifferent to the line that Chandra was pursuing.

## Chandra's Contentious Presentation

On 11 January 1935 the 24-year-old Chandra gave a presentation to the Royal Astronomical Society in London. His audience comprised 100 distinguished scientists in tiered rows, the leading lights seated at the front. This wasn't Chandra's first presentation to that august body, but on earlier occasions he had been plugging holes in the work of others. This time he was to discuss a topic of his own, namely the mass limit on white dwarfs.

Chandra explained his reasoning since the epiphany in 1930 through to his latest calculations, and then concluded: "The life of a star of small mass must be essentially different from that of a star of large mass. ... For a star of small mass the natural white dwarf stage is an initial step towards complete extinction. A star of large mass cannot pass into the white dwarf stage and one is left speculating on other possibilities."

The next speaker was Eddington. Chandra knew his topic was to be white dwarfs but had no idea what the great man intended to say.

In fact, Eddington vociferously rejected Chandra's assertion that the electrons in the degenerate core of a white dwarf were relativistic. If there was no maximum mass for a white dwarf then there could be no possibility of a collapse. Fowler's results were sufficient and Chandra's calculations had no physical significance.

---

14. It is only the interior of a white dwarf that is degenerate, the outer layers remain a perfect gas. Even as the electrons in the core cease to emit radiation, the hot surface remains luminous.
15. Chandrasekhar, S., (1931), 'The Maximum Mass of Ideal White Dwarfs', *The Astrophysical Journal*, **74**, 81–82.
16. Chandrasekhar, S., (1934), 'Stellar Configurations with Degenerate Cores', *The Observatory*, **57**, 373–377.

Chandra was devastated by this forthright condemnation because at no time at Cambridge had the world's leading astrophysicist privately expressed his doubts.

But that was typical of Eddington. Aware of his high standing in the astrophysics community he was arrogant, domineering, and intolerant of anyone who disagreed with him, irrespective of their standing. And of course the best knock downs were delivered in public. Attendees at the Society meetings were eager to hear how Eddington would react to a presentation. For the 'old hands' it was fun, but for a youngster it was vicious.

## Postscript

After finishing at Cambridge, Chandra took a train to Liverpool on 29 November 1935 and the following day sailed for America aboard the *White Star Britannica*. On arriving in Boston on 8 December he was met by Harlow Shapley, Director of the Harvard College Observatory, who had invited him to spend some time there.

In February 1936 Otto Struve, Director of the Yerkes Observatory of the University of Chicago, offered Chandrasekhar a research position, pointing out: "It seems to me your brilliant theoretical work could be made even more valuable to astronomers if it could be combined with practical investigations by observers."

As events transpired, Chandrasekhar spent the rest of his long career at the University of Chicago. He wrote research papers and definitive text books. He received many awards, including sharing the 1983 Nobel Prize for Physics "for his theoretical studied of the physical processes of importance to the structure and evolution of the stars". He died in 1995 of a sudden heart attack.

Of course, by that time Chandra's assertion that there was an upper limit to the mass of a white dwarf had been proven. It was difficult to calculate the precise value because it depended on many factors. The accepted value of 1.4 solar masses first appeared in a paper by Soviet-American theoretical physicist George Gamow.[17]

So stellar collapse does indeed occur, and because not even light can escape we call the result a 'black hole'; a term introduced into popular lexicon in 1967 by Princeton physicist John Archibald Wheeler. The first black hole to be located is contained in the binary system Cygnus X-1.[18] Furthermore, we have found black holes of several billion solar masses lurking in the cores of galaxies.[19]

---

17. Gamow, G., (1939), 'Physical Possibilities of Stellar Evolution', *Physical Review*, **55**, 718–725.
18. See my article 'The First Known Black Hole' in *Yearbook of Astronomy 2020*.
19. See my article 'Supermassive Black Holes' in *Yearbook of Astronomy 2018*.

In his excellent account of Chandra's 1935 presentation to the Royal Astronomical Society, Arthur I. Miller suggests that Eddington ought to have stood up, said that in his 1926 book he had identified the prospect of a star collapsing but lacked a mathematical proof. Now this brilliant student from his own university had devised the proof. But to his death in 1944, Eddington refused to accept that a star could collapse. We can only wonder how our understanding of astrophysics might have been accelerated if Eddington had backed Chandra.

## Further Reading
*Empire of the Stars: Friendship, Obsession and Betrayal in the Quest for Black Holes* by Arthur I. Miller, Little, Brown Co., 2005.

# Planetary Protection
## Keeping the Planets Safe from Earthly Bacteria

### Peter Rea

**Introduction**

"The Martians had no resistance to the bacteria in our atmosphere to which we have long since become immune. Once they had breathed our air, germs, which no longer affect us, began to kill them. The end came swiftly. All over the world, their machines began to stop and fall. After all that men could do had failed, the Martians were destroyed and humanity was saved by the littlest things, which God, in His wisdom, had put upon this Earth."

These words come from the novel *War of the Worlds* written by Herbert George (H.G.) Wells in 1897. It nicely summarises the need for planetary protection on all bodies in the solar system. We have built up resistance to most Earthly bacteria or viruses, but as recent events have shown, a new strain of virus can wreak havoc.

In 1898 H. G. Wells published his famous book *The War of the Worlds* about a Martian invasion of the Earth. The Martians were ultimately killed by the Earth's bacteria to which they had no immunity. It is perhaps ironic that when we send spacecraft to Mars, we go to great lengths to ensure that we do not contaminate Mars, the home of Wells' imaginary Martians, with our bacteria. (John McCue)

Our lack of resistance to Severe Acute Respiratory Syndrome Coronavirus 2 (SARS CoV 2) the virus that causes COVID-19 produced a pandemic with resulting chaos across the globe. In 1918 as The Great War came to an inglorious end, a Spanish flu pandemic caused by the H1N1 virus infected at least 500 million people with an estimated 50 million succumbing to the disease.

## Forward and Backward Contamination

The last few years has seen increasing efforts to look for current or past signs of life on Mars. There is increasing interest in Jupiter's natural satellites Europa, Ganymede and Callisto and Saturn's moon Enceladus. If we are to preserve these areas, then the appropriate precautions against contamination from Earth must

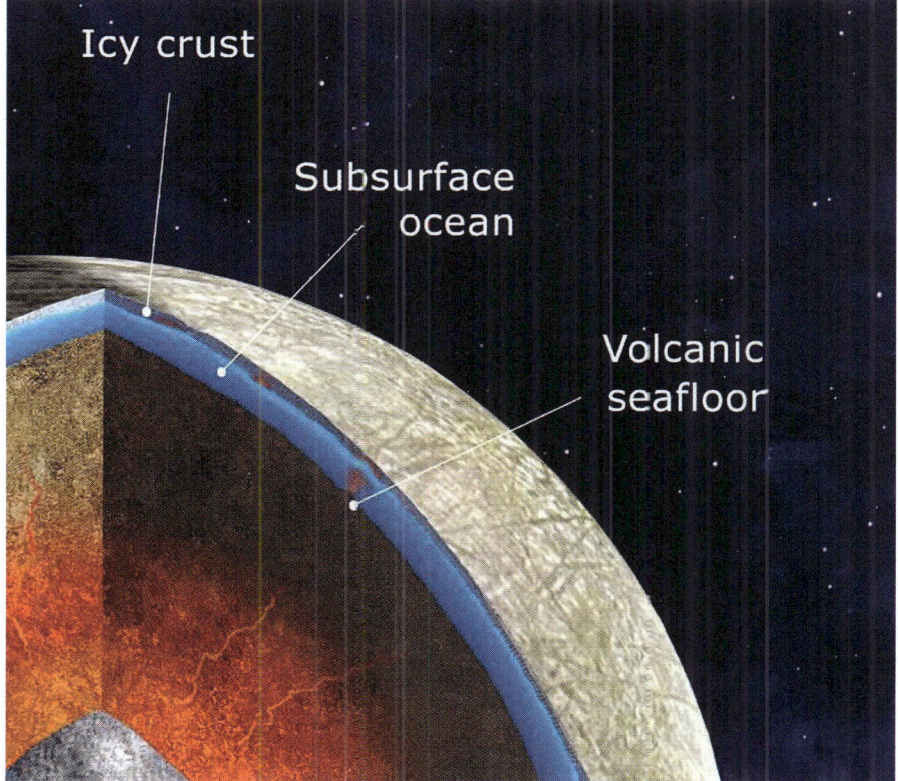

This artists rendering shows how Europa may look below the crust. It shows a subsurface ocean with ocean floor, a potential source of hydrothermal vents perhaps. (NASA/JPL-Caltech/Michael Carroll)

be in place. Forward contamination, sending our bacteria or viruses to other solar system bodies, could upset the natural environment and give false readings in any search for life. The reverse can also happen, Backward Contamination, bringing foreign bacteria or viruses to Earth.

## Planetwide Agreement on Planetary Protection

On 3 October 1958 – a whole decade before the first lunar rocks were returned to Earth and a day short of the first anniversary of the beginning of the space age – the Committee on Space Research (COSPAR) was established. Its purpose was to promote international collaboration in space research with a free exchange of ideas and results. The first meeting was held in London in 1958 and meetings have taken place regularly since then. COSPAR is made up of various scientific commissions looking at all aspects of Earth, planetary, solar and astrophysical research with international agreement. It also set up various panels looking more at space policy than research. One of these is the Panel on Planetary Protection, set up to agree an international protocol on how to deal with samples returned to Earth and how best to prevent contamination of solar system bodies from Earth's micro organisms.

## Different Target, Different Requirements

The COSPAR Panel on Planetary Protection lists the following categories of missions which reflect the diversity of bodies within the solar system and also the varying mission types, i.e. flyby, orbiter, or lander.

Category I: All types of mission to a target body which is not of direct interest for understanding the process of chemical evolution or the origin of life.

Category II: All types of missions (gravity assist, orbiter, lander) to a target body where there is significant interest relative to the process of chemical evolution and the origin of life, but where there is only a remote chance that contamination carried by a spacecraft could compromise future investigations.

Category III: Flyby (i.e. gravity assist) and orbiter missions to a target body of chemical evolution and/or origin of life interest and for which scientific opinion provides a significant chance of contamination which could compromise future investigations.

Category IV: Lander (and potentially orbiter) missions to a target body of chemical evolution and/or origin of life interest and for which scientific opinion provides a significant chance of contamination which could compromise future investigations.

# Planetary Protection: Keeping the Planets Safe from Earthly Bacteria

Engineers work on *Mars Exploration Rover Opportunity* (in its cruise configuration) in a clean room at Kennedy Space Center. A very important part of planetary protection is keeping contaminants from humans from riding aboard spacecraft. The pictured engineers have donned clean room suits that only allow their eyes to be exposed. (NASA)

Category V: Two subcategories exist: unrestricted Earth return for solar system bodies deemed by scientific opinion to have no indigenous life forms and restricted Earth return for all others.

The different planetary protection categories reflect the level of concern that contamination could compromise future investigations of solar system bodies or the safety of the Earth and its indigenous life forms.

## Launch Bias

I hope it follows that any spacecraft wanting to come into direct contact with a category IV body must undergo some form of sterilisation before leaving Earth. That spacecraft is not the only piece of hardware heading to the planet or planetary moon. The launch vehicle (rocket) will have an upper stage to which the spacecraft was attached. Having reached Earth escape velocity the spacecraft separates from the upper stage and over time will gradually drift further apart, but the upper stage will follow the spacecraft along its journey. It is not possible to sterilise the upper

Launch of the *Mars Perseverance Rover* from Cape Canaveral Space Force Station in Florida on 30 July 2020. The *Atlas 5* rocket would initially target the upper stage with the rover slightly away from Mars to avoid any possibility of contact. A week or two after launch the capsule containing the rover would use its thrusters to move it to the desired trajectory. The upper stage would fly past Mars at a safe distance and go into orbit around the Sun. (NASA/Joel Kowsky)

stage as this part of the rocket is exposed to our atmosphere whilst on the launch pad. For this reason the launch bias was introduced. Instead of aiming directly at the planet the guidance system will bias the trajectory away from the planet slightly so the chances of the upper stage coming under the gravitational influence of the planet and potentially crashing on it are remote. This means that shortly after launch, often between 10 and 30 days the spacecraft must make a small trajectory correction manoeuvre to remove the launch bias and put the spacecraft back onto the desired trajectory. With no means (and no need) to adjust its trajectory, the upper stage will miss the target object by many tens of thousands of kilometres.

## Early Lessons

Back in the 1960s, as America prepared to send *Apollo 11* to the Moon to attempt the first crewed moon landing, concern grew about the possibility of returning "lunar bugs" to Earth. The Moon is a harsh mistress. Its surface is baked by sunlight for 14 days, during which time temperatures can reach 120°C. During the lunar night,

# Planetary Protection: Keeping the Planets Safe from Earthly Bacteria

The crew of *Apollo 11* being recovered from their Command Module after splashdown. Note that the crew and a navy diver are wearing Biological Isolation Garments to prevent any lunar bugs from entering our atmosphere. (NASA)

also 14 days, temperatures can drop to as low as $-130\,°C$, the temperature at the poles perhaps getting down to lower than $-200\,°C$. Coupled with the fact that the Moon's surface is essentially a vacuum and therefore takes the full force of solar radiation, this makes the Moon a seemingly unsuitable place for micro-organisms. Yet, discoveries in the last decade or two of *extremophiles* – micro-organisms that exist on Earth in severe conditions like boiling water, locked in ice, acidic water or in extreme dry environments – suggest that micro organisms could exist on other planetary bodies with extreme conditions. This was not known to *Apollo 11* mission planners and, as there was concern about the possibility of lunar bugs, the first Apollo astronauts donned *biological isolation garments* inside the *Apollo 11* command module on their return to Earth. The crew were then placed inside a purpose built *mobile quarantine facility* (MQF) which had been placed onboard the recovery ship the *USS Hornet*. Once inside, the biological isolation garments could be removed whilst they were taken to a new facility, the Lunar Receiving Laboratory at the Manned Spacecraft Center (later the Johnson Space Center) outside of Houston in Texas. It was here that the crew were placed in isolation for three weeks including time spent in the MQF. Lunar samples collected by the crew had been placed in

The first *Apollo 11* sample return container, with lunar surface material inside, is unloaded at the Lunar Receiving Laboratory at the Manned Spacecraft Center (later the Johnson Space Center). The rock box had arrived only minutes earlier at Ellington Air Force Base by air from the Pacific recovery area. The lunar samples were collected by astronauts Neil Armstrong and Buzz Aldrin during their lunar surface extravehicular activity (a moon walk to you and me). (NASA)

special hermetically sealed boxes whilst on the lunar surface. These containers would prevent both forward and backward contamination and were only opened inside specially sealed laboratories. No contaminants were found, and after three weeks the crew were allowed out of quarantine. This isolation of crews was not used for later moonwalkers, but lunar samples were only sent to the Lunar Receiving Laboratory where most samples still reside even today.

# Planetary Protection: Keeping the Planets Safe from Earthly Bacteria  277

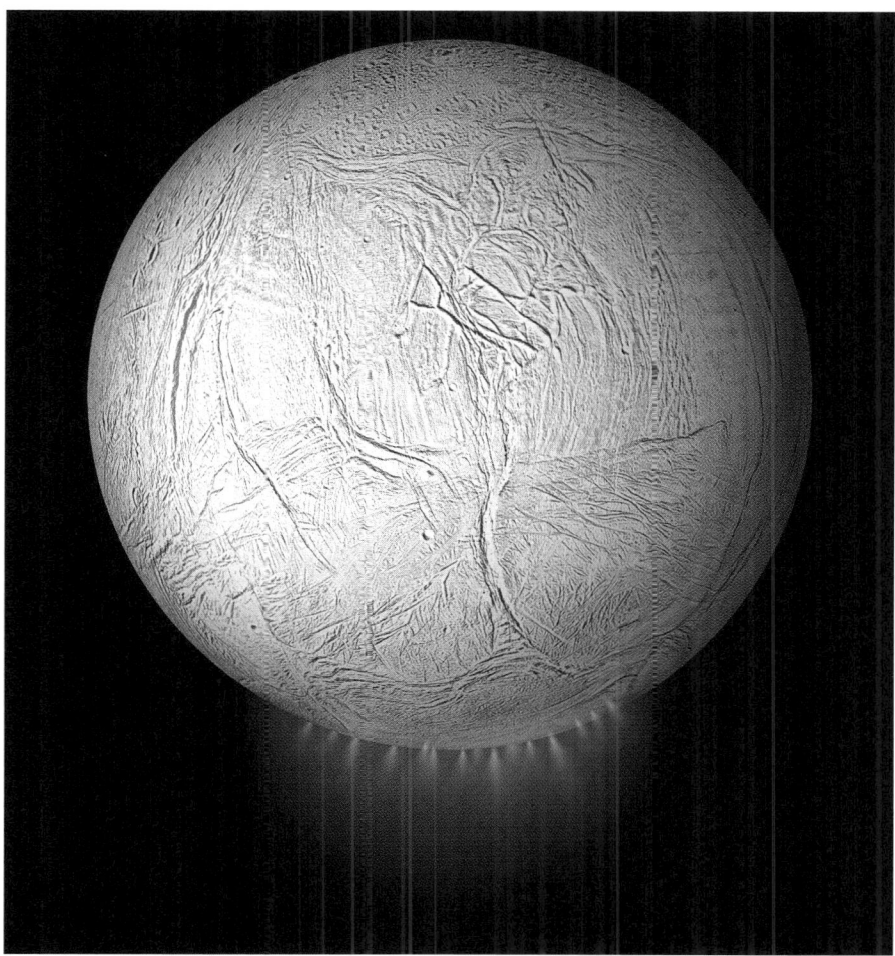

This artist's rendering shows Saturn's icy moon Enceladus with the plume of ice particles, water vapour and organic molecules that sprays from fractures in the moon's south polar region. Could life exist below the frozen surface of Enceladus? (NASA/JPL-Caltech)

To prevent Earthly bugs from being taken to solar system bodies of COSPAR Category II and higher, a process of spacecraft sterilisation is undertaken. This involves heating parts of the spacecraft to around 120°C. Some areas are subject to ultraviolet (UV) radiation whilst sensitive areas may just have alcohol or hydrogen peroxide applied. This can never totally sterilise the spacecraft but significantly reduces the risk. This begs one question: if even small amounts of micro organisms can reproduce and stay viable for long periods of time, what long term effect will this have on the planetary bodies to which we have sent – and are planning to send – landers and rovers?

## Possible Targets in the Search for Life

Mars remains the number one choice of planets likely to have some signs of past life. H. G. Wells thought so too as the introduction to this article shows. Recent discoveries by the *Galileo* and *Juno* spacecraft would indicate that the three largest icy moons of Jupiter – Europa, Ganymede and Callisto – may have sub-surface oceans, and are the target of the European Space Agency *Jupiter Icy Moons Explorer* (JUICE) mission launched in 2023. They are also of interest for the NASA *Europa Clipper* mission which, at time of writing, was due for launch in 2024. The Saturn orbiter *Cassini* saw evidence in camera images of something spewing out from its moon Enceladus, later confirmed as water when Cassini flew through this cloud of material. The *Huygens* probe that landed on the surface of Saturn's largest moon Titan saw glimpses of a world that reminded scientists of what an early Earth could possibly have looked like. When I was a young amateur astronomer in the 1960s I read books that talked about a "Goldilocks Zone", a habitable zone around stars where temperatures were neither too hot nor too cold and where water would remain liquid. In our solar system this would stretch from just beyond Venus to around the orbit of Mars, with Earth right in the middle. With the discovery of extremophiles (mentioned earlier) and black smokers – hydrothermal vents located at the bottoms of some oceans – a new range of possibilities opens up. These vents emit jets of particle-laden fluids from the Earth's crust. The particles are predominantly very fine-grained sulphide minerals formed when the hot

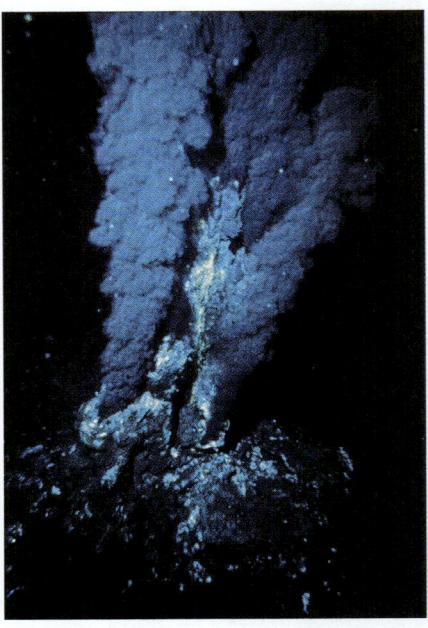

A black smoker at a mid-ocean ridge. These hydrothermal vents support life of different kinds in a dark, oxygen poor environment. Could these exist on other solar system bodies? (P.Rona / OAR / National Undersea Research Program (NURP); NOAA)

# Planetary Protection: Keeping the Planets Safe from Earthly Bacteria

The *Cassini/Huygens* spacecraft arrives at Saturn on 1 July 2004, where it would remain operational around the planet for just over 13 years. Of the many observations and discoveries made was a plume of water emanating from the Saturnian moon Enceladus indicating a sub-surface ocean, and identifying Enceladus as a potential target for a future lander. (NASA)

hydrothermal fluids mix with near-freezing seawater. These minerals solidify as they cool, forming chimney-like structures. Yet even here, down on the ocean floor, these vents are teaming with life where no light from the Sun can reach and oxygen and nutrition levels are very poor. Could such vents exist below the surface of the icy moons of Jupiter or Enceladus? Perhaps we need to rethink the range of the habitable zone around our Sun and other star systems.

## In Conclusion

An interesting point was made at a lecture I recently attended. Will we ever be able to land and live on other planets within our galaxy? In the opening paragraph of this article there is a quote from *War of the Worlds* by H. G. Wells. He cites that Earthly bacteria "no longer affect us". This is not entirely true. In past centuries missionaries and explorers who went to remote regions like the rain forests of South America, Africa or Borneo carried with them measles, influenza and the like to which local tribes had no resistance. The reverse was also true, as various viral

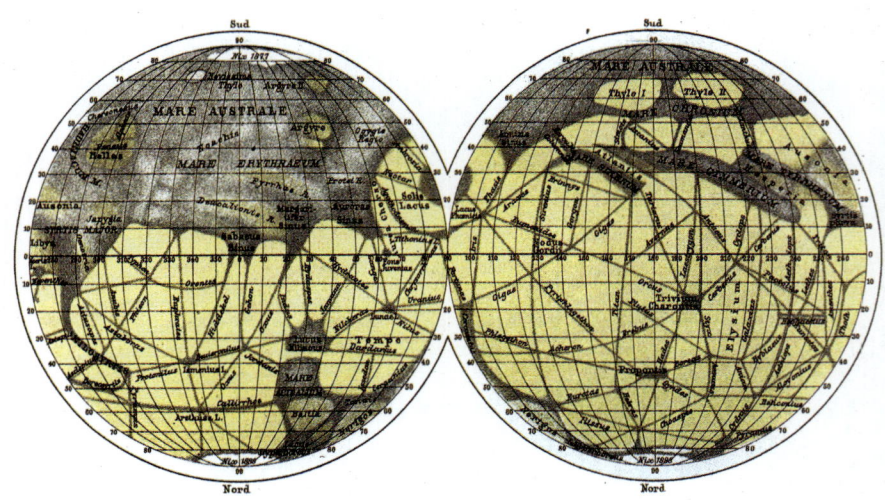

Carte d'ensemble de la planète Mars
avec ses lignes sombres non doublées
observées pendant les six oppositions de 1877-1888
par J.V. Schiaparelli.

Map of Mars by Giovanni Schiaparelli dated 1888 with south at the top. Early observations of the dark or albedo features on Mars made some astronomers believe it could be vegetation. The straight lines shown here were made famous (perhaps infamous) by Percival Lowell who referred to them as canals rather than channels (Italian *canali*) as Schiaparelli called them. Canals would suggest construction by intelligent life. (Wikimedia Commons / Giovanni Schiaparelli)

diseases made their way back to Europe and America and to which the populations of those countries had never previously been exposed and built up resistance. New viral diseases or mutated versions of existing diseases continue to spread across the globe, as the recent COVID pandemic has shown. One thing that is absolutely certain is that when humans set foot upon Earth-like planets in future centuries, we will have absolutely no resistance to that planet's bacteria or viruses, nor them to ours. Colonising the galaxy will not be as easy as the Sci-Fi films may portray. The biggest threat may come not from sentient life but the microscopic and almost invisible world of the virus. H. G. Wells knew that. Life is precious: it can also be deadly.

As H. G. Wells puts it in the opening of *War of the Worlds*: "No one would have believed in the last years of the nineteenth century that this world was being watched keenly and closely by intelligences greater than man's and yet as mortal as his own; that as men busied themselves about their various concerns they were scrutinised and studied, perhaps almost as narrowly as a man with a microscope might scrutinise the transient creatures that swarm and multiply in a drop of water."

## Acknowledgements

I am grateful to Sallie Keith of NASA for providing advice and some artwork and to David Todd for his insight and suggestions into the section order of this article. The *Yearbook of Astronomy* editor Brian Jones provided helpful advice and assistance in contacting key people within NASA.

## Further Information

NASA Office of Safety and Mission Assurance – Planetary Protection
**sma.nasa.gov/sma-disciplines/planetary-protection**

NASA Planetary Protection Center of Excellence at the Jet Propulsion Laboratory
**planetaryprotection.jpl.nasa.gov**

ESA Planetary Protection
**technology.esa.int/page/planetary-protection**

COSPAR Panel on Planetary Protection
**cosparhq.cnes.fr/scientific-structure/panels/panel-on-planetary-protection-ppp**

# Nearby Worlds Out There
## The Many Kinds of Exoplanet

## John McCue

Those of us who are a certain age will know that, when we were at school, finding planets elsewhere in our Milky Way galaxy was pure fantasy, especially any of them orbiting stars that are near enough to be seen in our own night sky without a telescope. Other worlds out there – exoplanets – were described in the adventures of Dan Dare and Flash Gordon, heroes of the comics we devoured when young.

Now, amateur astronomers can detect them with their back garden telescopes. If we are lucky enough that an exoplanet's orbit around its star lies flat to our view then it will cross the star's face regularly and dim its light, albeit not by very much! The immediate property of the exoplanet that we can measure from this transit is its size. It is natural therefore to classify an exoplanet according to its radius, especially as the majority of the 6,000 exoplanets that we know have been found by this transit method. Four groups have thus been alighted upon;[1] terrestrial-sized planets (less than 1.25 times earth's radius), Super-Earths (from 1.25 to 2 times), Neptune-size (from 2 to 6 times); and Jupiters (from 6 to 15 times the Earth's radius).

There are many exoplanets which appear near the boundary of the Super-Earths and the Neptunes – these are often seen with the name mini-Neptunes. Likewise, Jupiter exoplanets orbiting close to their star are understandably called Hot Jupiters.

Exoplanets detected by radial velocity changes of the host star, and by gravitational microlensing observations, allow the planet's mass to be estimated, giving a means of grouping according to mass. Radial velocity changes occur because the star and exoplanet perform a merry-go-round with the common centre-of-gravity as the turning point. As the star moves alternately towards and away from us under this swivelling action, we can observe changes in its radial velocity by the Doppler effect on the star's spectral lines.

The light from a very distant star will be focussed by the gravitational field of a closer star as this star passes between us and the distant one – microlensing.

---

1. Borucki, William J. et al., (2011). 'Characteristics of Planetary Candidates Observed by Kepler. II. Analysis of the First Four Months of Data'. *The Astrophysical Journal*, **736:19** (1). ui.adsabs.harvard.edu/abs/2011ApJ...736...19B/abstract

Many of us will remember the trick (to be discouraged!) of focussing sunlight with a magnifying glass on to a piece of paper; this is the well-known analogy, with the glass acting as the closer star and the sun as the more distant star. From an observational point of view, astronomers see a sudden flaring of the brightness of the distant star; if it has an exoplanet, that also will produce a spike in brightness, but not as much as its parent star. The terrestrial planet group can have a mass range of 0.1 to 2 earth-masses, Super-Earths being 2 to 10 times, Neptunes with 10 to 100 earth-masses and Jupiters having 100 to 1,000 times the mass of the Earth.[2] There are well over twice the number of exoplanets with known radii than known mass.

In April 2024, three Russians, Mikhail Korniyenko, Alexander Lynnik and Denis Efremov, in a spectacular experiment, jumped from a height of over 10,500 metres to land near their Barneo polar base, breaking the world parachuting record as they left the plane. On landing, they used a diesel generator to fire up a server that had been dropped earlier (from a lower altitude!) and connected with a satellite. Their aim of testing a low-cost communication system had been achieved.

They suffered minor frost-bite on their cheeks, despite heated masks, as they faced a wind chill of $-70°$ C falling from that lower part of the stratosphere. Surprisingly, though, at the much greater height of around 75,000 metres, the temperature can rise to above $2,000°$ C in a well-studied level known as the thermosphere. At this altitude, ozone is exposed to the sun's ultra-violet radiation, which absorbs it and warms it up. The density of the atmosphere though is so low that, despite the high temperature, very little heat energy develops. It doesn't feel like an oven!

In the upper atmospheres of Hot Jupiters, a similar process has been detected.[3] We know that metal oxides strongly absorb infra-red radiation and warm up, and indeed, eclipse spectra (when the exoplanet passes behind the host star) taken in the infra-red, show that these oxides, such as those of titanium and vanadium in gaseous form, are heating the upper atmospheres of these fiery exoplanets. Hot Jupiters are believed to form far from their parent stars, subsequently migrating inwards until very close to their star. This is asking questions of the accretion theory of the formation of our own solar system, which had been proposed by the French astronomer and mathematician Pierre-Simon, Marquis de Laplace (1749–1827) in

---

2. Stevens, Daniel J. and Gaudi, B Scott (2013). 'A Posteriori Transit Probabilities', *Publications of the Astronomical Society of the Pacific*, **125** (930), 933–950. **ui.adsabs.harvard.edu/abs/2013PASP..125..933S/abstract**

3. Changeat, Q. et al., (2022). 'Five Key Exoplanet Questions Answered via the Analysis of 25 Hot-Jupiter Atmospheres in Eclipse', *The Astrophysical Journal Supplement Series*, **260** (1). **iopscience.iop.org/article/10.3847/1538-4365/ac5cc2**

1796.[4] Why, for example, did our own Jupiter stay at its present location in the solar system?

Exoplanet HD 189733 b, located at a distance of 64.5 light years, is very similar to our own Jupiter in size and mass. However, it orbits its red dwarf star at a distance of ten times closer than Mercury does around our Sun, making it a Hot Jupiter. This planet's blue colour suggests an atmosphere like the Earth's, but any suggestion that life forms may exist there is easily discounted by the dangerous conditions at its surface, where temperatures of 1000° C, winds of around 8,000 kilometres per hour and glass rains of silicate particles make for a nightmare environment. This star, in the northern summer constellation of Vulpecula (the Fox), is just below naked eye visibility to us, although Tau (τ) Boötis, with a visual magnitude of 4.50, is just above that limit and can be seen fairly easily in a dark sky. Tau Boötis is host to another Hot Jupiter nearly six times the mass of our own giant planet Jupiter, and racing around its star (which is a little hotter than HD 189733) in just 3.3 days. Intriguingly, this star is a binary, its two component stars – 51 light years distant from us – named A and B, and the exoplanet orbiting the brighter star. Any imaginary resident creatures of Tau Boötis A b will be able to see the faint companion star B tracking around the background of stars over a period of 2,400 years. From Earth, the two stars appear very close in a small telescope, at 1.5 arc seconds, and given that A looks over 200 times brighter than B, the latter is difficult to detect visually.

Pierre-Simon, Marquis de Laplace, astronomer and mathematician, developed the accretion theory (or nebular hypothesis) for the creation of our solar system. (John McCue)

A typical Neptune is the exoplanet HAT-P-11 b, so named after its discovery in 2009 by the Hungarian Automated Telescope Network (HATNet) project. HAT-P-11 b lies at a distance of 122 light years in the constellation of Cygnus which, coincidentally, was the part of the sky constantly monitored by the *Kepler* space telescope – NASA's first planet-hunting mission – in the first major, and hugely successful, search for exoplanets by the transit method. This exoplanet was first spotted in thousands of transit images taken five years before Kepler was launched, then confirmed much later by radial velocity measurements almost as *Kepler* was

---

4. Woolfson, Michael (2000). 'The Origin and Evolution of the Solar System', *Astronomy & Geophysics*, 41 (1), 1.12–1.19. **academic.oup.com/astrogeo/article/41/1/1.12/182262**

leaving the launch pad in March 2009! Of course, it was immediately spotted by the *Kepler* mission and so can often be seen with the name Kepler-3 b.

The planet again revolves closely around its star, another red dwarf, making it a hot Neptune, but of curious note is the obliquity of its orbit to the star's axis of spin – about 100 degrees – meaning the planet almost has a polar orbit around its parent. An explanatory clue lies in the discovery nine years later, by radial velocity measurements, of a second exoplanet, HAT-P-11 c, one-and-a-half times the mass of our own Jupiter. It revolves around its star at a 'normal' Jupiter distance of 4.1 au, taking 93 years to complete one orbit, but crucially with a high eccentricity of 0.6, making its orbital path very cigar-shaped, and also at a significant tilt to the orbit of planet b.[5] These planetary movements are very different to those of our own planets – are there ways of building a system other than the accretion theory? The star's visual magnitude is 9.5, making it well below visibility level for the unaided eye.

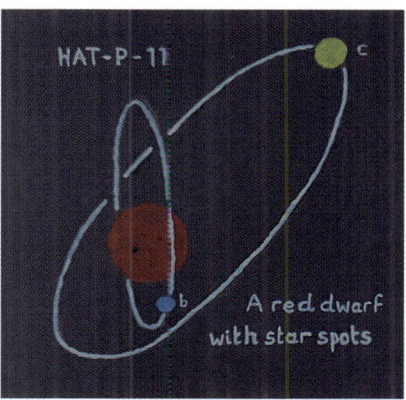

Exoplanets b and c in eccentric and tilted orbits around the star-spotted red dwarf, HAT-P-11. (John McCue)

The star HD 23472 (named in the Henry Draper catalogue of spectroscopic classifications) has a swarm of exoplanets of the smaller variety – terrestrials and super-earths. With a visual magnitude of 9.7, and distance from us 127 light years, this red dwarf lies in the southern constellation of Reticulum (the Net) and has a big family to look after. Not only is it a large family, but it stays close to home! All five planets reside within the constraints of an orbit the size of that of our own Mercury, the outer two, b and c, being Super-Earths with masses eight and three times Earth's mass respectively, and the inner three, d, e, and f, being terrestrials with masses 0.6, 0.7, and 0.8 likewise. Exoplanets that are so similar to our home planet beget thoughts of comfortable environments, even living creatures, but not so these. The outermost, c, has a calculated temperature of 470 K with those closer

---

5. Yee, Samuel W. et al., (2018), 'HAT-P-11: Discovery of a Second Planet and a Clue to Understanding Exoplanet Obliquities', *The Astronomical Journal*, **155** (6).
ui.adsabs.harvard.edu/abs/2018AJ....155..255Y/abstract

to the star gradually getting warmer until d, the closest, swelters at 900 K, all well above the boiling point of water.

Clearly, a habitable zone for earth-like creatures must allow the existence of liquid water under everyday circumstances, memorably called the Goldilocks zone. Considering the distances of planetary systems quoted so far, they seem incredible to our notions of distance on Earth. For example, if the reader finds Tau Boötis in a telescope, the light from that star has been journeying towards us for over 50 years. However, our Milky Way galaxy is around 100,000 light years in diameter, placing all exoplanets discovered so far in our immediate galactic neighbourhood. Even closer is the final planetary system we look at in this article, and which is right next door!

Alpha Centauri A is well-known as a nearby star, at a distance of 4.39 light years – navigators know it as Rigil Kent – and it has a close companion, B. These could claim to be the closest stars to Earth but for a third companion, C, better known as Proxima Centauri. It is not clear whether or not this star is gravitationally bound to A and B, but in any case it is a little closer at 4.22 light years, thereby snatching the record. With an apparent magnitude of 11.1, Proxima is not easy to see in amateur telescopes. However, it has three exoplanets – b, c and d – with Proxima Centauri b attracting huge attention since its discovery in 2016 by the radial velocity method. Proxima Centauri itself is a red dwarf star, smaller than the Sun, around which exoplanet b orbits every 11 days at a distance of just 0.04856 au (7,264,000 kilometres). With a mass only a little more than that of the Earth, it made exciting newspaper headlines, but having a temperature range suitable for liquid water as well sent public interest into overdrive. Proxima Centauri b does not pass across the face of its star, so astronomers cannot currently tell whether it has an atmosphere, but if it does, and life exists on its surface, it will be facing the dangers of ultra-violet radiation from the active star that provides its warmth.

# Comets and Literature in the Nineteenth Century

## Randall Stevenson

'The comet ... in all probability saved his life.' Careless with his health, and disappointed with the outcome of his observations, the young astronomer Swithin St Cleeve, in Thomas Hardy's 1882 novel *Two on a Tower*, languishes into potentially mortal illness. Yet he retains some of his interest in the sky, in comets particularly. 'Of all phenomena that he had longed to witness during his short astronomical career', Hardy explains, 'those appertaining to comets had excited him most. That the magnificent comet of 1811 would not return again for thirty centuries had been quite a permanent regret with him'. When Swithin hears that a new comet is approaching, '... the limitless and complex wonders of the sky resumed their old power over his imagination ... it seemed to lend him sufficient resolution to complete his own cure forthwith.'

The 'magnificent comet of 1811', discovered on 25 March 1811 by French astronomer Honoré Flaugergues, was the first of many bright comets appearing

The Great Comet of 1811 (Flaugergues) as depicted by the English engraver Henry Richard Cook from an observation made on 15 October 1811 from Otterbourne Hill in Winchester, Hampshire. (Wikimedia Commons/Henry Richard Cook/Jean-Michel Faidit)

throughout the nineteenth century, exercising extensive power over the imagination of artists, authors – Thomas Hardy in particular – and the public generally. Visible to the naked eye from around May 1811, and remaining in the skies for more than eight months, Flaugergues was often referred to as 'Napoleon's Comet'. Ever the opportunist, Napoleon proclaimed it a good omen, promising the success of his Russian campaign, begun in June 1811. For the Russians, it naturally took on an opposite significance, as a portent of doom – a role eventually ascribed to it by the French themselves, after their long retreat from Moscow. It figures in this way in Leo Tolstoy's epic novel of the Russian campaign, *War and Peace* (1869), his hero Pierre Bezukhov witnessing '… the enormous and brilliant comet of the year 1812 – the comet which was said to portend all kinds of woes and the end of the world.'

By the time Tolstoy's novel was published in the 1860s, Flaugergues had been followed – and often outshone – by several other dazzling comets, as bright as any appearing in the past millennium. Born in 1840, Hardy may have been just too young to observe the Great March Comet of 1843, which shone at magnitude −10, as bright as a Full Moon, and easily visible in daylight. He certainly witnessed Donati's Comet, appearing in 1858: one of the most beautiful of the century, and

The Great Comet of 1843, as depicted in *Das Wissen der Gegenwart, 27. Band: Kometen und Meteore* (The Knowledge of Today, Volume 27: Comets and Meteors) written by German astronomer Karl Wilhelm Valentiner and published in 1884. (Wikimedia Commons / Karl Wilhelm Valentiner)

Donati's Comet seen close to Arcturus, above Paris, 1858, from *Le Ciel* by French science writer Amédée Guillemin, published in 1877. (Wikimedia Commons / Amédée Guillemin)

widely represented in art and literature. Hardy probably refers to it in *The Comet at Yell'ham*, his poem contrasting ephemeral human life with the unending cycles of the heavens. It also appealed to Alfred, Lord Tennyson, whose son Hallam recalled him interrupting a dinner party in the autumn of 1858 to observe it, with the star Arcturus seeming to '… dance as if mad when it passed out of the comet's tail.' Tennyson later continued observing the comet from a platform on the roof of his house, long remembering the impression it made. Published sixteen years later, his verse play *Harold* begins with characters looking uneasily at a comet which '… glares in heaven …' and at a '… star / That dances in it as mad.'

Donati's Comet was followed three years later by a still brighter one – the Great Comet of 1861 – perhaps further focusing, for Hardy, astronomical interests begun in childhood. Comets, at any rate, provide a convenient metaphor in his 1874 novel *Far From the Madding Crowd*, helping characterise a gauche farmer who views women as '... remote phenomena ... comets of such uncertain aspect, movement and permanence that whether their orbits were as geometrical, unchangeable, and as subject to laws as his own, or as absolutely erratic as they superficially appeared, he had not deemed it his duty to consider.' Hardy's own awareness of remote astronomical phenomena was extended during the 1870s through reading Richard A. Proctor's *Essays on Astronomy* (1872), and by discussions with his doctor, Sir Henry Thompson, a distinguished amateur astronomer.

Perhaps the key moment in his deepening interests in comets and astronomy occurred at a key moment in his private life: on one of his first evenings in a house he had moved into, with his wife Emma, at Wimborne in Dorset. Walking out into their new garden, late on 25 June 1881, they found hanging in the northern sky yet another brilliant comet – discovered, like its predecessor in 1861, by the Australian amateur John Tebbutt. Hardy went on that summer to plan a novel on an astronomical theme, also drawing on the transit of Venus, due in December 1882, and on advice he had acquired from a visit to Greenwich Observatory. The result was *Two on a Tower*, in which the new comet more than fulfils Swithin's excited expectations, appearing 'visible in broad day', and flaunting at night a 'fiery plume' which 'filled so large a space of the sky as to completely dominate it.' As well as rescuing him from deadly illness, it also contributes – among other key roles – to a crucial moment in the novel's romantic plot, when a liaison with his patroness, Lady Viviette Constantine, develops while Swithin observes the comet one night in her company.

Reflecting the extent of Hardy's astronomical interests – later extended in his epic poetic drama *The Dynasts* (1904–8) – *Two on a Tower* also illustrates the shifting, sometimes-disquieting roles comets occupied in nineteenth-century imagination more generally. Swithin suggests the survival of one form of uneasiness about comets – fears of their collision with the Earth – when he reflects that their fascination accrues partly from '... the sinister suspicion attaching to them of being possibly the ultimate destroyers of the human race.' One of the novel's farm workers also indicates some continuing belief in comets as portents – though only in a rather dilute, almost comic form – when he asks '... what do this comet mean? ... That some great tumult is going to happen, or that we shall die of a famine?' A fellow farm labourer lightly dismisses his concern, reassuring him that '... it isn't to be supposed that a strange fiery lantern like that would be lighted up for folks with ten or a dozen shillings a week.' Even by the beginning of the century, as Tolstoy suggests in *War and Peace*, old superstitions about comets were

The Great Comet of 1881, also known as Tebbutt's Comet in honour of Australian astronomer John Tebbutt who discovered the comet in May 1881. The image seen here is a chromolithograph from *The Trouvelot Astronomical Drawings* (1882), a set of illustrations created from the astronomical observations of French astronomer and artist Étienne Léopold Trouvelot. (Wikimedia Commons / Étienne Léopold Trouvelot / Crystal Bridges Museum of American Art)

fading. Although Pierre Bezukhov acknowledges that Comet Flaugergues is said to 'portend all kinds of woes', he finds that it arouses in him '… no feeling of fear. On the contrary he gazed joyfully … at this bright comet … having travelled in its orbit with inconceivable velocity through immeasurable space.'

Focused on orbital velocities rather than omens of doom, Pierre's interests share late eighteenth- and early nineteenth-century movements from superstition towards rational enquiry. Especially after the object later named after Edmond Halley returned, as he had predicted, in 1758, comets had come to be conceived as celestial bodies whose nature and movements were subject to the understanding of scientists. By the end of the eighteenth century, this understanding had been further improved by the work of contemporary astronomers and mathematicians – in particular, Pierre-Simon Laplace. From orbital calculations, Laplace deduced that comets exert minimal gravitational influence on other celestial bodies, and are therefore of relatively insignificant mass. Previously, it had often been supposed that they were as massive as planets, and that any collision with one would be catastrophic for the Earth. After Laplace's work, comets seemed much less likely to be 'ultimate destroyers of the human race', and rather the 'desirable visitors' Swithin considers them, despite lingering sinister suspicions. For much of the nineteenth century, popular accounts of astronomy continued to counter such suspicions, generally describing comets as flimsy, lightweight objects, gaseous or nebulous in nature. In 1881, for example, newspapers reassured their readers that cometary collisions threatened the Earth no more than wind-blown mist might menace a mountain range.

Yet rather than altogether disappearing, the threatening aura comets had long maintained in popular imagination sometimes mutated into new forms, even encompassing recent scientific opinion in doing so. The gaseous nature now widely ascribed to comets, rather than innocent, was occasionally thought to threaten remote, malign influences over the weather, or even a risk of poisoning the atmosphere altogether. In the early Edgar Allan Poe story *The Conversation of Eiros and Charmion* – published in 1839, a few years after another passage of Halley's Comet in 1835 – characters acknowledge that scientific advances since the late eighteenth century ensured that '… reason had … hurled superstition from her throne'. They note that '… the very moderate density' of comets had been well established, allowing them to be 'regarded as vapory creations of inconceivable tenuity, and as altogether incapable of doing injury to our substantial globe.' Yet the story goes on to recount how '… that tenuity … which had previously inspired us with hope, was now the source of the bitterness of despair.' As a huge comet

A huge comet rises over Paris: illustration from the second chapter 'La Comète' in *La fin du monde* by French astronomer Camille Flammarion. (Wikimedia Commons/Camille Flammarion/F0x1/Creative Commons International Licence CC BY-SA 4.0)

approaches, its 'impalpable gaseous character' progressively damages the Earth's atmosphere – eventually, sufficiently seriously to set the air itself on fire, in a global conflagration utterly destroying humanity.

Half a century or more later, in *La fin du monde* (The End of the World, 1893), the French astronomer and author Camille Flammarion likewise describes an immense approaching comet entirely removing nitrogen from the Earth's atmosphere and threatening '… the end of the world … by fire.' Flammarion's scenario is in a way even grimmer than Poe's, reflecting developments in cometary science towards the end of the nineteenth century. By the later 1870s, astronomers had begun to consider that comets might not be altogether flimsy, but contain some solid elements. Published in 1877, Jules Verne's science fiction novel *Hector Servadac* (Off on a Comet) envisages a thoroughly solid body colliding with the Earth. Flammarion similarly adds to fears of poisoning and conflagration the possibility that '… the nucleus of the comet … contained within its mass of incandescent gas a certain number of solid uranolites', threatening appalling damage on colliding with the Earth.

Mention of spectroscopy in *La fin du monde* reflects another recent development in astronomy, also anticipating Flammarion's awkward influence on popular opinion about Halley's Comet. First successfully applied to comets, including Tebbutt's, in the 1880s, spectroscopy suggested the presence of poisonous cyanogen gas in their tails, and orbital calculations indicated that the Earth would pass through the tail of Halley's Comet when it returned in 1910. Articles Flammarion contributed to the Press at the time stressed that no danger was at all likely to result, but did mention eventualities which *might* lead to hugely damaging effects. Though tentative and speculative, these suggestions were amplified far enough in the newspapers to create some panic, particularly in the United States, where 'comet pills' and oxygen supplies were soon in demand.

No harm, of course, was caused by Halley's passage in 1910. The public might in any case have drawn some comfort from another comet novel, published a few years earlier in 1906. Written by H.G. Wells, *In the Days of the Comet* was also apocalyptic in much of its narrative, but eventually entirely affirmative in its vision. Possibly inspired, like Hardy, by the appearance of Tebbutt's Comet in 1881, Wells envisages – like Poe and Flammarion – an approaching comet's radical influence on the

Observable with the unaided eye for around 260 days before dropping to below naked-eye visibility in mid-January 1812, the Great Comet of 1811 was thought to be responsible for the long, hot summer that led to a particularly good year for wine. Merchants were selling 'Comet Wine' at inflated prices for several years after, in bottles that had an embossed seal featuring a comet. (Mary McIntyre)

Earth's atmosphere, but one which is very far from being damagingly poisonous or incendiary. It reproduces instead, on a global scale, effects rather like those often attributed to comets by nineteenth-century Oenophiles (lovers of wine). Ever since the appearance of Flaugergues, comets were widely supposed responsible for outstanding vintages: for helping produce what was celebrated – in years such as 1811, 1858 and 1881 – as 'Comet Wine'. Wells comparably envisages comet-induced, world-wide, euphoric intoxication, leading to the miraculous transformation of politics and institutions into the kind of utopian society he regularly advocates elsewhere in his writing.

Comets, in other words, though sometimes retaining their old threatening aura, also figured in nineteenth-century literature in ways entirely affirmative or salvational – for the ailing Swithin, in *Two on a Tower*; or, globally, for Wells, redeeming society from the 'desolating night of flaming industrialism' his novel begins by lamenting. The range of these roles indicates how widely the period's unusually brilliant comets impacted on literary imagination, figuring as perhaps the most influential, on nineteenth-century writers, 'of all phenomena... [among] the limitless and complex wonders of the sky.'

# On the Origin of NASA Names by Means of Imaginative Selection
# or
# The Preservation of Favoured Names in the Struggle for Exploration
(With apologies to Charles Darwin)

## Peter Rea

### Introduction
We only need to look at the names of some recent lunar and planetary missions to realise there must be an unpublished competition somewhere for the most absurd mission name. Acronyms seem to be in favour now. The *OSIRIS REx* mission to asteroid 101955 Bennu launched in 2016 is a classic example. *OSIRIS-REx (Origins, Spectral Interpretation, Resource Identification, Security-Regolith Explorer)* does not quite have the ring about it that *Voyager 2* did. Similarly the mission to Mercury called *MESSENGER (MErcury Surface, Space ENvironment, GEochemistry, and Ranging)* seems equally contrived. I would like to take you back to the early days of lunar and planetary exploration when mission names had a more poetic perhaps romantic ring to them that captured the imagination of this writer.

### Lunar Missions
The use of the word 'probe' to mean a spacecraft is in common usage these days. Its origin can be traced back to the year 1952. Two authors, E. Burgess and C. A. Cross delivered a paper to the British Interplanetary Society entitled 'A Martian

Edgar Maurice Cortright. (NASA)

The Ford Ranger. I wonder how many Ranger owners realise that this vehicle, albeit a much earlier model, gave rise to the name of a series of American spacecraft that returned the first close-up pictures of the Moon. (Wikimedia Commons/SsmIntrigue/Elise240SX)

Probe'. It would not be long before 'probes' were being launched regularly to the Moon and planets. In 1960 the then Assistant Director of NASA's Lunar and Planetary Programs Edgar Maurice Cortright (1923–2014) proposed a naming system that was adopted for the next two decades. Names for lunar probes were given 'land exploration activities'. This gave us mission names like *Surveyor* and *Prospector*. The Program Director at the Jet Propulsion Laboratory Clifford Cummings proposed the name *Ranger* after seeing a Ford Ranger pick-up truck (still in use today). Though perhaps stretching things a little, Ranger still refers to 'land exploration activities'.

## Planetary Missions

Whilst lunar mission were given 'land exploration activities', Cortright proposed that planetary mission names should have a nautical theme to convey, "… the impression of travel to great distances and remote lands." This way of thinking gave us mission names like *Mariner*, *Viking* and *Voyager*.

## Pioneer

Many *Pioneer* spacecraft were launched to the moon and planets. Although classed as 'land exploration activities', the name was already in use by the United States

Air Force who considered themselves as Pioneers in Space in a rivalry with the US Army who had launched America's first satellite *Explorer 1*. The first *Pioneer* was launched in 1958, two years before the introduction of the Cortright naming system. The early *Pioneers* were launched toward the Moon between 1958 and 1959. Pioneers 6 to 9 were sent to orbit the Sun as a space weather network between 1965 and 1968. In 1972 and 1973 *Pioneer 10* and *Pioneer 11* were sent to Jupiter and Saturn (*Pioneer 11* only) then out of the solar system. The last *Pioneers* went to Venus in 1978 and, although not officially *Pioneers*, they were named *Pioneer Venus Orbiter* and *Pioneer Venus Multiprobe*.

## Ranger

All *Ranger* probes were launched toward the Moon between 1961 and 1965, although the first six were not successful. Only the last three returned close-up images before they crashed onto the lunar surface. Due to the early failures of *Ranger*, the Americans had managed to run a successful planetary mission – *Mariner 2* in 1962 – before they eventually had a fruitful lunar mission, this being *Ranger 7* in 1964.

Astronaut Charles Conrad examines Surveyor 3 during the Apollo 12 mission in 1969. (NASA/Al Bean/Apollo 12 Image Library/AS12-48-7133)

## Surveyor

Seven soft landing spacecraft were launched between 1966 and 1968. Although *Surveyor 2* and *Surveyor 4* crashed, the other five were highly successful. *Surveyor 3* was visited by the crew of *Apollo 12* and some parts, including the camera, were returned to Earth for analysis.

## Lunar Orbiter

Perhaps the most unimaginative name given to a lunar mission, it fell outside the Cortright naming system and was purely descriptive of its mission. Despite numerous lines of enquiry, I have never understood why this series was not given a 'land exploration activities' name. *Prospector* had not been used up to that point. Five missions to photograph the Moon in detail were launched between 1966 and 1967 and were all successful.

Designed to give high resolution images of the Moon in support of Project Apollo, five Lunar Orbiters were launched in 1966/67. (NASA/Smithsonian/CC0 1.0 Universal)

## Prospector

Although named as part of the 'land exploration activities' system for naming lunar missions, no mission was launched during that Golden Age of lunar and planetary exploration, the 1960s and 1970s. However, the name was resurrected in 1998. The *Lunar Prospector* successfully entered orbit around the Moon on 11 January 1998, the mission lasting until 31 July 1999.

## Mariner

The early *Mariners* were based on the *Ranger* design. Ten *Mariners* were launched between 1962 and 1973. *Mariners 1, 2* and *5* were launched to Venus. *Mariner 1* failed. *Mariners 3, 4, 6, 7, 8* and *9* were launched to Mars. *Mariner 3* failed. *Mariner 10* was launched to Mercury with a flyby and gravity assist from Venus.

Mariner 9 was the second to last in the Mariner series and the first to orbit Mars. (NASA)

## Viking

The name was adopted quite early in the mission. The *Vikings* after all were truly great explorers of the sea. Had the *Viking* name not been used, the two spacecraft would have been *Mariners 11* and *12* as they were derived from earlier *Mariner* designs.

Vikings 1 and 2 showed their Ranger/Mariner heritage, being similar in size to Mariner 9 but carrying a lander. (NASA)

## Voyager

Showing its *Mariner* heritage, *Voyager* was arguably one of the most successful planetary missions. The name *Voyager* was adopted just before launch. During design and construction it was referred to as *Mariner Jupiter-Saturn 1977 (MJS '77)*. The name *Voyager* was adopted in March 1977, and is a designation resurrected from a defunct and very expensive Mars lander.

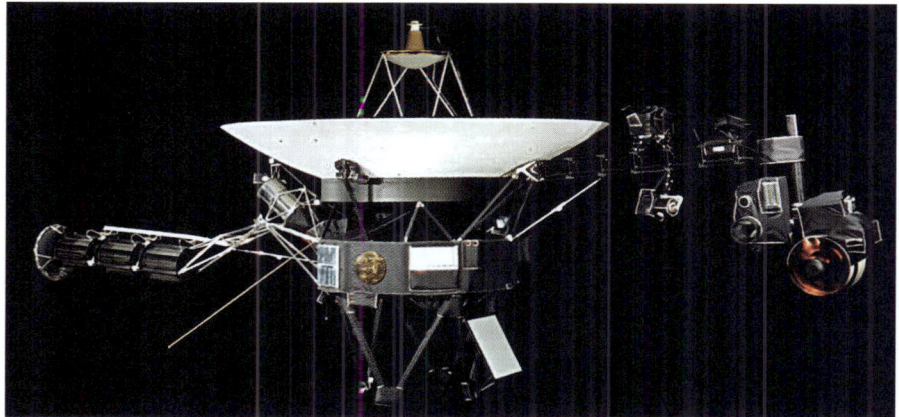

Recognised by the large high gain antenna, Voyager still shows its Ranger/Mariner heritage but was powered by Radioisotope Thermoelectric Generators rather than solar panels. (NASA/JPL-Caltech/PD-NASA)

## Imaginative Selection?

Project names are important; they can capture the imagination and inspire the younger generation. Perhaps they should not be left to those seeking the most unusual acronym. I miss the days of the *Mariners*, *Surveyors* and *Voyagers*. Names have evolved over the years, perhaps not for the best. I would expect imaginative selection to weed out the weak names but that does not appear to be the case. To paraphrase Charles Darwin, "Whilst this planet has gone cycling on according to the fixed law of gravity, from so simple a beginning endless names most beautiful and wonderful have been and are being evolved." Though how derived acronyms can be beautiful and wonderful is beyond me.

## Further Reading

Wells, Helen T., Whiteley, Susan H., Karegeannes, Carrie E., 1976, 'Origin of NASA Names', *The NASA History Series*, NASA SP-4402
    nasa.gov/wp-content/uploads/2023/03/sp-4402.pdf

# Mission to Mars
## Countdown to Building a Brave New World
## Pausing for Thought

### Martin Braddock

**Introduction**
With the announcement from NASA in early 2024 that the *Mars Sample Return (MSR)* project is to be put on hold, it is now appropriate for us to pause our series of articles describing a future for Mars colonisation. Between 2021 and 2025 this series of articles so far has explored the basic thinking behind establishing a settlement on Mars. Starting in 2021, with 'It All Starts With a Journey' we explored some of aspects of the space environment that mission planners will need to contend with for long, as in years, duration missions off Earth. In 2022's 'Laying the Foundations', we reviewed our understanding of Martian regolith and how 3D printing may be used to pre-assemble modular-like structures for habitats taking into account the growing influence 3D printing has on design and construction of Earth-based buildings. In 2023 in 'The Bare Necessities of Life', we started to dive into what it may take both physically and mentally to live on Mars, drawing on some of the experiences 'colonists' faced during Mars Analog Missions based in various countries around the world. We started to think about the mental and psychological requirements which we addressed in more detail in 2024's 'The Right Stuff at the Right Time', noting that we have a role for ourselves as educators to help the general public appreciate, understand and value the importance of people leadership whether in space or on Earth. In 2025 we discussed in 'It's Life But Not as We Know It' some of the incredible advances in humanoid robotics and how robots are likely to become our partners in space, in the same way that they are already assisting humans in tasks where it is either unsafe for humans to work, or where the repetitive nature does not require human inventiveness and thinking. Finally, at least for now, in 2026 and 'Pausing for Thought' we will review the current situation with Mars exploration as of July 2024 and attempt to reimagine the future and project a vision, taking into account any external factors which may lead to the need to delay or to accelerate delivering our ambition.

## What is MSR and Why is it Important?

The MSR mission is an ambitious NASA-ESA joint programme to return rock, dust, and atmospheric samples from Mars to Earth for detailed physical, chemical and biological analyses. MSR aims to fulfil, in part the objectives of *Mars2020*, and address key questions about the potential for Mars to have hosted life forms, to harbour life for the future and to further our knowledge in planetary evolution and conditions for habitability. It is the conclusion of the activities of the *Perseverance* rover which has been collecting and caching rock samples in the Jezero crater since landing in February 2021. An excellent review summarises the progress made and shows the route followed by the rover together with photographs of the Martian landscape in astonishing detail (Sun et al 2023). A successful *MSR* mission will demonstrate our capability to package samples on Mars and return them safely to Earth in a fully autonomous manner, which in turn will add enormous credibility to future efforts to support astronaut missions to the planet. The *MSR* mission is also an essential component of our ability to understand the capacity for life as it will enable analyses of material too complex for in situ robotic investigation and, as with the collection of lunar samples in previous missions, provide material to research laboratories around the world. As of the time of writing this article, *Perseverance* has collected 24 samples out of a total of 38 to be assembled for return to Earth and NASA publishes live updates for the general public on mission progress.[1]

Indeed, one of the first hints that microbial life could have been supported on Mars came from findings from NASA's *Curiosity* rover in 2017. *Curiosity* recorded elevated levels of manganese in rocks from the lake bed within Gale crater. The rocks which are sedimentary in nature are comprised of particles of larger grain sizes which may be indicative of their formation in a river, a river delta, or near the shoreline surrounding the ancient lake. A recent publication (Gasda et al 2024) considers and proposes mechanisms explaining how manganese could have been enriched in these rocks in the crater. Hypotheses include the percolation of groundwater through original sediments or through the rock after its formation and the identity of the oxidant which could have caused the precipitation of manganese.

Noteworthy is that Earth's geological history recorded abundant levels of manganese in rocks and oceans before the earliest forms of life were estimated to emerge approximately four billion years ago. Indeed it is believed that this element

---

1. For more information, see 'Perseverance Rover Updates' at **science.nasa.gov/mission/mars-2020-perseverance/science-updates**

# Mission to Mars: Countdown to Building a Brave New World 305

Exciting findings from Cheyava Falls! Could there have been life on Mars? (NASA/JPL-Caltech/MSSS)

was a key indicator for oxygen. Manganese becomes enriched due to the presence of high levels of atmospheric oxygen, a process which is often accelerated by the presence of microbial populations. Interestingly microbes on Earth can use the many different forms of oxidated manganese, call oxidation states, as energy sources for their metabolism (e.g. LaRowe et al 2021). It may be speculated that if microbial life was present on ancient Mars, elevated levels of manganese in rocks found at Gale crater could have been used as an energy source and to maintain and grow a population of microbes.

In July 2024, NASA announced that a rock on the surface contained signs of the possible past existence of microbial life.[2] A sample has been drilled from a rock, dubbed Cheyava Falls, located from a site believed to be a river delta that flowed into the Jezero crater. The limited analyses that can be formed in situ report minerals that precipitate from water and organic compounds. Intriguingly the

---

2. Could there have been life on Mars? **science.org/content/article/nasa-says-it-found-possible-signs-life-mars-there-are-lot-maybes**

sample also shows 'leopard spots' (indicated with a red circle on the accompanying image) which appear to contain iron and phosphate. This type of colouration on Earth is indicative of organic molecules reacting with haematite which can fuel microbial life forms. Our confirmation, or not, of the possible past existence of life can be only be ensured if the samples are returned to Earth for more extensive analyses. In addition, the veins in the rock are filled with crystals of olivine (red circle), a mineral that forms from magma. The olivine might be related to rocks that were formed farther up the rim of the river valley and that may have been produced by crystallization of magma.

Knowing whether Mars supported microbial life and whether it could do so in future would revolutionise our thinking on Mars habitability. For example, the potential for microbial life to be deliberately introduced to bioremediation soils of toxic perchlorate is a likely essential requirement for habitation (Keaney et al 2024) supporting both plant growth and a healthy human lifestyle.

## So What is the Problem with MSR?

It all comes down to cost.[3] In late 2023 an independent review board determined that the *MSR* schedule and budget were no longer viable.[4] With the project planned to launch in 2026, management and delays in program design and lack of detailed plans for both *Mars Ascent Vehicle (MAV)* and Mars orbiter reprojected, at earliest, a 2030 launch date and possibly as late as 2040, together with a cost increase from an estimated $3.8bn to $4.4bn in 2020 to a range of between $8bn and $11bn today (Witze 2024). This decision is a significant setback as operationally, it would be a huge advantage and vote of confidence is our technology and the capability of bringing home human beings safely.

It does, however, lay down the gauntlet to scientists and engineers.[5] The collection of 24 core samples in tubes by *Perseverance* at Jezero Crater, an ancient river delta with at least 30 samples planned to be cached remains important in the search for possible signs of Martian life. NASA's aim and challenge is to return some of the samples to Earth in the 2030s but for no more than a total budget of $7bn. In order to achieve this goal, engineers will need to design a space craft capable of going to

---

3. Further details can be found at **cbc.ca/news/science/nasa-mars-sample-return-1.7174976**
4. The 'Mars Sample Return (MSR) Independent Review Board-2 Final Report' can be downloaded at **nasa.gov/wp-content/uploads/2023/09/mars-sample-return-independent-review-board-report.pdf**
5. Coordinating a delivery an autonomous lift-off from Mars is a new challenge for scientists and engineers at **spectrum.ieee.org/mars-sample-return-mission**

Beautiful Mars and we will have to be patient! (Pixabay)

Pressing the pause button on *Mars Sample Return*. (Pixabay)

Mars, collecting the samples and then taking off from the surface to return them to Earth via an orbiting spacecraft as was conducted with the lunar missions. It is expected that analysis of the samples will inform mission planners where to send astronauts, now estimated to be in the 2040s.

Despite the concerns regarding both cost and time for completion, it is widely accepted that MSR provides a unique opportunity to sample materials from another planet in Earth-based laboratories. In addition to, perhaps, providing us with information which informs on planetary evolution and the minimal conditions required to support life, as a collaborative project across the two major space agencies NASA and ESA it will enhance our capabilities in robotic sample handling, Mars exploration operations including design considerations to support remote take off.

## Lifting off from Mars – How Will We Do It?

First we need think through the challenges and the various pros and cons relating to our past experiences; lifting off from Earth and the Moon. The gravitational field strength of Mars is ~38% of Earth, whereas for the Moon it is ~17%. The escape velocity – that is, the velocity needed for a spacecraft to escape from a gravitational centre of attraction without undergoing any further acceleration – is ~11.2 km/s for Earth ~5 km/s for Mars and ~2.4 km/s for the Moon. The Martian atmosphere provides some surface pressure, although only ~1% that of Earth.

At least today, the *MSR* mission is to be a collaboration of fully autonomous robotic missions combining the expertise of NASA and ESA scientists and engineers. In the first stage NASA will design a lander and *MAV* for delivery to the planet's surface. Design considerations for the *MAV* are under consideration (e.g. Renault et al 2023, Rowntree 2023)[6] and will be crucial component to the project. *Perseverance* will deliver the sample tubes in a canister to the *MAV*. In parallel, an Earth return orbiter craft will be orbiting the planet to receive the samples from the *MAV* once it has completed lift off and docking. With the samples safely onboard the return orbiter, they will be brought back to Earth via an entry capsule to a receiving facility for storage and analyses. The Northrop Grumman Corporation, which has a history of supporting space exploration and the lunar missions, will develop the *MAV's* solid rocket propulsion system. There are numerous technical challenges; the primary obstacle to overcome being that the craft function is 100% robotic and there are no opportunities for human intervention as there are on Earth. Secondly, rocket launches have been postponed for various reasons at short notice including late identification of technical faults and changes in weather conditions. As the one way signal latency, that is the time it takes for communication between Earth-Mars-Earth is approximately 20 minutes (Cahill and Braddock 2022), once the countdown has started there is no opportunity to abort. Lastly, there are a number of physical parameters to consider. On average the temperature on Mars ranges between $+20°C$ and $-60°C$ and there is a thin atmosphere. The designed

---

6. Designing the *MAV* components requires a Systems Requirements Cycle at ntrs.nasa.gov/api/citations/20210022133/downloads/MAV_SRC_IEEE_2022.pdf

propulsion system and rocket engines will need to be able to withstand those extremes and be resilient to dust storms should they arise.

## When is it Going to Happen?

With all dates firmly 'TBD', it is worth us considering the current proposed timeline for *MSR* while acknowledging the complexity of multiple programmes working both in series and in parallel. A pause in activity may make the path to Mars colonisation straight and clear and although obstacles may present, they can be overcome with human ingenuity and innovation. There is no better lever to pull than that of public opinion to make massive projects happen and it is important for us to consider the prize that could await us, perhaps not in our generation or even the next but in generations to come. The world has always been in a volatile state with nations in conflict and as technology becomes evermore able to wreak destruction on a huge scale, we need to reflect on our ability and readiness to preserve our species, to accept our differences and to realise that perhaps one day we will be able to colonise other worlds. The potential for climate change to dramatically alter the ability of life to be supported, even exist on Earth is a likely driver for urgency, not only to protect Earth, but to accelerate our knowledge of the feasibility of developing a 'plan B' (eg Kemp et al 2022).

The first phase of the mission will be the NASA launch of both the *Sample Retrieval Lander* and the MAV in 2028 to land as close to the *Perseverance* rover as

Pausing now will make the road to human missions clear and straight. (Pixabay)

practicable. The second phase of the mission in 2030 will be the delivery of the *MAV* rocket and two helicopters to the site by the lander. The role of the helicopters will be to transport the tubes containing the cached sample to the lander. At this stage of the planning, a further option is to use *Perseverance* to transport the sample tubes directly using its robotic arm functions and this level of detailed will be clarified in due course. Having safely transferred the samples onto the *MAV*, it will lift off and take them into Mars orbit. The third phase of the joint NASA-ESA mission uses an *Earth Return Orbiter (ERO)* built by ESA. This will have launched in 2027, will dock with the *MAV* in 2031 and then launch the sample container for return to Earth via an entry vehicle to be released on a course for Earth in 2033. The fourth and final stage of the flight mission is for the *ERO* vehicle to enter the Earth's atmosphere and land in Utah, USA.

Having secured the samples safely back on Earth, they will then be transported to secure containment facilities for storage, cataloguing and the plethora of analyses which will follow. It is anticipated that as for lunar samples from the *Apollo* missions, housed at lunar sample building at the Johnson Space Centre, there will be a central repository, location to be determined, for samples to be made available to scientists around the world.

In summary, for now it is business as usual. *Perseverance* is still collecting and storing samples with the planning assumption that one day they will be returned to Earth. Mission planners remain resolute in considering the technical and medical challenges that present themselves for both lunar and eventual Martian colonisation and continue to make breakthroughs devising and developing new technology that drives innovation for all on Earth as well as in space. We continue to dream of understanding what Mars was once, whether it has lessons for us on Earth with regards to sustainability and what prospects we have of making it our second home. The *MSR* project is a key stepping stone to demonstrating the viability of this vision and this mission together with other fully autonomous launches for the future will pave the way for humans landing on the first of likely many new worlds.

NASA's science mission chief, Nicky Fox said: "We've never launched from another planet, and that's actually what makes Mars sample return such a challenging and interesting mission."[7]

---

7. See 'NASA Seeks New Way to Bring Mars Rocks to Earth' at **learningenglish. voanews.com/a/nasa-facing-budget-issues-seeks-new-way-to-bring-mars-rocks-to-earth/7575867.html**

## References

Cahill, B., Braddock, M., 2022, 'A Crowd-Sourcing Project to Understand, Prevent and Manage Incidences of Injury and Wounding to Astronauts and Off-Earth Colonists' *Journal of Human, Earth and Future*, **3**, 299–321.

Gasda, P. J., Lanza, N. L., Meslin, P.-Y., Lamm, S. N., Cousin, A., Anderson, R., Forni, O., Swanner, E., L'Haridon, J., Frydenvang, J., Thomas, N., Gwizd, S., Stein, N., Fischer, W.W., Hurowitz, J., Sumner, D., Rivera-Hernández, F., Crossey, L., Ollila, A., Essunfeld, A., Newsom, H.E., Clark, B., Wiens, R.C., Gasnault, O., Clegg, S.M., Maurice, S., Delapp, D., Reyes-Newell, A., 2024, 'Manganese-rich Sandstones as an Indicator of Ancient Toxic Lake Water Conditions in Gale Crater, Mars', *Journal of Geophysical Research: Planets*, **129**, #5 **agupubs.onlinelibrary.wiley.com/doi/10.1029/2023JE007923**

Keaney, D., Lucey, B., Finn, K., 2024, 'A Review of Environmental Challenges Facing Martian Colonisation and the Potential for Terrestrial Microbes to Transform a Toxic Extraterrestrial Environment'. *Challenges*, **15** (1), 5 **mdpi.com/2078-1547/15/1/5**

Kemp, L., Xu, C., Depledge, J., Ebi, K. L., Gibbins, G., Kohler, T. A., Rockström, J., Scheffer, M., Schellnhuber, H. J., Steffen, W., Lenton, T. M., 2022, 'Climate Endgame: Exploring catastrophic climate change scenarios', *Proceedings of the National Academy of Science (PNAS)*, **doi.org/10.1073/pnas.2108146119**

LaRowe, D. E., Carlson, H. K., Amend, J. P., 2021, 'The Energetic Potential for Undiscovered Manganese Metabolisms in Nature', *Frontiers in Microbiology*, **12** **doi.org/10.3389/fmicb.2021.636145**

Renault, M., Lappas, V., 2023, 'Design of a Mars Ascent Vehicle Using HyImpulse's Hybrid Propulsion', *Aerospace*, 10. **mdpi.com/2226-4310/10/12/1030**

Rountree, I., 2022. 'MBSE Applications for the MSR SRC Mars Ascent Vehicle', *2022 IEEE Aerospace Conference (AERO)*, Big Sky, MT, USA, 1–14 **doi:10.1109/AERO53065.2022.9843502**

Sun, V. Z., Hand, K. P., Stack, K. M., Farley, K. A., Simon, J. I., Newman, C., et al., 2023. 'Overview and Results From the Mars 2020 Perseverance Rover's First Science Campaign on the Jezero Crater Floor'. *Journal of Geophysical Research: Planets*, **128**, #6 **agupubs.onlinelibrary.wiley.com/doi/full/10.1029/2022JE007613**

Witze, A., 2024. 'Is the Mars rover's rock collection worth $11 billion?' *Nature* **doi.org/10.1038/d41586-024-00831-0**

# A History of Observatory Designs
## The Telescope Age from the Seventeenth to Nineteenth Centuries

### Katrin Raynor

In the *Yearbook of Astronomy 2025*, I introduced the history of early observatory designs – early observatories in the sense that they were built before the modern structures we know today. We looked briefly at examples of stone circle observatories made by our ancient ancestors before making acquaintance with the gigantic sextant constructed by Ulugh Begh in Uzbekistan and the impressive observatory constructed by Danish astronomer Tycho Brahe, the latter being the first observatory to be custom built on mainland Europe and the last without the presence of a telescope. From around three decades after Brahe completed Uraniborg, the invention of the telescope began to substantially advance our understanding of the universe and our place within it.

The first telescope is credited to the Dutch spectacle-maker, Hans Lippershey (c.1570–1619) who tried, and failed, to patent his idea of a refracting telescope in 1608. Regardless of this set back, his invention drew great interest from the scientific community, in particular the Italian polymath Galileo Galilei (1562–1642), whose name is synonymous with the invention of the telescope and who is often referred to as the Father of Astronomy.

The telescopes made by Galileo, and the important telescopic observations he carried out, paved the way for new ideas in telescope making. During the centuries that followed, telescope design underwent significant modifications and refinements, and this observing tool is now one of the most important and key scientific instruments used by astronomers.

The following article provides a select few interesting examples of ground-based observatories housing telescopes and other astronomical observing instruments, ranging from the seventeenth to the nineteenth century.

**Leiden Observatory**
Located at the astronomical institute of Leiden University in the Netherlands, Leiden Observatory was one of the first purpose-built observatories and holds

the title of being the world's oldest operating facility of its kind. Albert Einstein (1879–1955) frequently visited Leiden Observatory, and Ejnar Hertzsprung (1873–1967) and Jan Oort (1900–1992) worked there.

Constructed in 1633 at the request of the Dutch Orientalist and mathematician Jacobus Golius (1596–1667), who worked at Leiden University, the observatory was needed to house and use a large quadrant which had been built by his colleague, the astronomer and mathematician Willebrord Snellius (1580–1626). Nearly 40 years after its construction, the observatory was expanded and a second turret built, each turret having a rotating roof to facilitate observing the sky. The observatory stood until 1817, when it was sadly torn down, rebuilt and renovated again in 1837.

By 1853, thanks to the growing popularity of the subject of astronomy attributed to the work of lecturer and director of the observatory Frederik Kaiser (1808–1872), a new observatory was commissioned to be built, this time at the university's botanical gardens in a quieter area of the city. The new observatory included domes, offices and accommodation for astronomers, the building being subsequently extended numerous times to include additional rooms for housing instruments and a new dome to contain a photographic telescope.

Leiden Observatory was the first purpose-built observatory and is one of the oldest operating observatories in the world. Constructed in 1633, it underwent reconstruction in 1837 before being relocated to the university's botanical gardens. (Wikimedia Commons / AWossink / CC BY-SA 4.0)

Astronomical observations are unfortunately no longer carried out at the old Leiden Observatory. In 1974, the astronomy department was moved to the Leiden Bio Science Park – the Jan-Hendrik Oortgebouw and the Huygens Laboratory on the Wassenaarseweg. Between 2008 and 2012, the old observatory under went restoration and visitors can now make use of the visitor centre located in the basement which hosts exhibitions and a telescope that is over 120 years old. The old Observatory domes still house four historical telescopes which, unfortunately, are no longer used.

## Rundetaarn Observatory

Constructed in Denmark just a few years after the Leiden Observatory, Rundetaarn Observatory is still standing today in the city of Copenhagen which, as you can imagine, is an unusual location to build a facility of this kind. King Christian IV started construction of the tower in 1637 by way of continuing Tycho Brahe's (1546–1601) research; Brahe had died in 1601 and was one of Denmark's most notable astronomers. Rundetaarn Observatory was, and remains, one of the most iconic buildings in the city. This is not necessarily due to any particular observing equipment but because of its remarkable spiral ramp. Known as an equestrian staircase, this is a very gently sloping flight of steps constructed in order to allow it to be negotiated by horses. A little over 200 metres in length, the staircase leads to the observing platform, located 34.8 metres above street level. To reach the platform, there are 7.3 turns along the impressive ramp, each turn having an elevation of 3.74 metres. The observatory that currently stands on the roof of the tower was built in 1929 and houses a powerful refracting telescope.

An unusual location to build an observatory, the 35-metre tall Rundetaarn Observatory, or Round Tower, still stands today in the city of Copenhagen. Built in the seventeenth century, it is one of the most iconic buildings within the city centre. The observatory's internal equestrian staircase that leads to the observing platform is a highlight for modern-day visitors. (Wikimedia Commons/Jay Galvin/CC BY 2.0)

## The Observatories of Johannes Hevelius

Johannes Hevelius (1611–1687) is one of the best-known astronomers of the seventeenth century, and his observatories are certainly noted for ingenuity in design and construction. The most famous example he built was actually his second observatory, constructed in 1650 and which spanned the roofs of three connected houses that he owned. Known as *Sternenburg* or 'Star Castle', the observatory was a 7.5 × 15 metre wooden platform housing a huge refracting telescope with a focal length of 46 metres. The observatory was well thought out, having a rotatable shelter for housing a brass horizontal quadrant and a hut for a large metal sextant and octant. The observatory was destroyed by fire in 1679, although prior to this tragic event, Hevelius had carried out a lot of useful work from *Sternenburg*, including the observation of sunspots and carrying out a four year programme charting the lunar surface. He had made some of the most incredible discoveries from the observatory, including four comets and the Moon's libration in longitude.

## Royal Greenwich Observatory

A UNESCO World Heritage site, Royal Greenwich Observatory (RGO) is the home of Greenwich Mean Time (GMT) and a location through which the Prime Meridian of the world – longitude 0° – passes, dividing the world into Eastern and Western hemispheres. It is perhaps the best-known observatory in Great Britain today. Since 1884, the Prime Meridian has been used to measure time internationally.

Located at Greenwich Park, London, the Royal Greenwich Observatory was established in 1675. It was designed by Sir Christopher Wren (1632–1723) and built on the ruins of Greenwich Castle, with Flamsteed House being the first part of the observatory to be constructed. The first Astronomer Royal, John Flamsteed (1646–1719) took up residency at the observatory just under a year after it had been built, and would remain there for 40 years. The Greenwich Time Ball, installed in 1833 by the sixth Astronomer Royal, John Pond (1767–1836) and still in operation today, was the first public time signal in Great Britain. At a pre-determined time each day, a bright red ball at the top of Flamsteed House was dropped down a mast to signal the time to onlookers and, most importantly to ships on the River Thames. Since its construction, the RGO has made many significant contributions to astronomy, time keeping and the determination of star positions. Astronomical observations were always a key part of the observatory. One of its significant instruments was the Great Equatorial Telescope, a 28-inch refractor produced by the Irish telescope maker Howard Grubb (1844–1931) and installed in an iconic onion-shaped dome at the GRO in 1893.

The onion shaped dome at the Royal Observatory Greenwich, London houses one of the largest refracting telescopes in the world. Installed in 1893, the Great Equatorial Telescope was to be used for astrophotography purposes but became pertinent to the observatory's research into double stars. The Onion Dome was originally constructed using a riveted iron frame and covered with papier mâché. At its widest point, the dome bulges five feet from the observatory's walls. (Royal Observatory, Greenwich / Elliott Brown / CC BY 2.0)

## MacFarlane Observatory

This was the first purpose-built university observatory, not only in Scotland but in Great Britain. It was named after Scottish merchant and astronomer Alexander MacFarlane (1702–1755) who had bequeathed his astronomical instruments to the University of Glasgow following his death. During his time living and working in Jamaica, MacFarlane had bought a selection of astronomical instruments from fellow astronomer Colin Campbell (d.1752), including a 5-foot transit telescope, a 5-foot zenith sector and a one-month regulator clock. When the instruments arrived in Glasgow in 1756 following MacFarlane's death, the Scottish chemist, inventor and mechanical engineer James Watt (1736–1819) was tasked with restoring the instruments to a useable condition. A location was needed to house the instruments, and construction of what became the MacFarlane Observatory began the following year. Upon its completion, the observatory boasted a frontage of 60 feet. In 1760, Alexander Wilson (1714–1786) was appointed as professor of practical astronomy, his most notable work being that in solar physics. The MacFarlane Observatory has since long gone – the construction of taller buildings near the site during the nineteenth century rendered it obsolete and flats now stand in its place.

## Observatories at the Vatican

The Observatory of the Vatican in Vatican City, Rome, Italy was established in 1774, at which time was known as the Observatory of the Roman College. The Vatican has a long and rich history in astronomy which can be traced back to the constitution by Pope Gregory XIII which was, according to the Vatican Observatory website, '… a committee to study the scientific data and implications involved in the reform of the calendar which occurred in 1582.' Later observatories founded by the papacy included the *Specula Vaticana* (Vatican Observatory) established in 1789 in the Tower of the Winds in the Vatican, and the Observatory of the Capitol, operational from 1827 to 1870). The closure of the *Specula Vaticana* in 1321 was primarily due

The Vatican has a long and rich history with astronomy. This photograph of the Pontifical palace and Vatican Observatory, located at Castel Gandolfo, clearly shows two of the observatory domes. The observatory was re-located here in the 1930s following the increase in light pollution. The larger dome has a diameter of 8.5 metres whilst the smaller dome has a diameter of 8 metres. Both were constructed in 1935. (Wikimedia Commons/H.Raab (User Vesta)/CC BY-SA 4.0)

to the obstruction of viewing caused by the dome of St Peter's Basilica. All the instruments were subsequently transferred to the College Observatory. In 1878, the College Observatory was nationalised by the Italian government and re-named the *Regio Osservatorio al Collegio Romano* (Royal Observatory at the Roman College). This change signalled the end of astronomical research at the Vatican. Thirteen years later however, Pope Leo XIII re-founded the Vatican Observatory, locating it on a hill behind St Peter's Basilica where it operated for 40 years. In the latter part of the nineteenth century, the Vatican Observatory was involved in creating a photographic celestial map and an astrographic catalogue. Due to the increase in light pollution, a further observatory was established in the 1930s, 25 kilometres south-east of Rome at the Papal Summer Residence at Castel Gandolfo. As the decades passed by, recurring light pollution issues meant that the observatory had to be re-located yet again. In 1981, the Vatican Observatory Research Group founded a second research centre at Steward Observatory, University of Arizona. The Vatican Advanced Technology Telescope was installed at Mount Graham International Observatory, Arizona (a division of Steward Observatory) in 1993.

## The Leviathan of Parsonstown

Also known as the Rosse six-foot telescope, the Leviathan of Parsonstown certainly lives up to its name. Built between 1842 and 1846 it was the largest telescope in operation until the construction of the 100-inch Hooker Telescope for Mount Wilson Observatory, California in 1917. The 72-inch reflecting telescope was built

Taken from the Grubb Parsons Ltd collection at Tyne & Wear Archives, the photograph shows the ginormous mirror that was used for the Leviathan of Parsonstown telescope. (Wikimedia Commons / Grubb Parsons Ltd / Tyne & Wear Archives & Museums)

A History of Observatory Designs 319

The Leviathan of Parsonstown certainly lived up to its name. This Victorian colour-cotton wall hanging shows the scale of the six-foot telescope which, between 1845 and 1917, was the largest telescope in operation. Construction of the 72-inch reflector took place between 1842 and 1846 and was carried out by William Parsons, 3rd Earle of Rosse at Birr Castle, Ireland. The telescope was re-constructed during the 1990s following the renewed interest inspired by Patrick Moore. (Wikimedia Commons/Working Men's Educational Union/National Maritime Museum, Greenwich, London/CC BY-NC-SA 3.0)

by William Parsons, 3rd Earle of Rosse (1800–1867), who was an engineer and astronomer, on his estate at Birr Castle in King's County (now County Offaly), Ireland. The construction of the telescope was no mean feat; the mirror alone weighed three tons and was five inches thick! Combined with the length of the tube and mirror box of 54 feet, the total weight of the Leviathan was approximately 12 tons. An observatory of sorts was built between 1843 and 1844 whereby the telescope was housed in walls approximately 71 feet (21 metres) long, 40 feet (12 metres) high and 23 feet (7 metres) apart. These walls restricted the telescope's movement, which were controlled using a cable chain and pulley system, meaning that only one hour of observing time of an object on the celestial equator could be undertaken each night.

William Parsons spent much time observing nebulae and galaxies through the telescope. When he died in 1867, his son Lawrence Parsons, 4th Earl of Rosse (1840–1908), took to using the telescope, his most notable work including confirmation of the discovery of the satellites of Mars in 1877 and pioneering research relating to the infrared emission of the Moon and estimations of the temperature of the lunar surface. After Lawrence Parsons' death the Leviathan was not used again; it was partly dismantled and its mirror sent to the Science Museum in London where it still resides today. During the 1970s, following a television programme and publication of a book by Patrick Moore,[1] interest in the Leviathan of Parsonstown was renewed, the telescope and observatory eventually being restored during the 1990s.

Lick Observatory – the first permanently occupied mountain top observatory – is situated at Mount Hamilton, California and is now operated by the University of California. Constructed between 1876 and 1887, the observatory was built at the bequest of James Lick. This photograph was taken in 1902 and illustrates the neo-classical design of the building and the two domes, the larger of which houses the Great Lick Refractor telescope. (Wikimedia Commons/Lick Observatory/Detroit Publishing Company/Trialsanderrors/US Library of Congress/ppmsca.17974)

---

1. *The Astronomy of Birr Castle*, Patrick Moore, 1971. Mitchell Beazley, ISBN 0 85533 004 X

## Lick Observatory

The world's first permanently occupied mountain top observatory, Lick Observatory was constructed between 1876 and 1887 on the summit of Mount Hamilton, just east of San Jose, California, United States of America. The observatory was built at the bequest of the American carpenter, piano maker and patron of the sciences James Lick (1796–1876). An elevation of 1,283 metres above sea level, understandably, proved to be problematic when construction started. Materials were transported by horse and cart, and entailed long journeys up and along an inclined path to the top of the mountain. The observatory was built in neoclassical style architecture and housed many impressive telescopes during the nineteenth century. These included the famous 36-inch Great Lick Refractor – at the time the world's largest refracting telescope – and the 36-inch Crossley Reflector. Two domes were built at the observatory; a smaller-dome in 1881 where a 12-inch telescope was installed and, several years later, and after facing various obstacles, the great dome which, when completed, would house the Great Lick Refractor. During the nineteenth century much notable work was carried out from Lick Observatory including measurements of the angular diameters of the four Galilean satellites of Jupiter Albert A. Michelson in 1891 and the discovery of Amalthea, the fifth moon of Jupiter, by Edward Emerson Barnard in 1892.

## Further Reading

Glasgow University Observatories
**www.astro.gla.ac.uk/observatory/history/obs-hist.shtml**

Portal to the Heritage of Astronomy
Hevelius Observatory, Danzig, Poland
**web.astronomicalheritage.net/show-entity?identity=101&idsubentity=1**

The Leviathan's Legacy: the Story of the Birr Castle Telescope
**skyatnightmagazine.com/space-science/the-leviathans-legacy**

History of the Vatican Observatory
**www.vaticanobservatory.va/en/history**

Historical Resources: Lick Observatory
**lickobservatory.org/explore/historical-resources**

Royal Museums, Greenwich
**rmg.co.uk**

# Sidewalk Astronomy
## Cosmos to Kerbside

## Jonathan Powell

Many scientists and philosophers have committed their entire lives to a chosen discipline in the knowledge that in the event of any discovery, or worldly philosophical insight, the fruits of their endeavours will have in some way helped to unravel humankind's understanding of the universe in which we live. Whilst written words of explanation can readily inform and enlighten, there remains an unquestionable premise that the imparting of such gained insight and wisdom is, if possible, best delivered in person, first hand, as a tutor would engage with a student.

It would be a travesty if all those concerted efforts across the centuries had no audience to realize why an individual would invest their life on such a journey, for all that mastery gained to simply gather dust in a book on a library shelf. To educate and enthuse others not only signifies an acknowledgement to the generations of people who spent their time attempting to unravel the great cosmic questions, but to also act as a catalyst to those

This image, which featured in the March 1921 issue of *Popular Science* magazine, accompanying the article 'Adventures in Street Corner Astronomy' by artist, photographer and amateur astronomer Latimer J. Wilson (1878–1948), depicts sidewalk astronomer Joseph G. White showing '... the new comet or the planets ...' to the public through his 4½-inch refracting telescope from a location on 42nd Street, New York. (Wikimedia Commons/Latimer J. Wilson/*Popular Science*)

who would take up such a mantle and work toward making that next significant breakthrough or insightful determination.

Outside of academic institutions, and for those intentionally seeking to find out more about space and astronomy, one of the first points of contact has for decades been a local astronomical society, where the experiences of other likeminded people can nurture and encourage the eager mind. Along with observing sessions, the practical and progressive input one can glean from workshops shows astronomy to be not only a science, but in the likes of such topics as telescope making, an art and a skill.

However, for some, the pursuit of astronomy may not be found within an astronomical society, with a fair proportion wishing their interest to be a solitary one. There are also those that harbour less of a fascination with the science but still possess an inkling that once further inspired, could see more of an involvement whereby a dormant desire is subsequently ignited. Without that stimulus of being engaged outside of a school, college, or university, their curiosity and potential may never be fulfilled.

It would therefore seem prudent that bringing science and astronomy into a wider public domain outside of a lecture hall could well meet the needs of a good many people, and perhaps in-turn encourage that initial spark into a burning flame?

## Early Beginnings of Sidewalk Astronomy

Prior to sidewalk astronomy, and before its presence in urban life became a recognized part of modern times, the want to share such images of our Moon or the stars and planets through a telescope did exist, but in a much more isolated set-up. Dating back to the eighteenth century, with possible earlier references even pre-dating that period, such a practice has been documented. Amongst those who probably couldn't really be classed as a sidewalk astronomer, but certainly engaged in doing so, was Sir William Herschel, (1738–1822). Whilst observing the Moon from the street (possibly in Bath) Herschel is said to have engaged with many passers-by, a selection of whom being curious as to his endeavours. References to one such passer-by would suggest that it was one Sir William Watson (1744–1824) who, after engaging with Herschel, were to become good friends. Indeed, as their friendship grew in later years, Herschel was to build Watson his own telescope.

During the nineteenth century, sidewalk astronomy had taken on the more popularized guise of 'street corner astronomy', with the sight of telescopes as well as other instruments such as microscopes, bringing science to the public as a first-hand experience outside of the observatory and laboratory. Whilst some would charge for the 'service' of allowing a person to gaze through a telescope or look down a microscope, others offered the opportunity for free.

In later years, and entering the realms of local folklore, people like self-proclaimed 'star hustler' Herman Heyn (1930–2021) would, for three decades, erect his telescope on a street corner in Baltimore, Maryland, USA, and for just a monetary 'tip', share images of the night sky with the public. Heyn was to intermittently upgrade his optics via a succession of telescopes that he was to acquire, forever enhancing the experience. His enthusiasm for astronomy inspired many, some who were to buy their own telescopes or even venture into the world of astronomy as a career.

## The New Era of Sidewalk Astronomy

American amateur astronomer John Lowry Dobson (1915–2014) built upon those early foundations of sidewalk astronomy, extensively widening the possible catchment area of those who could be reached. Like the expansion of the universe and under Dobson's mentoring, the seeds were embedded for what was to eventually become a global entity. Dobson, whose legacy of achievements echoes around the world to this day, moved sidewalk astronomy into a different realm of possibility and encapsulated his own incredible drive, passion, and vision to motivate and involve others in astronomy.

Dobson was born in Beijing, China, relocating with his parents to San Francisco, California, in the mid 1920s. Here, Dobson was to gain a degree in Chemistry from Berkeley, and after attending a lecture by a Vedanta *swamiji* in 1944 (Vedanta – based on Indian (Hindu) philosophy) duly joined the Vedanta Society in a San Francisco monastery where, as a monk of the Ramakrishna Order, reconciled his passion for astronomy with its teachings. Perhaps unknown to Dobson at that time, the kernel of what he was to later deliver to the world as sidewalk astronomy was slowly manifesting itself. For, with his closely guarded interest in telescope making shielded from those around him at the monastery, Dobson found himself venturing outside of its confines, randomly engaging with people to enlighten them about the wonders of the night sky through his newly furnished optical device.

As telescope making was not part of the curriculum at the monastery his antics were frowned upon, with Dobson resorting to using coded messages in any of his correspondence to conceal the fact that such construction was taking place, referring to a telescope as a 'geranium'. Eventually his superiors, suspicious of his doings, could tolerate his behaviour no more and Dobson was offered the straight choice of either terminating the making of telescopes or leaving the monastery. Dobson chose the former but was expelled regardless in 1967 when, as Dobson

John Dobson dedicated his life to fulfilling an inner passion and resolve to bring astronomy to those who may never, without his concept of sidewalk astronomy, have witnessed the wonders of the night sky. Classing himself as an amateur astronomer and offering reluctance to accept the credit bestowed upon him for the innovative Dobsonian telescope, Dobson's commitment has seen the evolution of a network of sidewalk astronomers that spans the globe, all providing the mission statement of "Bringing astronomy to the Public". His legacy was to provide a means to educate, inform, and connect astronomy with the public like never before. (Richard Berry)

himself documents, a misunderstanding took place which led to the head *swamiji* believing Dobson had rejected his teachings by contradicting the reconciliation of science with Vedanta.

Following his exit, and just one year later in 1968, San Francisco Sidewalk Astronomers was created by Dobson and two students, Bruce Sams and Jeff Roloff. Ironically, Dobson was to later address the Vedanta Society of California, Hollywood, where he was to spend time in the years that followed teaching telescope making and conducting cosmology classes, where his own personal views on the cosmos saw a distinct lean toward the Steady State model rather than the Big Bang theory.

## The Birth of the Dobsonian Telescope

Dobson's endeavours in the area of telescope making have afforded him a rather special place in astronomy. There are very few who can claim to have their name associated with a telescope which, alongside sidewalk astronomy, has heralded a

John Dobson's revolutionary creation introduced the astronomical community to the first large, portable, low-cost telescope, whose wide appeal in the 'light bucket' category of instruments would allow distant objects to be viewed from the more challenging and polluted skies, such as witnessed in urban areas, providing the perfect scope for those engaged with sidewalk astronomy. The 'truss tube' derivative of the Dobsonian pictured here affords the user the ability to both quickly assemble and de-assemble the scope. Globally recognized, the Dobsonian has woven itself into astronomy folklore. (Wikimedia Commons/ECeDee/ CC BY-SA 3.0)

celestial marriage of two concepts that have propelled Dobson to near legendary status.

Whilst Dobson was reluctant to accept any credit for his creation, the Dobsonian boasts a combination of many aspects that was to ultimately offer the astronomical community, a large, portable, low-cost telescope. Whilst other designs of a similar nature existed, Dobson was able to meld a telescope that to the present day remains widely used by amateur and professional astronomers alike.

The basis of the design is an altazimuth mounted Newtonian telescope built on the premise that its simplified design has made it easy to construct and relatively inexpensive. The Dobsonian telescope name embedded itself into the world of astronomy, existing in a range of telescopes commonly referred to as 'light buckets', with the ability to target fainter objects such as galaxies. Dobson's idea was to use a mirror rather than a lens, working on the design of a Newtonian telescope. The mechanics operate by reflecting the light travelling down the tube of the telescope from a curved primary mirror to a flat secondary mirror, which then directs the light into the focuser from which an image is produced.

Dobson knew that in order to gather light from dim objects, the telescope would require a large objective mirror, and whereas observing such night sky phenomena as nebulae and star clusters was generally restricted to areas offering 'dark sky' conditions, the Dobsonian would afford the best chance of sighting such objects under more challenging skies where light pollution would be the dominant feature.

Dobson had found the perfect match; sidewalk astronomy and a telescope that could deliver results under conditions that were poor and, indeed, that other astronomers would seek to avoid.

## Bringing Astronomy to the Public

Sidewalk astronomy has become a global affair with presence in Australia, Brazil, Canada, Chile, Germany, Greece, India, Ireland, Italy, Mexico, New Zealand, Russia, Ukraine, and the United Kingdom, plus many other countries. Around the world there are likeminded astronomers taking the night sky, and indeed the daytime sky, to the streets just as Dobson did himself, providing the casual passer-by with an insight into the world of astronomy. With a mission statement of 'Bringing Astronomy to the Public', sidewalk astronomy has taken on a global guise that surely would have made Dobson proud as the catalyst that made it all possible.

An insight into the life of a Sidewalk Astronomer' is best gleaned from the experiences of those who've taken their telescopes out into the streets to engage with the public.

Recalling his days as a sidewalk astronomer in Waikiki, Honolulu, Hawaii, is Barry Peckham:

> "Sidewalk astronomy in Waikiki was immensely rewarding. I set up a homemade (not by me), 8-inch Dobsonian telescope at 'The Waikiki Wall' for nearly 20 years before the advent of COVID".

Pictured on the grounds of a hotel near Diamond Head in Honolulu, Barry Peckham proudly stands by a 15-inch telescope, one of his many scopes. During his sidewalk astronomy years, Barry has visited a number of place including Molokai, Lanai, Hawaii Island, Kauai and Maui, as well as Rhode Island, Tucson and much further afield in Australia. Kapiolani Park was utilized as the location for monthly public star parties in Honolulu with monthly events hosted in Kahala Community Park, on the opposite side of the Diamond Head crater. (Barry Peckham)

Now a resident of Rhode Island, Barry recalls many fine moments on the Waikiki sidewalk:

"Other scope wranglers joined me over the years, and I started the habit with John Dobson on hand, trying to maximize foot traffic. The best part for me was coaxing a collection of strolling strangers, locals and tourists, surfers and party girls, nuns, and hookers, to assemble an instant astronomy appreciation group there on the edge of this highly illuminated tourist Mecca. The international flavours of exclamation were especially fun for me as each person put eye to the eyepiece and saw what they had never seen before. Among the sights, the Moon, Saturn, Jupiter, and sometimes a low, crescent Venus which fuelled this instant community. Reliably stable air, plus excellent optics, gave most of these folks the best views they will ever have of celestial sights."

Like Barry, many sidewalk astronomers found great inspiration from Dobson himself. Donna Smith from Burbank, California, travelled with John Dobson to many locations:

"The social aspect of sidewalk astronomy is what has kept me active in astronomy for so long. John Dobson had this great need to understand our place in the Universe and to get everyone else to 'know the neighbourhood

Sidewalk astronomers in Burbank, California, offering passers-by a glimpse of the wonders of the cosmos under heavily polluted skies. Seen here next to the Chandler Bike Path, John Dobson's vision of bringing astronomy to urban areas under challenging conditions is being dutifully upheld by a band dedicated enthusiasts. One member, Donna Smith, spent much of her time travelling with Dobson, gaining a true insight into the man whose astronomy ethic inspired millions. (Donna Smith)

into which they were born'. Maybe I spent too much time around John, but sidewalk astronomy is pretty much our social life, and we get as much out of it as the public does. Maybe we spark an interest in science or widen someone's views, but I know for sure that I made several people's day a little better by letting them see some amazingly beautiful sights through the eyepiece. This is especially true in areas where the public have had fewer opportunities and are looking through a telescope for the first time. Such joy, and how lucky am I to experience that with them?"

Andy Poniros is a science correspondent and NASA/JPL Solar System Ambassador. Andy's journey into sidewalk astronomy started with the encouragement of a colleague who, knowing Andy's love for astronomy, literally demanded that he make a telescope. A local salvage yard saw the twosome collect several items including a stove pipe, toilet bowl mount flanges, plywood, plus various pieces of aluminium, PVC, and steel which, together with the necessary optics, saw the creation of their own an 8-inch Dobsonian telescope.

"A few months after building my telescope I was watching a lunar eclipse in my front yard with Saturn sporting its ring nearby. As I was viewing, a young girl in the neighbourhood came by on her bicycle and asked what I was

Sidewalk astronomer Andy Poniros at Dick Parker's mirror making class in Connecticut, USA, with a piece of 16½-inch original porthole glass. The glass, which was given to Andy by John Dobson, was to be used as part of a project to create a downscaled replica of Dobson's original 24-inch mirror telescope. Working alongside Dobson the pair brought the cosmos to the kerbside across the northeast of America for over a decade. (Jim Hendrickson@ skyscraperinc)

doing? Of course, I asked her if she wanted take look through the telescope. After looking for a few seconds she said, "Everyone has to see this". The girl seemingly rounded up the entire neighbourhood with everyone looking through my Dobsonian. The chatter and excitement shared by all reached into everyone's soul. For me, sidewalk astronomy has been a love affair ever since."

Whilst Andy had no intention of performing sidewalk astronomy that night, the Dobsonian telescope provided the perfect platform to enable him to do so. Later, Andy was to meet and become friends with John Dobson. Along with Dobson, the pair brought sidewalk astronomy to the northeast of America for ten consecutive years.

Andy recalls a memorable quote from Dobson, "The importance of a telescope is not on how big it is, it's not on how well made it is – it's how many people, less fortunate than you, got to look through it."

Marilyn Roberts and her partner Eric Moon are based in Kaslo, British Columbia, Canada.

As a teacher at Selkirk College for 38 years, Marilyn recalls one summer when they received a grant for a 'Have You Seen the Light' project, where two large rooms full of interactive demonstrations were created to explain the various properties of light. It was suggested that the addition of a telescope would make for a perfect talking point, so John Dobson was duly contacted. Marilyn explains; "John volunteered to come to Kaslo and give a telescope making workshop and lectures on astronomy. He stayed with me and Eric for a week, and we learned so much. The telescope mirrors were ground in my yard and the telescopes were made. The last night of his stay we took them to town and showed people the night sky."

Marilyn and Eric have taken their telescopes to star parties all over north-western America, conducting astronomy lectures and demonstrations in various places. "So many people don't have any idea about how the seasons, moon and stars work, or what those bright points in the sky really are. It's fun to see the lights go on in their heads when they start to see the big picture and how small we really are.

Eric recalls, "John was very abrasive and if you cared to please him, you woke up and started to think. He also laughed at himself. A few of his sayings stuck with me like calling a person a 'silly goose' and 'if you break your telescope it won't lead to a life of sorrow'.

Seasoned Canadian sidewalk astronomers Marilyn Roberts and her partner Eric Moon – pictured here with a selection of their self-made telescopes – have dedicated much of their lives to engaging with the public at star parties, lectures, exhibitions, and upon the sidewalks across North-western America. Their teaching background and passion for astronomy was enhanced following a visit from John Dobson who was to work alongside them. Their dedication to sidewalk astronomy and quest to present the cosmos to others via the eyepiece has spanned many decades. (Marilyn Roberts and Eric Moon)

> "When you meet the universe in person, look down the eye piece tube and see the rings of Saturn or the dust lanes on The Andromeda Galaxy or the glow of the Orion Nebula, it changes you. When you know what these things are and realize how big and complex the Universe is, you realize that your problems are very, very small. The 'how' and 'why' things are the way they are becomes real."

Chris Woodcock is founder of Sussex Sidewalk Astronomers based in the United Kingdom.

> "I started this group in October 2021, initially on Facebook, to connect with fellow amateur and professional astronomers from all walks of life. In just over two years our group has grown to over 1,200 members. Although we are primarily based in (east) Sussex, we have had the pleasure of travelling across the UK, offering members and newcomers the chance to view celestial wonders through our variety of telescopes. Best of all, we do not

Members of Sussex Sidewalk Astronomers not only engage with the public in the own locality but travel the UK to inspire others with images of the wonders of the night sky. Founded in 2021, Chris Woodcock has brought professional and amateur astronomers together with the motto of 'The universe doesn't charge us to look at it, so why should we?' With an array of telescopes, members actively seek to encourage and nurture those who yearn to learn more about astronomy. (Chris Woodcock)

charge for any of our viewings. Our motto is, 'The universe doesn't charge us to look at it, so why should we?' The most rewarding part of our work is seeing the expressions of awe and wonder on the faces of those experiencing the sky's marvels for the first time. It's incredibly fulfilling to leave such a lasting impression on people's lives through our shared passion. One of my favourite success stories is that of Josh Lee, our current Chairman. Josh has navigated the challenges of ADHD and autism throughout his life, but since discovering a passion for astronomy through Sussex Sidewalk Astronomers, he has blossomed into a confident adult with a zest for space facts and astrophotography."

## John Dobson's Legacy

The testimonies of individuals across the world resonate the mission statement of John Dobson's will to bring astronomy to those who would perhaps have otherwise missed out on seeing such wonders as the shape and form of lunar craters and the shadows that are created, or the glorious rings of Saturn, or perhaps the Orion Nebula with its stellar nursery. Some of those who have been stopped by sidewalk astronomers and offered a glimpse through a telescope eyepiece may have, for just a short time, been shown a different world and with it perhaps pondered for just a little while about their own place within it.

Dobson's legacy is remarkable, and a testimony to one man's vision to bring the cosmos to the kerbside.

# Miscellaneous

# Some Interesting Variable Stars

## Tracie Heywood

You may have considered taking up variable star observing but how should you choose which stars to observe? There are so many variable stars in the night sky and you don't want to waste your time attempting to follow the "boring" ones. Your choice of stars will, of course, depend on the equipment that you have available, but also needs to be influenced by how much time you can set aside for observing.

This article splits some of the more interesting variable stars into three groups. The group that is most suited to you will depend on how often you can observe each month and for how long you can observe on a clear night. The light curves included have been constructed from observations stored in the Photometry Database of the British Astronomical Association Variable Star Section. Comparison charts for most of these stars can be found on the BAA Variable Star Section website at **britastro.org/vss**

### One-Nighters

These are stars that can go through most of their brightness variations in the course of a single (reasonably long) night and would suit *people who can only observe occasionally, but can then observe well into the night.*

| STAR | TYPE | RA | | DEC | | MAX / MIN | PERIOD |
|---|---|---|---|---|---|---|---|
| | | H | M | ° | ' | | |
| TT Aurigae | EB | 05 | 10 | +39 | 35 | 8.6 / 9.5 | 1.332735 days (~32 hours) |
| TV Cassiopeiae | EA | 19 | 19 | +59 | 08 | 7.3 / 8.2 | 1.8125919 days (~44 hours) |
| RR Lyrae | RR Lyr | 19 | 25 | +42 | 47 | 7.2 / 8.1 | 0.566777 days (~13.6 hours) |
| Beta (β) Persei (Algol) | EA | 03 | 08 | +40 | 57 | 2.1 / 3.4 | 2.86734 days (~69 hours) |

**TT Aurigae** is a Beta Lyrae type eclipsing variable. Primary eclipses last for just less than 5 hours and are 'V-shaped', indicating that they are partial eclipses. Predictions for upcoming eclipses can be found at **www.as.up.krakow.pl/minicalc/AURTT.HTM**

**TV Cassiopeiae** is an Algol type eclipsing variable. Predictions for upcoming eclipses can be found at **www.as.up.krakow.pl/minicalc/CASTV.HTM**

**RR Lyrae** is a pulsating variable similar in some ways to Cepheid variables, but its variations do not always repeat exactly from one cycle to the next.

**Beta Persei (Algol)** is an eclipsing variable that shows deep primary eclipses but only very shallow secondary eclipses. Although circumpolar from the UK, it is too low in the sky for observation from April to June.

## Most Clear Nights

These are stars that vary a bit more slowly, but which can display significant changes over a week or two. They would suit *people who can observe for a short while on (nearly) every clear night.*

| STAR | TYPE | RA | | DEC | | MAX / MIN | PERIOD |
|---|---|---|---|---|---|---|---|
| | | H | M | ° | ' | | |
| Eta Aquilae | Cepheid | 19 | 52 | +01 | 00 | 3.5 / 4.3 | 7.17679 days |
| BM Cassiopeiae | EA | 00 | 55 | +64 | 05 | 8.8 / 9.4 | 197.28 days |
| V482 Cygni | R CrB | 20 | 00 | +33 | 59 | 10.7 / 15.9 | None |
| AB Draconis | UGZ | 19 | 49 | +77 | 44 | 12.3 / 16.2 | 3.65 hours (orbital) |

**Eta Aquilae** is a Cepheid variable. Its brightness variations repeat exactly from one cycle to the next.

**BM Cassiopeiae** is an Algol-type eclipsing variable. Primary eclipses in 2026 are predicted for mid-February and late August.

**V482 Cygni** is a R Coronae Borealis type variable. It spends most of its time near maximum brightness, but sometimes shows dramatic deep fades which can last for weeks or many months.

**AB Draconis** is a dwarf nova system that produces frequent outbursts, often two or three times a month. Sometimes, however, it enters 'standstill' phases, going for several months without outbursts.

## Several Times per Month

Slower variables whose brightness will change significantly over several months or a year. These variables would suit *people who can observe several times per month, but not necessarily on every clear night.*

| STAR | TYPE | RA | | DEC | | MAX / MIN | AVERAGE PERIOD |
|---|---|---|---|---|---|---|---|
| | | H | M | ° | ' | | |
| UU Aurigae | Semi Reg | 06 | 37 | +38 | 27 | 5.1 / 6.6 | 235 days |
| X Camelopardalis | Mira | 04 | 46 | +75 | 06 | 7.1 / 14.3 | 143.6 days |
| AC Herculis | RV Tau | 18 | 30 | +21 | 52 | 6.9 / 9.0 | 75 days |
| Alpha Orionis | Semi Reg | 05 | 55 | +07 | 24 | 0.4 / 1.6 | 423 days, 2,100 days |
| R Scuti | RV Tau | 18 | 47 | −05 | 42 | 4.2 / 8.6 | 144 days |
| T Ursae Majoris | Mira | 12 | 36 | +59 | 29 | 6.6 / 13.8 | 256 days |

**UU Aurigae** and **Alpha Orionis (Betelgeuse)** are semi-regular variables. 'Semi-regular' indicates that the brightness variations repeat only roughly from one cycle to the next. The observed brightness range differs between cycles, sometimes covering the whole of the listed range (occasionally exceeding it), but more often being somewhat smaller.

**X Camelopardalis** is a Mira type variable. As for other Mira type variables, some maxima are brighter than others. The August 2021 maximum peaked at magnitude 7.1, while that of February 2022 only reached magnitude 9.5. Maxima in 2026 are predicted to occur in January, May and October.

**AC Herculis** and **R Scuti** are RV Tauri type variables. These usually show alternating deep and shallow minima. The depths of the deep minima can differ considerably from one cycle to the next.

**T Ursae Majoris** is a Mira type variable. Maxima are predicted to occur in late November 2025 and in early August 2026.

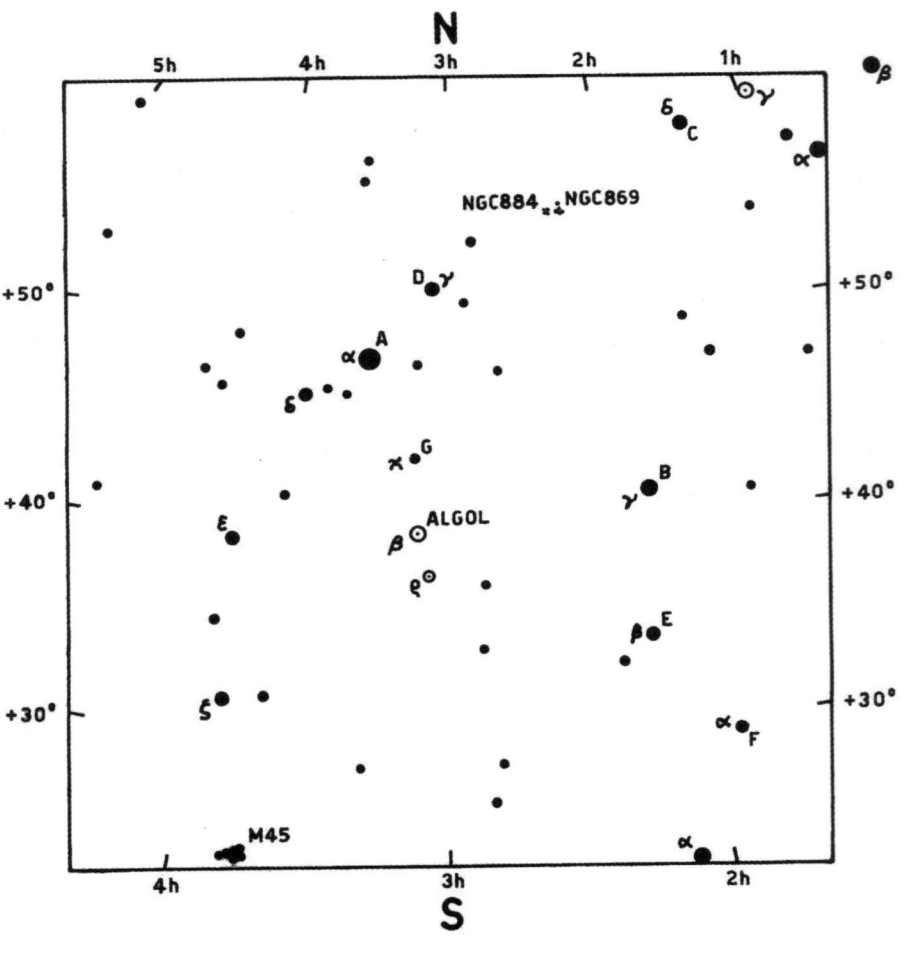

The BAA VSS finder chart for Beta (β) Persei (Algol). (BAA Variable Star Section)

# Minima of Algol in 2026

## Beta (β) Persei (Algol): Magnitude 2.1 to 3.4 / Duration 9.6 hours

|     |    | h    |     |     |    | h    |   |     |    | h    |   |     |    | h    |   |
|-----|----|------|-----|-----|----|------|---|-----|----|------|---|-----|----|------|---|
| Jan | 1  | 11.2 |     | Feb | 2  | 0.1  | * | Mar | 2  | 16.3 |   | Apr | 3  | 5.3  |   |
|     | 4  | 8.0  |     |     | 4  | 21.0 | * |     | 5  | 13.1 |   |     | 6  | 2.1  |   |
|     | 7  | 4.8  |     |     | 7  | 17.8 | * |     | 8  | 9.9  |   |     | 8  | 22.9 |   |
|     | 10 | 1.6  | *   |     | 10 | 14.6 |   |     | 11 | 6.8  |   |     | 11 | 19.7 |   |
|     | 12 | 22.4 | *   |     | 13 | 11.4 |   |     | 14 | 3.6  |   |     | 14 | 16.5 |   |
|     | 15 | 19.2 | *   |     | 16 | 8.2  |   |     | 17 | 0.4  |   |     | 17 | 13.4 |   |
|     | 18 | 16.1 |     |     | 19 | 5.0  |   |     | 19 | 21.2 | * |     | 20 | 10.2 |   |
|     | 21 | 12.9 |     |     | 22 | 1.9  |   |     | 22 | 18.0 |   |     | 23 | 7.0  |   |
|     | 24 | 9.7  |     |     | 24 | 22.7 | * |     | 25 | 14.8 |   |     | 26 | 3.8  |   |
|     | 27 | 6.5  |     |     | 27 | 19.5 | * |     | 28 | 11.6 |   |     | 29 | 0.6  |   |
|     | 30 | 3.3  |     |     |    |      |   |     | 31 | 8.5  |   |     |    |      |   |
| May | 1  | 21.4 |     | Jun | 2  | 10.4 |   | Jul | 1  | 2.6  |   | Aug | 1  | 15.6 |   |
|     | 4  | 18.3 |     |     | 5  | 7.2  |   |     | 3  | 23.4 | * |     | 4  | 12.4 |   |
|     | 7  | 15.1 |     |     | 8  | 4.1  |   |     | 6  | 20.2 |   |     | 7  | 9.2  |   |
|     | 10 | 11.9 |     |     | 11 | 0.9  |   |     | 9  | 17.0 |   |     | 10 | 6.0  |   |
|     | 13 | 8.7  |     |     | 13 | 21.7 |   |     | 12 | 13.8 |   |     | 13 | 2.8  | * |
|     | 16 | 5.5  |     |     | 16 | 18.5 |   |     | 15 | 10.7 |   |     | 15 | 23.6 | * |
|     | 19 | 2.3  |     |     | 19 | 15.3 |   |     | 18 | 7.5  |   |     | 18 | 20.5 |   |
|     | 21 | 23.2 |     |     | 22 | 12.1 |   |     | 21 | 4.3  |   |     | 21 | 17.3 |   |
|     | 24 | 20.0 |     |     | 25 | 8.9  |   |     | 24 | 1.1  | * |     | 24 | 14.1 |   |
|     | 27 | 16.8 |     |     | 28 | 5.8  |   |     | 26 | 21.9 |   |     | 27 | 10.9 |   |
|     | 30 | 13.6 |     |     |    |      |   |     | 29 | 18.7 |   |     | 30 | 7.7  |   |
| Sep | 2  | 4.5  |     | Oct | 3  | 17.5 |   | Nov | 1  | 9.7  |   | Dec | 2  | 22.7 | * |
|     | 5  | 1.4  | *   |     | 6  | 14.3 |   |     | 4  | 6.5  |   |     | 5  | 19.5 | * |
|     | 7  | 22.2 | *   |     | 9  | 11.1 |   |     | 7  | 3.3  | * |     | 8  | 16.3 |   |
|     | 10 | 19.0 |     |     | 12 | 8.0  |   |     | 10 | 0.1  | * |     | 11 | 13.1 |   |
|     | 13 | 15.8 |     |     | 15 | 4.8  |   |     | 12 | 20.9 | * |     | 14 | 9.9  |   |
|     | 16 | 12.6 |     |     | 18 | 1.6  | * |     | 15 | 17.8 | * |     | 17 | 6.7  |   |
|     | 19 | 9.4  |     |     | 20 | 22.4 | * |     | 18 | 14.6 |   |     | 20 | 3.5  |   |
|     | 22 | 6.2  |     |     | 23 | 19.2 | * |     | 21 | 11.4 |   |     | 23 | 0.4  | * |
|     | 25 | 3.1  | *   |     | 26 | 16.0 |   |     | 24 | 8.2  |   |     | 25 | 21.2 | * |
|     | 27 | 23.9 | *   |     | 29 | 12.9 |   |     | 27 | 5.0  |   |     | 28 | 18.0 | * |
|     | 30 | 20.7 | *   |     |    |      |   |     | 30 | 1.8  | * |     | 31 | 14.8 |   |

Eclipses marked with an asterisk (*) are favourable from the British Isles, taking into account the altitude of the variable and the distance of the Sun below the horizon (based on longitude 0° and latitude 52°N).

All times given in the above table are expressed in UT/GMT.

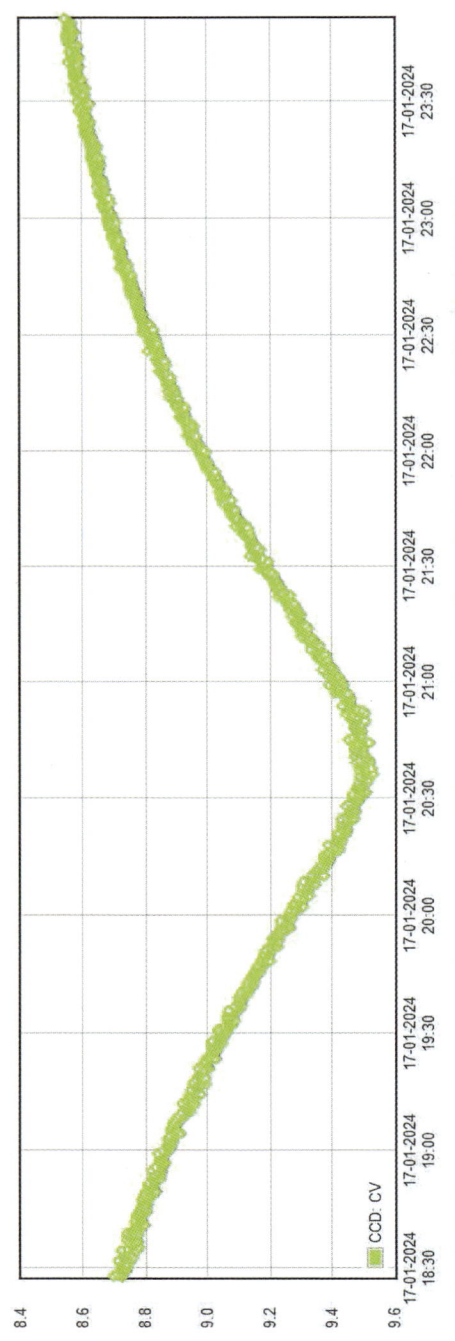

A light curve showing CCD observations by David Conner, of a primary eclipse of TT Aurigae during the evening of 17 January 2024. (David Conner)

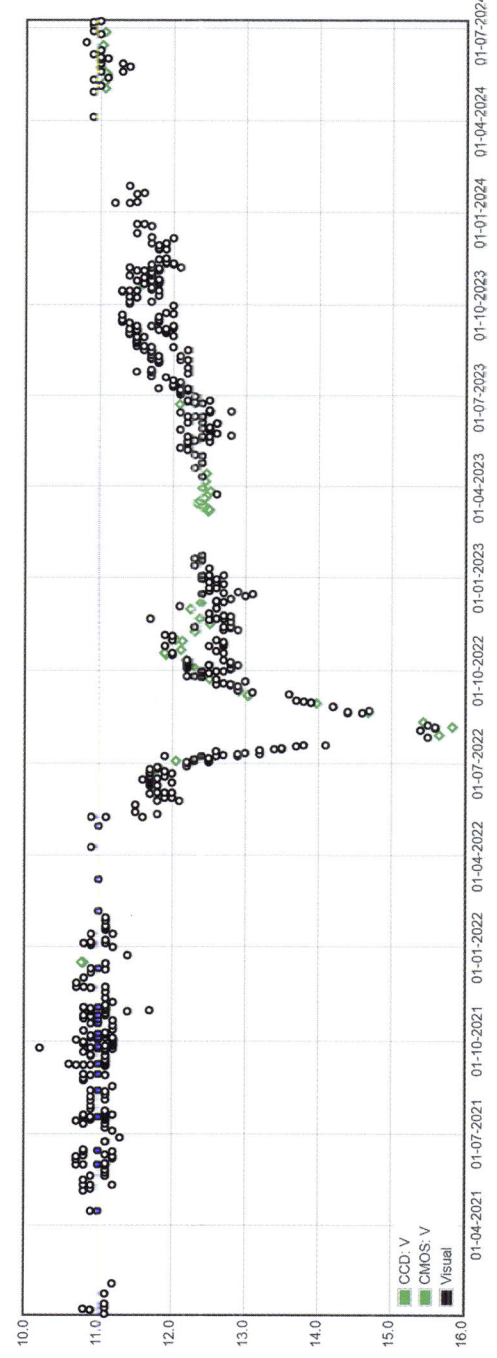

The BAA VSS light curve for V482 Cygni (R CrB type) between January 2021 and June 2024. (BAA Variable Star Section)

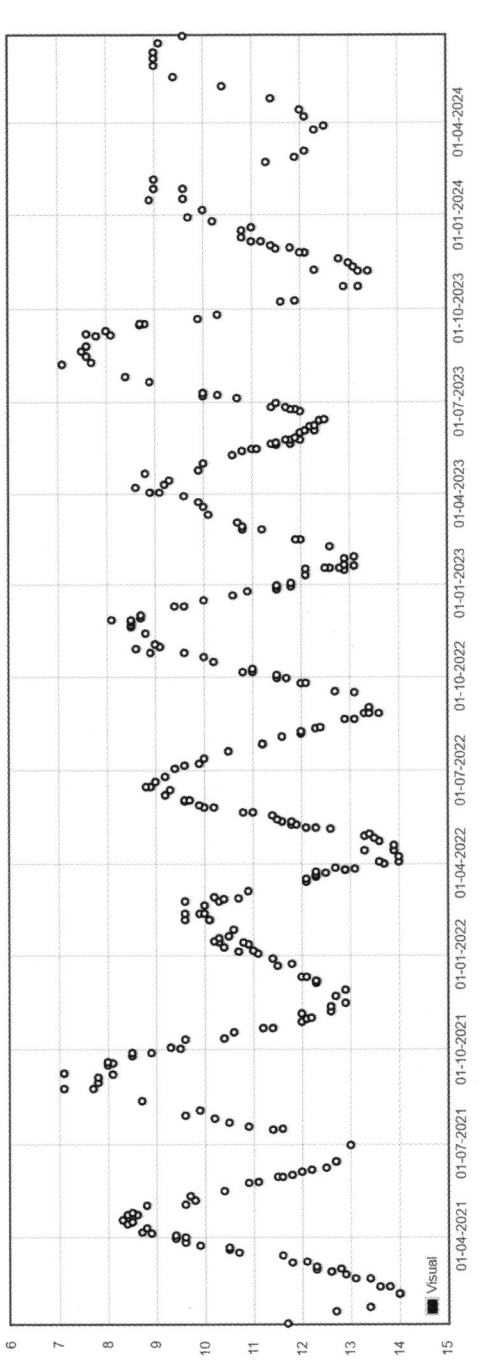

The BAA VSS light curve for the Mira-type variable X Camelopardalis between January 2021 and June 2024. (BAA Variable Star Section)

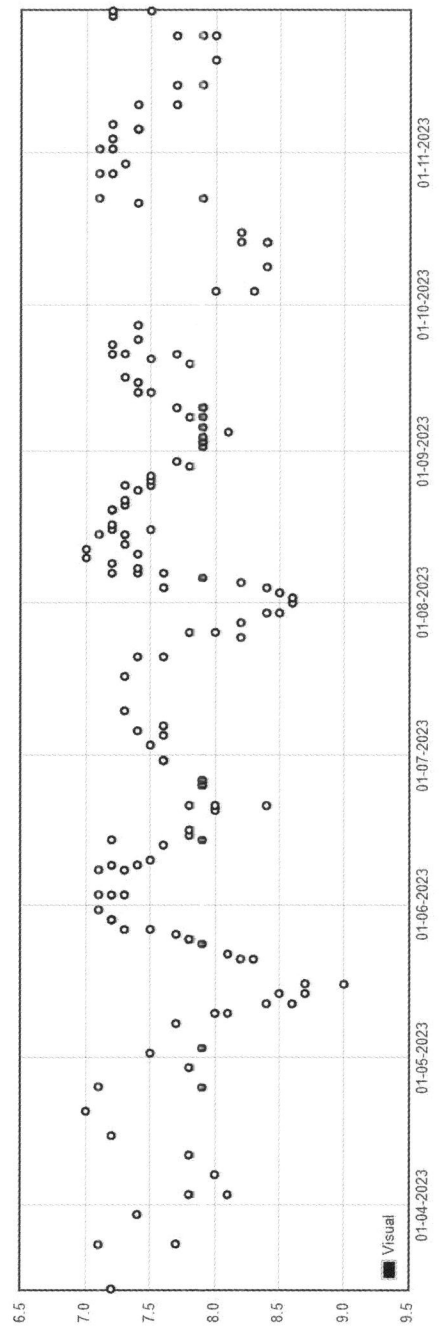

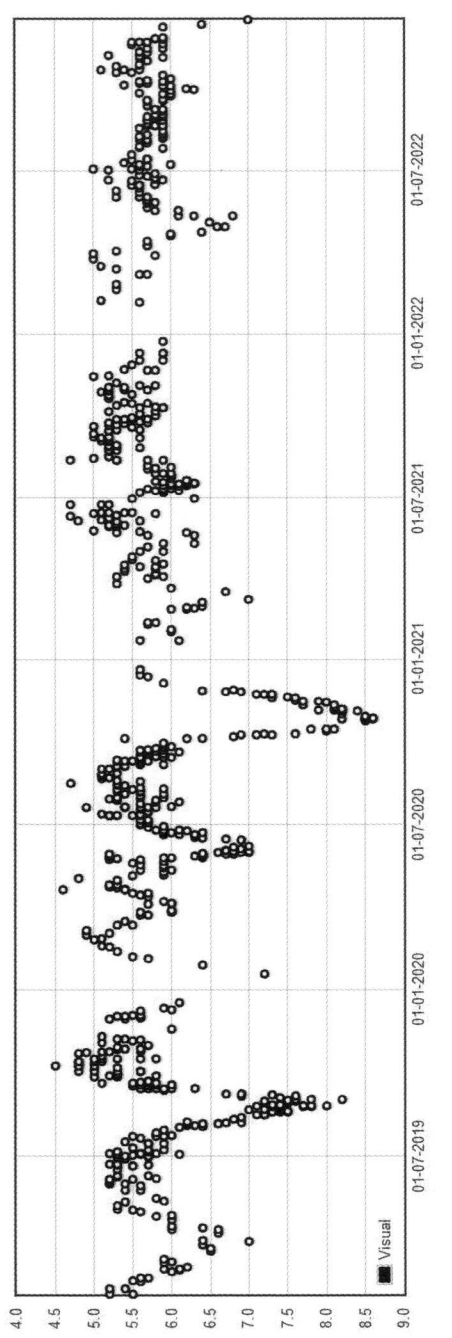

The BAA VSS light curve for R Scuti (RV Tauri type) from 2019 to 2022. (BAA Variable Star Section)

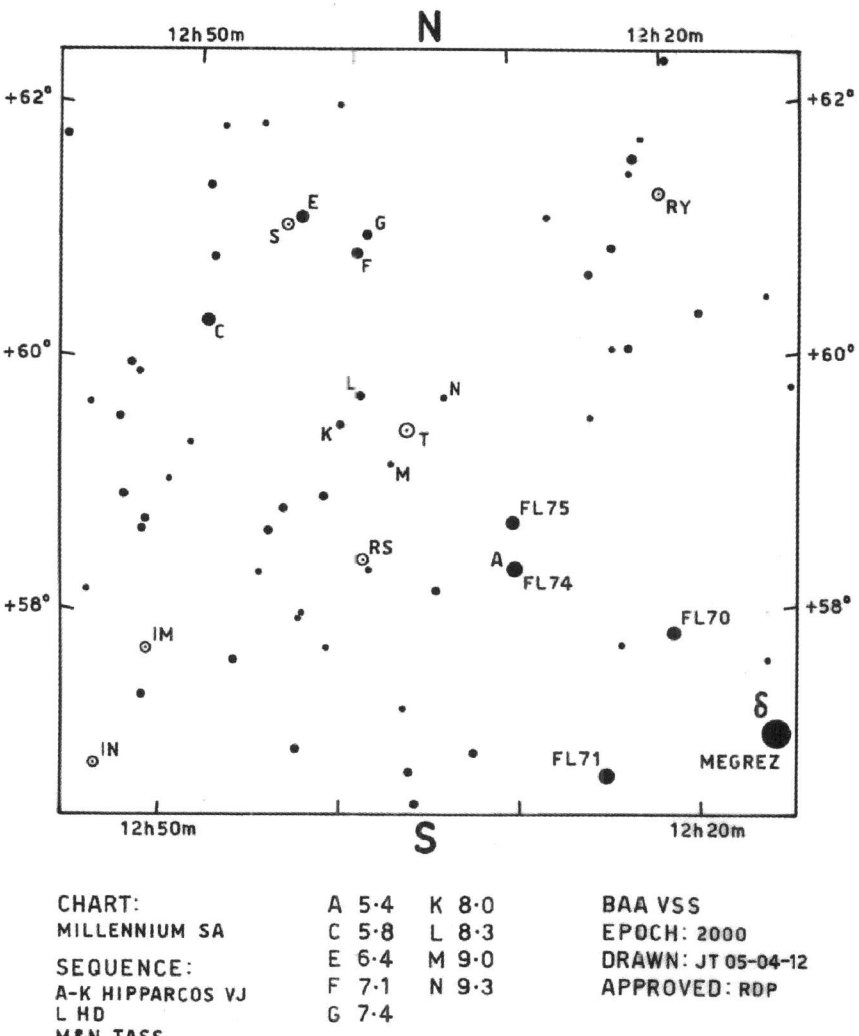

The BAA VSS finder chart for the Mira-type variable T Ursae Majoris, showing several useful comparison stars. (BAA Variable Star Section)

# Some Interesting Double Stars

## Brian Jones

The accompanying table describes the visual appearances of a selection of double stars. These may be optical doubles (which consist of two stars which happen to lie more or less in the same line of sight as seen from Earth and which therefore only appear to lie close to each other) or binary systems (which are made up of two stars which are gravitationally linked and which orbit their common centre of mass).

Other than the location on the celestial sphere and the magnitudes of the individual components, the list gives two other values for each of the double stars listed – the angular separation and position angle (PA). Further details of what these terms mean can be found in the article *Double and Multiple Stars* published in the 2018 edition of the Yearbook of Astronomy.

Double-star observing can be a very rewarding process, and even a small telescope will show most, if not all, the best doubles in the sky. You can enjoy looking at double stars simply for their beauty, such as Albireo (β Cygni) or Almach (γ Andromedae), although there is a challenge to be had in splitting very difficult (close) double stars, such as the demanding Sirius (α Canis Majoris) or the individual pairs forming the Epsilon (ε) Lyrae 'Double-Double' star system.

The accompanying list is a compilation of some of the prettiest double (and multiple) stars scattered across both the Northern and Southern heavens. Once you have managed to track these down, many others are out there awaiting your attention …

# Some Interesting Double Stars

| Star | RA h | RA m | Declination ° | Declination ′ | Magnitudes | Separation (arcsec) | PA ° | Comments |
|---|---|---|---|---|---|---|---|---|
| Beta[1,2] ($\beta^{1,2}$) Tucanae | 00 | 31.5 | −62 | 58 | 4.36 / 4.53 | 27.1 | 169 | Both stars again double, but difficult |
| Achird ($\eta$ Cassiopeiae) | 00 | 49.1 | +57 | 49 | 3.44 / 7.51 | 13.4 | 324 | Easy double |
| Mesarthim ($\gamma$ Arietis) | 01 | 53.5 | +19 | 18 | 4.58 / 4.64 | 7.6 | 1 | Easy pair of white stars |
| Almach ($\gamma$ Andromedae) | 02 | 03.9 | +42 | 20 | 2.26 / 4.84 | 9.6 | 63 | Yellow and blue-green components |
| 32 Eridani | 03 | 54.3 | −02 | 57 | 4.8 / 6.1 | 6.9 | 348 | Yellowish and bluish |
| Alnitak ($\zeta$ Orionis) | 05 | 40.7 | −01 | 57 | 2.0 / 4.3 | 2.3 | 167 | Difficult, can be resolved in 10cm telescopes |
| Gamma ($\gamma$) Leporis | 05 | 44.5 | −22 | 27 | 3.59 / 6.28 | 95.0 | 350 | White and yellow-orange components, easy pair |
| Sirius ($\alpha$ Canis Majoris) | 06 | 45.1 | −16 | 43 | −1.4 / 8.5 | | | Binary, period 50 years, difficult |
| Castor ($\alpha$ Geminorum) | 07 | 34.5 | +31 | 53 | 1.93 / 2.97 | 7.0 | 55 | Binary, 445 years, widening |
| Gamma ($\gamma$) Velorum | 08 | 09.5 | −47 | 20 | 1.83 / 4.27 | 41.2 | 220 | Pretty pair in nice field of stars |
| Upsilon ($\upsilon$) Carinae | 09 | 47.1 | −65 | 04 | 3.08 / 6.10 | 5.03 | 129 | Nice object in small telescopes |
| Algieba ($\gamma$ Leonis) | 10 | 20.0 | +19 | 50 | 2.28 / 3.51 | 4.6 | 126 | Binary, 510 years, orange-red and yellow |
| Acrux ($\alpha$ Crucis) | 12 | 26.4 | −63 | 06 | 1.40 / 1.90 | 4.0 | 114 | Glorious pair, third star visible in low power |
| Porrima ($\gamma$ Virginis) | 12 | 41.5 | −01 | 27 | 3.56 / 3.65 | | | Binary, 170 years, widening, visible in small telescopes |
| Cor Caroli ($\alpha$ Canum Venaticorum) | 12 | 56.0 | +38 | 19 | 2.90 / 5.60 | 19.6 | 229 | Easy, yellow and bluish |
| Mizar ($\zeta$ Ursae Majoris) | 13 | 24.0 | +54 | 56 | 2.3 / 4.0 | 14.4 | 152 | Easy, wide naked-eye pair with Alcor |
| Alpha ($\alpha$) Centauri | 14 | 39.6 | −60 | 50 | 0.0 / 1.2 | | | Binary, beautiful pair of stars |
| Izar ($\varepsilon$ Boötis) | 14 | 45.0 | +27 | 04 | 2.4 / 5.1 | 2.9 | 344 | Fine pair of yellow and blue stars |
| Omega[1,2] ($\omega^{1,2}$) Scorpii | 16 | 06.0 | −20 | 41 | 4.0 / 4.3 | 14.6 | 145 | Optical pair, easy |
| Epsilon[1] ($\varepsilon^1$) Lyrae | 18 | 44.3 | +39 | 40 | 4.7 / 6.2 | 2.6 | 346 | The Double-Double, quadruple system with $\varepsilon^2$ |
| Epsilon[2] ($\varepsilon^2$) Lyrae | 18 | 44.3 | +39 | 40 | 5.1 / 5.5 | 2.3 | 76 | Both individual pairs just visible in 80mm telescopes |
| Theta[1,2] ($\theta^{1,2}$) Serpentis | 18 | 56.2 | +04 | 12 | 4.6 / 5.0 | 22.4 | 104 | Easy pair, mag 6.7 yellow star 7 arc minutes from $\theta^2$ |

| Star | RA | | Declination | | Magnitudes | Separation (arcsec) | PA ° | Comments |
|---|---|---|---|---|---|---|---|---|
| | h | m | ° | ' | | | | |
| Albireo (β Cygni) | 19 | 30.7 | +27 | 58 | 3.1 / 5.1 | 34.3 | 54 | Glorious pair, yellow and blue-green |
| Algedi (α$^{1,2}$ Capricorni) | 20 | 18.0 | −12 | 32 | 3.7 / 4.3 | 6.3 | 292 | Optical pair, easy |
| Gamma (γ) Delphini | 20 | 46.7 | +16 | 07 | 5.14 / 4.27 | 9.2 | 265 | Easy, orange and yellow-white |
| 61 Cygni | 21 | 06.9 | +38 | 45 | 5.20 / 6.05 | 31.6 | 152 | Binary, 678 years, both orange |
| Theta (θ) Indi | 21 | 19.9 | −53 | 27 | 4.6 / 7.2 | 7.0 | 275 | Fine object for small telescopes |
| Delta (δ) Tucanae | 22 | 27.3 | −64 | 58 | 4.49 / 8.7 | 7.0 | 281 | Beautiful double, white and reddish |

# Some Interesting Nebulae, Star Clusters and Galaxies

## Brian Jones

| Object | RA | | Declination | | Remarks |
|---|---|---|---|---|---|
| | h | m | ° | ' | |
| 47 Tucanae (in Tucana) | 00 | 24.1 | −72 | 05 | Fine globular cluster, easy with naked eye |
| M31 (in Andromeda) | 00 | 40.7 | +41 | 05 | Andromeda Galaxy, visible to unaided eye |
| Small Magellanic Cloud | 00 | 52.6 | −72 | 49 | Satellite galaxy of the Milky Way |
| NGC 362 (in Tucana) | 01 | 03.3 | −70 | 51 | Globular cluster, impressive sight in telescopes |
| M33 (in Triangulum) | 01 | 31.8 | +30 | 28 | Triangulum Spiral Galaxy, quite faint |
| NGC 869 and NGC 884 | 02 | 20.0 | +57 | 08 | Sword Handle Double Cluster in Perseus |
| M34 (in Perseus) | 02 | 42.1 | +42 | 46 | Open star cluster near Algol |
| M45 (in Taurus) | 03 | 47.4 | +24 | 07 | Pleiades or Seven Sisters cluster, a fine object |
| Large Magellanic Cloud | 05 | 23.5 | −69 | 45 | Satellite galaxy of the Milky Way |
| 30 Doradus (in Dorado) | 05 | 38.6 | −69 | 06 | Star-forming region in Large Magellanic Cloud |
| M1 (in Taurus) | 05 | 32.3 | +22 | 00 | Crab Nebula, near Zeta ($\zeta$) Tauri |
| M38 (in Auriga) | 05 | 28.6 | +35 | 51 | Open star cluster |
| M42 (in Orion) | 05 | 33.4 | −05 | 24 | Orion Nebula |
| M36 (in Auriga) | 05 | 36.2 | +34 | 08 | Open star cluster |
| M37 (in Auriga) | 05 | 52.3 | +32 | 33 | Open star cluster |
| M35 (in Gemini) | 06 | 06.5 | +24 | 21 | Open star cluster near Eta ($\eta$) Geminorum |
| M41 (in Canis Major) | 06 | 46.0 | −20 | 46 | Open star cluster to south of Sirius |
| M44 (in Cancer) | 08 | 38.0 | +20 | 07 | Praesepe, visible to naked eye |
| M81 (in Ursa Major) | 09 | 55.5 | +69 | 04 | Bode's Galaxy |
| M82 (in Ursa Major) | 09 | 55.9 | +69 | 41 | Cigar Galaxy or Starburst Galaxy |
| Carina Nebula (in Carina) | 10 | 45.2 | −59 | 52 | NGC 3372, large area of bright and dark nebulosity |
| M104 (in Virgo) | 12 | 40.0 | −11 | 37 | Sombrero Hat Galaxy to south of Porrima |
| Coal Sack (in Crux) | 12 | 50.0 | −62 | 30 | Prominent dark nebula, visible to naked eye |
| NGC 4755 (in Crux) | 12 | 53.6 | −60 | 22 | Jewel Box open cluster, magnificent object |
| Omega ($\omega$) Centauri | 13 | 23.7 | −47 | 03 | Splendid globular in Centaurus, easy with naked eye |
| M51 (in Canes Venatici) | 13 | 29.9 | +47 | 12 | Whirlpool Galaxy |
| M3 (in Canes Venatici) | 13 | 40.6 | +28 | 34 | Bright Globular Cluster |

| Object | RA h | RA m | Declination ° | Declination ' | Remarks |
|---|---|---|---|---|---|
| M4 (in Scorpius) | 16 | 21.5 | −26 | 26 | Globular cluster, close to Antares |
| M12 (in Ophiuchus) | 16 | 47.2 | −01 | 57 | Globular cluster |
| M10 (in Ophiuchus) | 16 | 57.1 | −04 | 06 | Globular cluster |
| M13 (in Hercules) | 16 | 40.0 | +36 | 31 | Great Globular Cluster, just visible to naked eye |
| M92 (in Hercules) | 17 | 16.1 | +43 | 11 | Globular cluster |
| M6 (in Scorpius) | 17 | 36.8 | −32 | 11 | Open cluster |
| M7 (in Scorpius) | 17 | 50.6 | −34 | 48 | Bright open cluster |
| M20 (in Sagittarius) | 18 | 02.3 | −23 | 02 | Trifid Nebula |
| M8 (in Sagittarius) | 18 | 03.6 | −24 | 23 | Lagoon Nebula, just visible to naked eye |
| M16 (in Serpens) | 18 | 18.8 | −13 | 49 | Eagle Nebula and star cluster |
| M17 (in Sagittarius) | 18 | 20.2 | −16 | 11 | Omega Nebula |
| M11 (in Scutum) | 18 | 49.0 | −06 | 19 | Wild Duck open star cluster |
| M57 (in Lyra) | 18 | 52.6 | +32 | 59 | Ring Nebula, brightest planetary |
| M27 (in Vulpecula) | 19 | 58.1 | +22 | 37 | Dumbbell Nebula |
| M29 (in Cygnus) | 20 | 23.9 | +38 | 31 | Open cluster |
| M15 (in Pegasus) | 21 | 28.3 | +12 | 10 | Bright globular cluster near Epsilon (ε) Pegasi |
| M39 (in Cygnus) | 21 | 31.6 | +48 | 25 | Open cluster, good with low powers |
| M52 (in Cassiopeia) | 23 | 24.2 | +61 | 35 | Open star cluster near 4 Cassiopeiae |

M = Messier Catalogue Number    NGC = New General Catalogue Number

The positions in the sky of each of the objects contained in this list are given on the Monthly Star Charts printed elsewhere in this volume.

# Astronomical Organizations

### American Association of Variable Star Observers
49 Bay State Road, Cambridge, Massachusetts, 02138, USA
**aavso.org**
The AAVSO is an international non-profit organization of variable star observers whose mission is to enable anyone, anywhere, to participate in scientific discovery through variable star astronomy. We accomplish our mission by carrying out the following activities:

- observation and analysis of variable stars
- collecting and archiving observations for worldwide access
- forging strong collaborations between amateur and professional astronomers
- promoting scientific research, education and public outreach using variable star data

### American Astronomical Society
1667 K Street NW, Suite 800, Washington, DC 20006, USA
**aas.org**
Established in 1899, the American Astronomical Society (AAS) is the major organization of professional astronomers in North America. The mission of the AAS is to enhance and share humanity's scientific understanding of the universe, which it achieves through publishing, meeting organization, education and outreach, and training and professional development.

### Association of Lunar and Planetary Observers (ALPO)
Matthew L. Will (Secretary), P.O. Box 13456, Springfield, IL 62791-3456, USA
**alpo-astronomy.org**
Founded in 1947 by Walter Haas, the ALPO is an international non-profit organization that studies all natural bodies in our solar system. ALPO Sections include Lunar, Solar, Mercury, Venus, Mars, Minor Planets, Jupiter, Saturn, Remote Planets, Comets, Meteors, Meteorites, Eclipses, Exoplanets, Outreach and Online, many with separate "Studies Programs" within these Sections. Minimum

membership is very reasonable and includes the quarterly full colour digital *Journal of the ALPO*. Interested observers of any experience are welcome to join. Many members stand ready to improve the skills and abilities of novices.

## Astronomical Society of Australia
c/o A/Prof. J.W. O'Byrne, School of Physics, The University of Sydney, NSW 2006, Australia
**asa.astronomy.org.au**
The Astronomical Society of Australia (ASA) was formed in 1966 as the organisation of professional astronomers in Australia. Membership of the ASA is open to anyone contributing to the advancement of Australian astronomy or a closely related field. As well as publishing a refereed journal *Publications of the Astronomical Society of Australia*, the Society runs an annual conference and several workshops and schools.

## Astronomical Society of the Pacific
390 Ashton Avenue, San Francisco, CA 94112, USA
**astrosociety.org**
Formed in 1889, the Astronomical Society of the Pacific (ASP) is a non-profit membership organization which is international in scope. The mission of the ASP is to increase the understanding and appreciation of astronomy through the engagement of our many constituencies to advance science and science literacy. We invite you to explore our site to learn more about us; to check out our resources and education section for the researcher, the educator, and the backyard enthusiast; to get involved by becoming an ASP member; and to consider supporting our work for the benefit of a science literate world!

## Astrospeakers.org
astrospeakers.org
A website designed to help astronomical societies and clubs locate astronomy and space lecturers which is also designed to help people find their local astronomical society. It is completely free to register and use and, with over 50 speakers listed, is an excellent place to find lecturers for your astronomical society meetings and events. Speakers and astronomical societies are encouraged to use the online registration to be added to the lists.

## British Astronomical Association
Burlington House, Piccadilly, London, W1J 0DU, England
**britastro.org**
The British Astronomical Association is the UK's leading society for amateur astronomers catering for beginners to the most advanced observers who produce

scientifically useful observations. Our Observing Sections provide encouragement and advice about observing. We hold meetings around the country and publish a bi-monthly *Journal* plus an annual *Handbook*. For more details, including how to join the BAA or to contact us, please visit our website.

## British Interplanetary Society
Arthur C. Clarke House, 27/29 South Lambeth Road, London, SW8 1SZ, England
**bis-space.com**
The British Interplanetary Society is the world's longest-established space advocacy organisation, founded in 1933 by the pioneers of British astronautics. It is the first organisation in the world still in existence to design spaceships. Early members included Sir Arthur C Clarke and Sir Patrick Moore. The Society has created many original concepts, from a 1938 lunar lander and space suit designs, to geostationary orbits, space stations and the first engineering study of a starship, Project Daedalus. Today the BIS has a worldwide membership and welcomes all with an interest in Space, including enthusiasts, students, academics and professionals.

## Canadian Astronomical Society
**Société Canadienne D'astronomie (CASCA)**
100 Viaduct Avenue West, Victoria, British Columbia, V9E 1J3, Canada
**casca.ca**
CASCA is the national organization of professional astronomers in Canada. It seeks to promote and advance knowledge of the universe through research and education. Founded in 1979, members include university professors, observatory scientists, postdoctoral fellows, graduate students, instrumentalists, and public outreach specialists.

## Royal Astronomical Society of Canada
203-4920 Dundas St W, Etobicoke, Toronto, ON M9A 1B7, Canada
**rasc.ca**
Bringing together over 5,000 enthusiastic amateurs, educators and professionals RASC is a national, non-profit, charitable organization devoted to the advancement of astronomy and related sciences and is Canada's leading astronomy organization. Membership is open to everyone with an interest in astronomy. You may join through any one of our 29 RASC centres, located across Canada and all of which offer local programs. The majority of our events are free and open to the public.

## Federation of Astronomical Societies

The Secretary, 147 Queen Street, Swinton, Mexborough, S64 8NG

**fedastro.org.uk**

The Federation of Astronomical Societies (FAS) is an umbrella group for astronomical societies in the UK. It promotes cooperation, knowledge and information sharing and encourages best practice. The FAS aims to be a body of societies united in their attempts to help each other find the best ways of working for their common cause of creating a fully successful astronomical society. In this way it endeavours to be a true federation, rather than some remote central organization disseminating information only from its own limited experience. The FAS also provides a competitive Public Liability Insurance scheme for its members.

## International Dark-Sky Association

**darksky.org**

The International Dark-Sky Association (IDA) is the recognized authority on light pollution and the leading organization combating light pollution worldwide. The IDA works to protect the night skies for present and future generations, our public outreach efforts providing solutions, quality education and programs that inform audiences across the United States of America and throughout the world. At the local level, our mission is furthered through the work of our U.S. and international chapters representing five continents.

The goals of the IDA are:

- Advocate for the protection of the night sky
- Educate the public and policymakers about night sky conservation
- Promote environmentally responsible outdoor lighting
- Empower the public with the tools and resources to help bring back the night

## The Planetary Society

60 South Los Robles Avenue, Pasadena, CA 91101, USA

**planetary.org**

The Planetary Society was founded by Carl Sagan, Louis Friedman and Bruce Murray in 1980 in direct response to the enormous public interest in space, and with a mission to introduce people to the wonders of the cosmos. With a global membership in excess of 50,000 from over 100 countries, it is the largest and most influential non-profit space organization in the world. The Planetary Society bridges the gap between the scientific community and the general public, inspiring and educating people from all walks of life and empowering the world's citizens to advance space science and exploration.

## Royal Astronomical Society of New Zealand

PO Box 3181, Wellington, New Zealand
**rasnz.org.nz**
Founded in 1920, the object of The Royal Astronomical Society of New Zealand is the promotion and extension of knowledge of astronomy and related branches of science. It encourages interest in astronomy and is an association of observers and others for mutual help and advancement of science. Membership is open to all interested in astronomy. The RASNZ has about 180 individual members including both professional and amateur astronomers and many of the astronomical research and observing programmes carried out in New Zealand involve collaboration between the two. In addition the society has a number of groups or sections which cater for people who have interests in particular areas of astronomy.

## Astronomical Society of Southern Africa

Astronomical Society of Southern Africa, c/o SAAO, PO Box 9, Observatory, 7935, South Africa
**assa.saao.ac.za**
Formed in 1922, The Astronomical Society of Southern Africa comprises both amateur and professional astronomers. Membership is open to all interested persons. Regional Centres host regular meetings and conduct public outreach events, whilst national Sections coordinate special interest groups and observing programmes. The Society administers two Scholarships, and hosts occasional Symposia where papers are presented. For more details, or to contact us, please visit our website.

## Royal Astronomical Society

Burlington House, Piccadilly, London, W1J 0BQ, England
**ras.org.uk**
The Royal Astronomical Society, with around 4,000 members, is the leading UK body representing astronomy, space science and geophysics, with a membership including professional researchers, advanced amateur astronomers, historians of science, teachers, science writers, public engagement specialists and others.

## Society for the History of Astronomy

Birmingham and Midland Institute, 9 Margaret Street, Birmingham, B3 3BU
**shastro.org.uk**
The Society for the History of Astronomy was founded in 2002 to promote the study of the history of astronomy by hosting talks by members and publishing new research into the field. One of the main objectives was to encourage research

into past astronomers who have previously been neglected within the history of science. Some of its members are professional historians of science but most are amateur historians. The Society hosts several one-day conferences at venues across the United Kingdom each year. A *Bulletin* is published twice yearly containing articles and news items about astronomical history along with short reports of original research by members. The SHA also issues a quarterly electronic newsletter "e-News" which supplements the email messages from the society with updated events/meetings, and general news from council and SHA library. A library of publications of importance to the history of the science is maintained by the Society at the Birmingham and Midland Institute.

One of the Society's major activities is organising a Survey of Astronomical History in the form of lists of historical astronomers and observatories in every part of Britain and Ireland. This has been motivated by a desire to promote research into local astronomical activities that may have previously been neglected. The Society publishes annually a refereed journal called *The Antiquarian Astronomer* containing new research into the history of astronomy, particularly articles written by members. Published papers have discussed activities in major observatories, scientific research by individuals of particular note, scientific instrument makers, and the activities of prominent amateurs.

## Society for Popular Astronomy
Secretary: Guy Fennimore, 36 Fairway, Keyworth, Nottingham, NG12 5DU
**popastro.com**
The Society for Popular Astronomy is a national society that aims to present astronomy in a less technical manner. The bi-monthly society magazine Popular Astronomy is issued free to all members.

## Webb Deep-sky Society
Secretary: Steve Rayner, 11 Four Acres, Weston, Portland, Dorset, DT5 2JG
**webbdeepsky.com**
Founded in 1967 – and named after Thomas William Webb, author of *Celestial Objects for Common Telescopes* – the Webb Deep-Sky Society is one of the leading international deep sky organisations, and publishes a journal *The Deep-Sky Observer* together with a regular double star *Circular*. The original aim of the society was to update Webb's publications, and this was achieved through a series of eight handbooks. It still publishes material that it believes is relevant to deep sky observing. The society welcomes all levels of observers and has a number of sections dedicated to the observations of Double Stars, Nebulae and Clusters, and Galaxies.

# Our Contributors

**Martin Braddock** is a professional scientist and project and people leader in the fields of drug discovery and development with 40 years' experience of working in academic institutes and large multi-national corporate organizations. He holds a BSc in Biochemistry and a PhD in Radiation Biology and is a former Royal Society University Research Fellow at the University of Oxford. He was elected a Fellow of the Royal Society of Biology in 2010, and in 2012 received an Alumnus Achievement Award for distinction in science from the University of Salford. Martin has published over 210 peer-reviewed scientific papers, filed nine patents, and edited two books for the Royal Society of Chemistry. He also serves as a proposal evaluator for multiple international research agencies. Martin holds further qualifications from the University of Central Lancashire and Open University. He is a member of the Mansfield and Sutton Astronomical Society and was elected a Fellow of the Royal Astronomical Society in May 2015. An ambassador for science, technology, engineering and mathematics (STEM), Martin seeks to inspire the next generation of young scientists to aim high and be the best they can be. In September 2024 he started his own consulting business helping small, medium and large business progress their new medicine pipelines. Martin can be contacted via email at **martin.braddock@genixiconsulting.com** or via his LinkedIn address at **linkedin.com/in/martin-braddock**

**Matt Caplan** is a theoretical nuclear astrophysicist and currently a professor of physics at Illinois State University where he studies white dwarfs, neutron stars, and microscopic black holes. Dr. Caplan received his PhD from Indiana University and his BS from the University of Virginia. Before beginning at ISU he was a Canadian Institute for Theoretical Astrophysics National Fellow at McGill University, and is an inaugural fellow of the Physicists Coalition for Nuclear Threat Reduction. Beyond academia, he is a scriptwriter for PBS Digital Studios and Kurzgesagt.

**Neil Haggath** has a degree in astrophysics from Leeds University and has been a Fellow of the Royal Astronomical Society since 1993. A member of Cleveland and Darlington Astronomical Society since 1981, he has served on its committee since

1989. Neil is an avid umbraphile, clocking up seven total eclipse expeditions so far to locations as far flung as Australia and Hawai'i. Five of these were successful, the most recent being in Waco, Texas on 8 April 2024. In 2012, he may have set a somewhat unenviable record among British astronomers – for the greatest distance travelled (6,000 miles to Thailand) to NOT see the transit of Venus. He saw nothing on the day … and got very wet!

**David M. Harland** gained his BSc in astronomy in 1977 and his PhD in computer science in 1981. He has lectured in computer science, worked in industry and managed academic research. In 1995 he 'retired' in order to write on space themes.

**David Harper**, FRAS has had a varied career which includes teaching mathematics, astronomy and computing at Queen Mary University of London, astronomical software development at the Royal Greenwich Observatory, bioinformatics support at the Wellcome Trust Sanger Institute, and a research interest in the dynamics of planetary satellites, which began during his PhD at Liverpool University in the 1980s and continues in an occasional collaboration with colleagues in China. He is married to fellow contributor Lynne Marie Stockman.

**Tracie Heywood** is an amateur astronomer from Leek in Staffordshire and is one of the UK's leading variable star observers, using binoculars to monitor the brightness changes of several hundred variable stars. Tracie currently writes a monthly column about variable stars for *Astronomy Now* magazine. She has previously been the Eclipsing Binary coordinator for the Variable Star Section of the British Astronomical Association and the Director of the Variable Star Section of the Society for Popular Astronomy.

**Rod Hine** was aged around ten when he was given a copy of *The Boys Book of Space* by Patrick Moore. Already interested in anything to do with science and engineering he devoured the book from cover to cover. The launch of Sputnik I shortly afterwards clinched his interest in physics and space travel. He took physics, chemistry and mathematics at A-level and then studied Natural Sciences at Churchill College, Cambridge. He later switched to Electrical Sciences and subsequently joined Marconi at Chelmsford working on satellite communications in the UK, Middle East and Africa. This led to work in meteorological communications in Nairobi, Kenya and later a teaching post at the Kenya Polytechnic. There he met and married a Yorkshire lass and moved back to the UK in 1976. Since then he has had a variety of jobs in electronics and industrial controls, and until recently was lecturing part-time at the University of Bradford. Rod got fully back into astronomy in around 1992 when his wife bought him an astronomy book, at which time he joined Bradford Astronomical Society. He is currently working part-time at Leeds University providing engineering support for a project to convert redundant satellite dishes into radio telescopes in developing countries.

**Brian Jones** hails from Bradford in the West Riding of Yorkshire and was a founder member of the Bradford Astronomical Society. He developed a fascination with the night sky at the age of five when he first saw the stars through a pair of binoculars, his interest in astronomy eventually taking him into the realms of writing sky guides for local newspapers, appearing on local radio and television, teaching astronomy and space in schools and, in 1985, to becoming a full time astronomy and space writer. His books have covered a range of astronomy and space-related topics for both children and adults and his journalistic work includes writing articles and book reviews for several astronomy magazines as well as for many general interest magazines, newspapers and periodicals. His passion for bringing an appreciation of the universe to his readers is reflected in his writing, his books having covered a range of astronomy and space-related topics for both children and adults. The minor planet 45689 Brianjones is named after him.

**John McCue** graduated in astronomy from the University of St Andrews and began teaching. He gained a PhD from Teesside University studying the unusual rotation of Venus. In 1979 he and his colleague John Nichol founded the Cleveland and Darlington Astronomical Society, which then worked in partnership with the local authority to build the Wynyard Planetarium and Observatory in Stockton-on-Tees. John is currently double star advisor for the British Astronomical Association.

**Neil Norman**, FRAS first became fascinated with the night sky when he was five years of age and saw Patrick Moore on the television for the first time. It was the Sky at Night programme, broadcast in March 1986 and dedicated to the Giotto probe reaching Halley's Comet, which was to ignite his passion for these icy interlopers. As the years passed, he began writing astronomy articles for local news magazines before moving into internet radio where he initially guested on the Astronomyfm show 'Under British Skies', before becoming a co-host for a short time. In 2013 he created Comet Watch, a Facebook group dedicated to comets of the past, present and future. His involvement with Astronomyfm led to the creation of the monthly radio show Comet Watch, which is now in its fourth year. Neil lives in Suffolk with his partner and three children. Perhaps rather fittingly, given Neil's interest in asteroids, he has one named in his honour, this being the main belt asteroid 314650 Neilnorman, discovered in July 2006 by English amateur astronomer Matt Dawson.

**Jonathan Powell** is a contributor to the *Sky at Night* magazine and has also had articles published in *Astronomy Now*. He has written three books on astronomy; *Rare Astronomical Sights and Sounds* (which was selected by *Choice* magazine as an Outstanding Academic Title for 2019); *Cosmic Debris: What It Is and What We Can Do About It;* and *From Cave Art to Hubble: A History of Astronomical Record Keeping*, all of which are available from Amazon. Jonathan worked at BBC Radio Wales as their astronomy correspondent, and was astronomy and space correspondent for *The National*, (an online newspaper for Wales). He is currently a columnist at the *South Wales Argus*, and a contributor to *CAPCOM*, an online magazine which promotes astronomy and spaceflight to the public. He has also presented on commercial radio at Sunshine FM in Worcester, Brunel FM in Swindon, and Bath FM, and has been a presenter on the astronomy and space dedicated radio station Astro Radio (UK). In addition, he is a former panellist for the YouTube channel 'Space Oddities'. Jonathan is currently at 107.9 GTFM in South Wales. He has also written a book on castles, *Fortress Wales*, and was part of the writing team for the BBC Television programme, 'The Fast Show', which won a BAFTA.

**Katrin Raynor** lives in Pontypridd, Glamorgan, South Wales and is a Fellow of the Royal Astronomical Society and Royal Geographical Society. She is an amateur astronomer and philatelist, and in her spare time writes articles and interviews for popular astronomy magazines including *Sky at Night* magazine and the Society for Popular Astronomy magazine, *Popular Astronomy*. Katrin is also a co-presenter on the Sky at Night's podcast, Star Diary. The minor planet 446500 is named after her.

**Peter Rea** has had a keen interest in lunar and planetary exploration since the early 1960s and frequently lectures on the subject. He helped found the Cleethorpes and District Astronomical Society in 1969. In April of 1972 he was at the Kennedy Space Centre in Florida to see the launch of Apollo 16 to the moon and in October 1997 was at the southern end of Cape Canaveral to see the launch of Cassini to Saturn. He would still like to see a total solar eclipse as the expedition he was on to see the 1973 eclipse in Mali had vehicle trouble and the meteorologists decided he was not going to see the 1999 eclipse from Devon. He lives in Lincolnshire with his wife Anne and has a daughter who resides in Melbourne, Australia.

## Our Contributors

**Andrew Santarelli** is theoretical astrophysicist and a graduate student at Illinois State University studying black holes, stellar evolution and populations, and a combination of the two in Hawking stars and quasi-stars. Andy received his BS in Physics from Illinois State University, and prior to that worked as a music recording engineer in Los Angeles.

**Randall Stevenson** is the author of many studies of literature, most recently *Reading the Times: Temporality and History in Twentieth-Century Fiction* (2018). He is Emeritus Professor of English Literature in the University of Edinburgh – a position allowing more scope, these days, to study and photograph the night sky, and to recall distant days when he was briefly a student of astrophysics, before turning to literature.

**Lynne Marie Stockman** holds degrees in mathematics from Whitman College, the University of Washington and the University of London. She has studied astronomy at both undergraduate and postgraduate levels, and is a Friend of the Astronomical Society of the Pacific and a member of the Society for the History of Astronomy. A native of North Idaho, Lynne has lived in Britain since 1992. She was an early pioneer of the World Wide Web; with her husband and fellow Yearbook contributor David Harper, she created the web site **obliquity.com** in 1998 to share their interest in astronomy, computing, family history and cats.

# Society for the History of Astronomy

## *All are welcome*

Members receive the Society's publications: eNews, Bulletin and Antiquarian Astronomer. There are also meetings and visits to places of interest, use of our unique library and access to research grants.

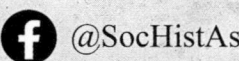 @SocHistAstro

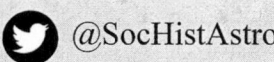 @SocHistAstro

Contact - general.secretary@shastro.org.uk

**societyforthehistoryofastronomy.com**

# Become a Friend of the Royal Astronomical Society

### Benefits include...

Friends of the RAS lecture programme
Visits to observatories and other places of interest
Discounted subscription to 'Astronomy & Geophysics' magazine
Use of the RAS library

For more information visit:

## www.ras.ac.uk/friends

# The Federation of Astronomical Societies

The FAS is a union of astronomical societies, groups and clubs, liaising together for their mutual benefit.

Formed in 1974, we celebrated our **50th anniversary** in 2024.

**Benefits of membership include:**

- Access to Public Liability Insurance for societies at extremely favourable rates
- A Resource Centre containing documents providing a common framework for the running of astronomical societies and their outreach activities
- A Newsletter where societies can feature their events and information and also discover other activities
- Public Conventions featuring prominent speakers from Academic and Industrial backgrounds, at sites around the UK, and also online webinars and lectures
- A website announcing events from societies and other organisations that would be of interest to its member societies
- Around 200 astronomical societies are FAS Members, not only in the United Kingdom but also overseas.

For details about the organisation, as well as other information, visit our website
**www.fedastro.org.uk**

## Magazine
### A Space Oddities Publication
#### Edited and produced by Michael Bryce

CAPCOM Magazine contains original articles by enthusiastic, knowledgable and published authors. Plus features on spaceflight and astronomy including the latest news from around the world. CAPCOM is published six times per year and is free to download. Contributions to CAPCOM are welcome. Original articles on any aspect of spaceflight and astronomy are considered.

For more information and to download CAPCOM please visit:
**youtube.com/@spaceoddieslive** and follow the link in the description for CAPCOM

# Space Oddities Live!

Interested in Astronomy or Space Exploration?
Do you use YouTube or Facebook?

Then check out:

# Space Oddities Live!

Streaming Live Every week
Visit our YouTube channel for details!
**youtube.com/@spaceoddieslive**

Astronomy and Space Exploration news
Discussion
Special Guests
Night Sky Notes
Viewers' Gallery
and more

We livestream every week on YouTube and Facebook. We are an international panel of amateur and professional astronomers. Panel Members are from the UK, Spain, the US and Canada. We chat about anything relating to the Universe and space exploration, keeping our audience up to date with the latest news. We also present interesting presentations on a huge variety of astronomical subjects and also create our own space-related videos for all levels of astronomical knowledge. As well as our weekly shows we also go live for important space launches or other special events. We have a lot of fun, so why not join us? For livestream details, please visit our YouTube Channel at

# youtube.com/@spaceoddieslive

For inclusion in our weekly viewers' gallery, please send your images to

**spaceoddieslive@gmail.com**
Please include your name, location, equipment, processing details etc
One image per email please, entitling it "Gallery Entry"

*Background Image Credit: Jonathan Wood and Rachael Wood: Doncaster Astronomical Society*

Advertisement

# Fully in the picture?

The British Astronomical Association (BAA) has been a driving force in amateur astronomy since 1890, and is one of the world's leading amateur groups. By joining the BAA, you will become part of a diverse community of enthusiasts of all levels of ability and with a varied array of interests and expertise.

**Planetary**

**Deep Sky**

**Comets**

**Equipment**

**Lunar**

## Benefits of membership

- Receive the bimonthly *BAA Journal*, the annual *Handbook*, and access online copies of current and historic issues.
- Receive our regular *BAA Newsletter* and *Section Circulars*, delivered by e-mail.
- Develop and share your expertise through the BAA's regular meetings, including Back to Basics workshops held around the country to encourage newcomers.
- Access online tutorials to improve your observing skills and get the most out of your equipment, or watch videos of talks by leading experts.
- Contribute to scientifically valuable observations, often in collaboration with professional colleagues. Our members include Tom Boles, who has discovered more than 150 supernovae from his Suffolk observatory, and Damian Peach, whose planetary images are among the world's best.
- Talk to fellow members on the BAA's online forum.
- Become involved in our observing programmes organised by the BAA Observing Sections.
- Present your work on your own BAA Member Page and contribute articles to the *BAA Journal*.
- For membership details scan the QR code or follow the Join the BAA Today link at britastro.org/join.

Proud to support Dark Skies for Everyone
**Join today - britastro.org**